# Species Problem: Issues and Challenges

# Species Problem:
# Issues and Challenges

Edited by **Shirley Striker**

New York

Published by Callisto Reference,
106 Park Avenue, Suite 200,
New York, NY 10016, USA
www.callistoreference.com

**Species Problem: Issues and Challenges**
Edited by Shirley Striker

International Standard Book Number: 978-1-63239-574-0 (Hardback)

# Contents

**Permissions**

**List of Contributors**

# Preface

The main aim of this book is to educate learners and enhance their research focus by presenting diverse topics covering this vast field. This is an advanced book which compiles significant studies by distinguished experts in the area of analysis. This book addresses successive solutions to the challenges arising in the area of application, along with it; the book provides scope for future developments.

An elucidative account regarding the topic of species problem and certain issues as well as challenges related to it have been provided in this book. It is a compilation of theoretical research works regarding the species problem, which is among the most elementary issues in biology. The primary topics covered in this book are: evolutionary emergence of the species as an individual unit of a specific level of generality; consideration of the species problem from the point of view of modern non-conventional science paradigm, with stress on its conceptual status presuming its characterization in the boundaries of specific conceptual framework; a review of the concept of biological species on the basis of the "friend-foe" identification system; considerations of the concepts of phylogenomics and evolution of species as candidates for the universal one; species delimitation approach employing multi-locus coalescent-based technique; epistemological view of the species as a specific explanatory hypotheses; review of species concept proposed by Darwin.

It was a great honour to edit this book, though there were challenges, as it involved a lot of communication and networking between me and the editorial team. However, the end result was this all-inclusive book covering diverse themes in the field.

Finally, it is important to acknowledge the efforts of the contributors for their excellent chapters, through which a wide variety of issues have been addressed. I would also like to thank my colleagues for their valuable feedback during the making of this book.

**Editor**

# Introductory

# The Species Problem, Why Again?

Igor Ya. Pavlinov

Additional information is available at the end of the chapter

## 1. Introduction

Every scientific discipline, in the semantic terms, is a set of theoretical constructs, i.e theories, concepts, ideas, etc., of different levels of generality. One of paradoxes of development of the science lies in the fact that the more general and fundamental are constructs of such kind, the less clearly can they be delineated at the level of common understanding and defined by means of formal language of the given discipline. As a result, the latter may be likened to a building with a very shaky foundation (basic concepts), with pretty loosely aligned walls (derived concepts), and with a roof just looking quite solidly (the facilities of solving technical problems).

The idea of *the* species belongs to such basic conceptions in the biological sciences, this idea has being been acknowledged repeatedly over the centuries. Accordingly, in the light of the above paradox, the species notion was and remains to be among the most disputed and controversial in biology, with a compass of viewpoints ranging from acknowledging the unconditional and self-evident objective reality of the species to denying it as an objective (natural) phenomenon. Despite the efforts of generations of theoreticians, it appeared impossible to reach a universal and all-suiting understanding and definition of what is *the* species of living organisms, i.e. the "biological species" in its most general (not particular Mayrian) sense.

The fundamental nature of the species notion in biology has led to an attempt to establish a particular biological discipline about it proposed to be called "eidology" or "eidonomy" (after the Greece term "eidos", see 2.2; not in sense of Husserl) [1-6]. Its focal point was declared to be development of some general theory of *the* species of living beings, which would explain both the existence and most general properties of the species as a natural phenomenon, along with variety of its manifestations in different groups of organisms reflected by particular species concepts.

Disputability and ambiguity of the basic notion of *the* species has generated the well-known "species problem", which appears to be of the same fundamental character to biology as that notion. It was explicitly highlighted in the early 20th century [7-8], but it is clearly much older; as a matter of fact, it had emerged, though without an official nomination, at the time when both natural philosophers and subsequently natural scientists had began to use the term "species" ("eidos") to describe the diversity of both organisms and other things. Current attitude toward this problem varies from its ignoring by practicing biologists to its explicit fixation in theoretical studies as a particular theoretical construct built upon the species notion. Not a once biologists and philosophers participating in the discussion of this problem tried to offer their understanding of the species as more or less radical and more or less general solutions of the species problem. But each of them appeared eventually proved to be more or less particular and not deciding but just supplementing the problem and thus making it far more "problematic". So the species problem in biology seems to be doomed to remain eternal as a consequence of fundamentally irremovable disputability and ambiguity of the very notion of *the* species.

In this chapter, I draw attention to some key issues of the species problem as it is seemed to me now. First, I shall try to delineate somehow what precisely might be called the "species problem" and to identify its origins, both historical and cognitive. Second, I shall present possible scientific and philosophical contexts of its analysis, with emphasis on the non-classical philosophy of science. Third, I shall consider, within the latter philosophy, a possible natural science context of the consideration of the species problem represented in the form of a "conceptual pyramid", a part of which is the species notion as a theoretical construct. At last, it will be shown that another "radical solution" of the species problem may be just to acknowledge objective multiplicity of the "kinds of species" of living beings, corresponding to which is subjective multiplicity of the species concepts.


## 2. Whence the species problem

Any cognitive problem is systemic by its nature, and the species problem provides no exception. It is structured, multifaceted, multi-component, with the issues of different levels of generality and significance interacting within it. These issues appear and disappear with the development of the problem, which, in its turn, is caused by development of the scientific discipline in which it has been subsisting. In particular, taxonomic aspect of consideration of the species problem was dominating previously, while its "detaxonomization" is noticeable at the present time, according to which the "species in classification" becomes separated from the "species in nature" and it is the latter that is now being considered as a focal point of the species problem [9-11]. Respectively, discussion of this problem should begin with consideration of the following key issues: *What* is the species problem? Why is it about just the *species*? Why is it just the *problem*? [10, 12-13].

## 2.1. What is the species problem

Generally speaking, any problem is generated by a cognitive issue that has no clear-cut single answer, and this is true for the species problem. The latter is a consequence of the above-stressed irremovable ambiguity of the species notion (in its general biological sense), which means impossibility to give an exhaustive comprehensive theoretical definition of the species as a biological phenomenon. This is referred to as the "species uncertainty" [14-15].

I think, however, that ambiguity of the species concept in itself is not the whole problem. Its important (maybe the core) part seems to be a contradiction between polysemy of the species notion and unsuccessful striving of discussants to reduce it to a single most general (or at least most appropriate) definition common to the entire biology.

An aspiration for a unified comprehension and definition of the species is quite understandable; every science must have some unified thesaurus, through which the subject area of that science is uniformly described. From such a perspective, usage of some common term for a certain natural phenomenon—in our particular case, for a manifestation of diversity of organisms—implies that the phenomenon in question is endowed with a unique property, which allows to recognize it among other phenomena of the "same kind". Therefore, the history of the species problem appeared to be largely a story of searches for such a fundamental overall property of *the* species ("specieshood", see 5), which could be adequately reflected in a single definition.

The species problem, in such a general meaning, emerged simultaneously with the very notion of species (= eidos) in the Ancient times, where it initially had quite different interpretations (see 2.2). In the scholastic period, this ambiguity has been reduced to a logical interpretation of the species. In modern times, however, dominated became biological understanding of *the* species as a group of organisms, which diverse interpretations are currently being tried to reduce to its evolutionary or genetic (reproductive) or operational meanings. Another contemporary attempt, if not to reduce but at least to put diverse treatments in some order, is to build a kind of "conceptual pyramid" of different levels of generality of these treatments (see 4.1).

One of the key issues that shapes contemporary understanding of the species problems concerns explanation of emergence of both *the* species (in the general sense) as a natural phenomenon and actual diversity of its manifestations. I think that there cannot be any properly developed theory of the species (whatever might it be) without putting and answering these fundamental questions.

## 2.2. Why the species

Fundamental status of the species concept has deep historical roots, without reference to which one can hardly understand the reasons for such a great attention paid both to the species proper and to the species problem under consideration.

In a very rough approximation (for details, see [2, 16-17], the history of the term "species" dominating nowadays in biology goes back to the Aristotelian notion of "eidos" denoting certain "form" through which the formless "matter" assumes its actual existence. So, the "species" (= "eidos" = "form") such treated was "external" with respect to the "matter", which is evident, for instance, from Theophrastos' concept of plants changing their "species" due to changes of conditions of their growth [18]. Under this naturphilosophical doctrine, the actual existence of any natural body is impossible without respective "eidos" making the thing what it is. This ontology had been supplemented by a cognitive construct called later "genus-species scheme" by neo-Platonists and scholastics, in which the "eidos"="species" got rather logical status of one of the universal categories of knowledge. According to this integrated onto-epistemological construct, the "eidos"="species" is universal and fundamental in both to the Nature itself and to the knowledge about the Nature. Therefore, nothing can exist without the species, be it a body in the objective world or its image inferred within the logical generic-species subjective scheme. This led to a strong belief of earlier Aristotle interpreters formulated explicitly by Boethius that "[if] we do not know what is the species, nothing would secure us from misunderstandings" (translated from the Russian edition [19]).

Strictly speaking, it is this Ancient historical and cognitive landmark from which it is reasonable to trace the above "eidology" with its presumption of universality and fundamentality of the species, whatever its particular interpretation may be, and all that is associated with it. Searching for a "final answer" to the question "What is the species?" gave birth to some "Boethian tradition". It was brought to biology by Aristotelian A. Cesalpino having first applied explicitly generic-species scheme to classification of botanical objects. Subsequently, it was filled in part with the biological content by J. Ray, and then fixed by Linnaeus, for whom it was the species that was the basic unit of the Natural System. So, past and present theoreticians, having tried and still trying to answer somehow the above question, were and still are "Boethians", as they were and still are believing this issue is one of the most fundamental in biology.

Possible answers to that "Boethian question" have been being traditionally sought most often in the framework of the dichotomy preset by neo-Platonists and early scholastics in the form of opposition of realism vs. nominalism [13, 20-22]. Commitment to the realism requires acknowledge of the species objectively and undoubtedly existing as a kind of fundamental and universal "unit of Nature". Nominalists deny objectivity (reality) of the species in the sense just indicated, or at least do not recognize its particular fundamental status in the hierarchy of the Nature (bionominalism, see [11]), though acknowledge necessity and universality of the species as a useful "unit of classification".

Discussants, even belonging to opposite research schools, can quite agree with each other in recognition of fundamental status of the above "Boethian question", whatever its particular answer might be. For instance, both "methodist" Linnaeus and "naturalist" Buffon (in his later years) believed in objective (real) status of the species as a universal and fundamental "unit of the Nature". On the other hand, evolutionist Darwin, rejecting alongside with logician J. Bentham distinctiveness of the species as a fundamental taxonomic and eventually natural category, called however his famous book just "The Origin of Species...", and not of races or of something like that.

One of peculiar manifestations of the "Boethian tradition", I think, is an exaggerated attention to the species category displayed by many biologists who use to pay too much attention to it. Due to this, other aspects of the biological diversity, both "vertical" (e.g. supraspecific groups) and "horizontal" (e.g. ecomorphs), are usually treated as of secondary importance. This standpoint seems to be obsolete with regard to modern understanding of biodiversity, but it nevertheless still persists in contemporary biology thus impoverishing the overall picture of the biodiversity [23].

### 2.3. Why the problem

A brief answer to this question was given above (see 2.1); the problem is that the notion of species, which has become fundamental for biology due to, among others, its "historical burden", cannot be filled with a single content [12, 17, 22]. It has many meanings, which cannot be reduced to a single, albeit rather complicated, formula such as " The species is...".

An ambiguity of the species notion has as deep historical roots as this notion itself. It has been originally used to refer to essentially different phenomena, some of which belonged to the actual diversity of organisms, while others to the ways this diversity was described. And this is one of the main sources of the species problem.

Thus, Aristotle understood the "eidos" as both the groups of organisms (e.g. "tetrapods") and the essential properties characterizing them (e.g. "tetrapodness"). Such "dual" (from the modern standpoint) usage of the same term "eidos" was quite natural to the Ancient understanding of the Nature as the "Physis" and understanding of the species (eidos) as the "form" shaping the matter [24] (see 2.2). This standpoint was partially preserved in the natural history at least until the 16th century (occurred in J. Ray's writings, see [25]. However, these two aspects of the Ancient understanding of *the* species (eidos), as a taxon or as a meron, are recognized in the modern biology as fundamentally different, so their joining under the same term became removed by separation of two aspects, taxonomic and meronomic, of the organismal diversity [26]. Accordingly, taxonomically treated "eidos" became fixed as *the* species, while its meronomic treatment provides the notion of homologue.

Further, although Aristotle distinguished terminologically between "natural" and "logical" groups and seemed to use the term "eidos" only to designate the second ones [16], scholasticism united them under the single Latin term "species". It has not probably been without influence of Thomism, as one of its key ideas related to the topic under discussion was assertion of the unity of three "hypostases" of the essences—before things (*ante res*), within things (*in rebus*), and after things (*post res*)—as different aspects of the same universal organizing principle of the world of both things and ideas. Modern natural science recognizes a necessity of handling the "natural species" separately from the "logical species" [10, 11, 12, 13, 27], but this is not yet reflected properly in the existing thesaurus of "eidology". And this also contributes to the problematic situation; obviously, any discussion of ontological status of the species becomes meaningless if it is not indicated explicitly what kind of "species", natural or logical, is referred to (see also 3.4).

An important source of the species problem, in its general sense, is the multidimensional nature of the "species in nature" understood also in its general sense. It means that the species a) is a member of different natural processes, and b) it possess its own internal structure of different kind. Every aspect of the species natural history (e.g. genealogical, ecological, reproductive, etc.) can be fixed in the form of its key (essential) property to be used for elaborating certain species concept, which is advocated by its authors and proponents as a "principal" one. An aspiration for ascribing a universal meaning of the species to such particular concepts and, accordingly, the belittling of other concepts leads to competition between them, which however can be inconsistent under certain circumstance (see 5.2).

A particular aspect of the species multidimensionality and thus of the species problem became apparent relatively recently; it is the necessity for separate consideration of "the species taxon problem" and "the species category problem" [16, 28-29]. In the terms adopted here, the species category is defined by the specieshood, while the species taxon is (quite roughly) defined by particular manifestation of the specieshood in particular groups of organisms.

One of the sources of the species problem is that biologists (and occasionally philosophers) put quite different questions analyzing the species concepts and their applications; this was noticed repeatedly by many authors [9, 10, 12, 13, 16, 22, 30-34]. Some of these questions are about essential properties of the species (i.e. about the above mentioned "specieshood"), others deal with the mechanisms of emergence and sustainable subsistence of the species, and more others consider how to recognize particular species in the empirical studies. In this regard, the species problem is quite comparable to the homology problem or to the gene problem; in each of them, respective unit, though uniformly called (the species, the homologue, the gene, respectively), are recognized and treated much differently in particular research programs.

Pretty curious seems to be a kind of "psychological" source of the species problem, i.e. conviction of the debate participants that this problem does actually exist [35]. Due to this, the species problem takes certain kind of independence and self-sufficiency as a particular conceptual construct interested mainly to some theoreticians.

It is important to bear in one's mind that the species problem is a dynamic construct. It has been developing in parallel with development of both the natural science and the philosophy of science, responding one or another way to the new ideas elaborated by them. Accordingly, the content of the problem has been changing with time; some of its aspects fallen away, some came as new ones to gain particular attention. One of the most important recent changes was due to completing the above rigid dichotomy between "realism *vs.* nominalism" to a trichotomy by adding a modern version of the conceptualism to them [10, 27, 36]. The latter brings its own focus to the general species problem, which allows to take a fresh look at the multidimensional nature of the species proper and to legitimizes the "species pluralism" (see 3.1).

# 3. Understanding species: Cognitive situation

One of the most important in the contemporary cognitive science is the notion of cognitive situation, within which object, subject, purpose, and means of knowledge are determined. Understanding of its content and structure was changing considerably with the evolution of philosophy of science. The most significant shift occurred in the second half of the 20th century in connection with transition from the classical to the non-classical scientific paradigm [37]. The latter evidently, albeit it is not fully acknowledged yet, affects understanding of the entire species problem [10].

## 3.1. Classical *vs.* non-classical views of the species

Classical science is based on the following key assumptions. The Universe is organized (structured) by a single principle; the structure of the Universe is therefore linear and admits a reduction of its diversity to a minimum ("atomic") level; the unity of the Universe as a global natural phenomenon is reflected in the unity of a "final theory" describing it; it is comprehended by means of a unified general method (in its broadest sense, i.e. *Organon*). This general idea, in its natural philosophy version, is rooted in the Biblical worldview, according to which the Universe arose as a result of realization of the unified plan of Divine creation, and none other that Linnaeus wrote that *"Natura est lex Dei"* (see [38]). In the positivist version of the classical science, emphasis is made not on the unity of the Universe origin, but just on the method of its cognition; it is acknowledged that "the world is simple and allows as a simple description" following some unified protocol (R. Carnap). This general position is known as the onto-epistemological monism.

With respect to the species issues, monistic position, in its extreme form, is expressed in the recognition of the species as a universal unit of organization of the living matter, which existence does not require any proof [2]. Accordingly, there can be only one "true" species concept (or theory) describing (and eventually explaining) this universal phenomenon by means of some universal theory. In a more moderate version, which recognizes validity of different concepts, it means a possibility to elaborate finally an "ideal" [39], or a "primary" [40-42], or a "universal" [43] species concept, in relation to which other concepts, though locally true, have a subordinate (secondary, derivative) status. But it turns out that different philosophical backgrounds leads to different understanding of which exactly species concept (theory) should be considered as "primary". An emphasis on ontology leads to aspiration for as broad as possible biologically meaningful definition of *the* species. An emphasis on epistemology presumes search of as wide as possible operational theory-neutral concept. So, in some broadest perspective, any such candidates for a "universal" species concept provide just some partial decisions of the overall species problem.

The non-classical scientific paradigm is based on acknowledging complexity of both the Universe and of any of its components (fragments, aspects, levels, etc.), which are endowed with some emergent properties and are ontologically irreducible to each other. This means a fundamental impossibility of any kind of "universal theory of everything"; instead, different components (fragments, aspects, etc.) of the Universe are described by different partial theories that do not compete with each other but are complementary [44]. A part of non-classical paradigm is the modern conceptualism, according to which no empirical knowledge can exist out of the context configured by an informal (content-wise) theory of certain level of generality. The same is thought to be true for the method; a unified "Organon" (except for the comparative method in its most general sense) is impossible, various mutually irreducible components (fragments, aspects, etc.) of the Universe are described by particular methods satisfying conditions of the relevant informal theories. Of essential importance is recognition of irremovable presence of an "observer" in the cognitive situation; it is the cognizing subject that chooses somehow what and how exactly should be investigated in the Universe. This means fundamental impossibility of any kind of "absolutely objective" knowledge. From this it follows the onto-epistemological pluralism, with respect to the species issues meaning the following.

It is acknowledged, as an initial condition for analysis of the species problem, that (a) the biota is objectively structured in multi-faceted and multi-level ways, (b) one of manifestations of this structuredness is the subsistence of certain structural units, and (c) one of these units is what is usually called *the* species. Further, it is recognized that, just like the biota itself, the "species in Nature" understood in such a very general sense is by itself a complex and multi-faceted phenomenon. Recognition of this "species unit" in its whatever manifestations at the theoretical level is based on an informal (biologically meaningful) theory, which provides some general criteria of what is *the* species as a natural phenomenon. Therefore, any kind of theorizing about *the* species involves, by necessity, explicit fixation of some biologically meaningful context within which this natural phenomenon with its properties (manifestations) should be considered. Different mutually irreducible manifestations of *the* species are reflected in different species concepts which describe it in various ways and thus are complementary to each other. Together, they constitute a kind of general conceptual space as an "existential domain" of the species problem as a theoretical construct (see 3.2). It is also acknowledged that any empirical species concepts (in particular, those based on the similarity as such) are biologically sound only if they are correlated with certain biologically meaningful (evolutionary, or ecological, or else) theoretical concept. And, at last, no empirical identification of *a* particular "species in Nature" is possible without the above informal concept defining *the* species at theoretical level, as it is just the meaningful theory that indicates to a researcher what and how to "see" (to research) in the Nature (A. Einstein).

### 3.2. Three-partitioned cognitive situation for the species problem

Cognitive situation [37] is, in general, three-partitioned; it includes objective (ontological), epistemic and subjective components. The first component defines *what* to study, the second defines *how* to study, and the third defines *who* studies. In the framework of classical and non-classical paradigms, interrelations between these components are interpreted in significantly different ways.

In the classical science seeking for an "absolutely objective" knowledge by an "absolutely objective" method, the mutual influence of the above three components is though to be minimized. With this, the learning subject is "excluded" from the cognitive situation in order to eliminate its influence on the results of the learning, so the entire situation is sup-posed to be two-partitioned, consisting of non-interrelated ontological and epistemic components.

In the non-classical science, an irremovable presence and interaction of all the above three components of cognitive situation is acknowledged, which means the following. The objec-tive component forming ontological basis of the species problem is construed taking into ac-count certain epistemological conditions (e.g. observability). Epistemic component, as a set of principles and standards of studying the species issues, is formed, on the one hand, by a subject of the cognitive activity and, on the other hand, should be adequate to the ontology of the object (e.g. to its probabilistic nature). Subjective component in its most general sense embraces the entire spectrum of the learning subject ranging from particular scholars to sci-entific communities formed around particular scientific paradigms (research programs). It is the subject that captures, in some or other way, certain aspect of the biotic structure, in which context it becomes meaningful to consider *the* species (in its general sense) as an ele-ment of that structure. This "capturing" is a kind of cognitive act that makes it possible to identify *the* species in the cognitive situation as something liable to a theoretical comprehen-sion and empirical identification. And it is the learning subject that, after all, decides how to define and to study that structure.

Each of these components exists in the cognitive situation by means of various concepts, def-initions and occasionally personal ideas fixing them some or other way. This means that each cognitive situation involves a kind of "conceptualizing the world" [45] and therefore is associated with certain "conceptual space" [46], outside of which it does not exist. Such a "space" should be outlined as explicitly as possible; as a matter of fact, if some phenomenon is not reflected in concepts and definitions (or at least does not appear as a part of personal knowledge), then it is absent in the cognitive situation and cannot be reasonably investigat-ed. One of such conceptual spaces is built around the species notion and eventually the spe-cies problem. This space can generally be regarded as three-dimensional; its "cognitive axes" correspond to the above three components of the cognitive situation. Such an under-standing of the latter allows to consider every partial species concept as a local area (sub-space) in that conceptual space, so its content can be properly and fully determined only by its projecting onto all three axes of that space. In particular, the latter means that, say, evolu-tionary species should be apprehended not in an "absolute" sense as something uncondi-tionally existing in the Nature but as a particular aspect of the biota's structure recognized by a particular research community based on a particular theoretical concept.

With this way of considering particular species concepts, it is to be taken into account that they can be "loaded" with each of the components in a different degree; or, in other words, they can be projected onto corresponding axes of the conceptual space in different ways. In this regard, it is important to emphasize that these axes, although intercorrelated because of interaction of respective components of the cognitive situation, can be considered as "or-thogonal" in some utmost sense. Therefore, the species concepts, to the extent that they are "loaded" with (projected onto) basically different axes, may have substantially different cog-nitive meaning, with some of them being primarily ontological (e.g. phylogenetic) while others being primarily epistemological (e.g. phenetic). Such way of viewing of the overall conceptual space allows to stress that only the species concepts basically "loaded" with (projected onto) the same "cognitive axis" may be considered as the items of the "same kind", and thus may compete with each other (for instance, evolutionary and phylogenetic concepts). Contrary to this, species concepts basically "loaded" with (projected onto) different "cognitive axes" are not of the "same kind" and cannot compete directly in the given conceptual space; the instances are theory-burden phylogenetic and theory-neutral phenetic concepts. What compete actually under such a circumstance are not particular species concept but respective "cognitive axes" which are given more or less significance within the frameworks of particular natural science philosophies.

Further structuring of the overall conceptual space of the species problem is an important issue involving each of its "cognitive axes". Thus, the object (ontological) axis includes, for instance, ecological and phylogenetic aspects of subsistence of the "species in Nature"; or its phenomenological (e.g. genealogy) and causal (e.g. reproductive mechanisms) aspects. The epistemological axis includes, for instance, logical or mathematical foundations of the researches concerning the species subject. At last, the subject axis includes personal (intuitive) or "collective" (paradigmal) attitude to the "species in Nature". All this has a significant relevance to consideration of certain conditions of comparability and "competibility" of the species concepts considered elsewhere (see 5.2).

### 3.3. Species concept as an onto-epistemological model

In considering structure of the cognitive situation of the species problem, it is fundamentally important to understand that its objective (ontological) component encompasses not infinite objective reality (the Universe itself), but its finite model (representation) suitable for its handling as a theoretical construct. This model is given in a form of fixed concepts and definitions, it emerges as a result of some reduction operation, which is based on certain ideas of what is essential and what is not for analysis of the species problem. First, the biota is "extracted" from the Universe by breaking off some of its relationship with other components of the Universe irrelevant to representation of the biota in terms of its own structure. Then some structural units of the biota are singled out, one of which is designated as the species. When considering these items, only those characteristics of the biotic structure become evidently included that are deemed relevant to the species problem. This sequential operation of reduction is resulted in an onto-epistemological "species model" as a part of the objective component of cognitive situation of the species problem.

Each such "species model" is a biologically meaningful theoretical construct, which in more conventional terms is usually called the "species concept". It provides an item that could be properly denoted as the "species in theory". As it can be seen from the foregoing, the latter exists in the form of certain verbal definitions, which allow to distinguish the species from other units in the biotic structure (e.g. macro-monophyletic groups, ecomorphs, discrete age and sex groups, etc.). The combination of these definitions, as noted above, outlines the conceptual space of the species problem, and each onto-epistemological species model (concept) can be regarded as a local area of that space. In the terms adopted here, the less reducing is a species model, the greater part of conceptual space is occupied by the respective area.

In Max Weber's terms (see [47]), such an ontological species model can be interpreted as an "ideal type" that fixes essential properties of what is perceived by a researcher as the "species in Nature" being an objective natural phenomenon. Various properties are regarded as essential or nonessential under some biologically meaningful theory, which defines simultaneously (a) particular consideration aspect of the biotic structure in general and (b) the candidates "species in Nature" in particular. It is such a theory that gives a reduction basis resulted into a particular ontological "species model" (this issue is considered in some detail in one of the following sections, see 4.2). It is clear that the more reducing a model is, i.e. the more supposedly "nonessential" properties are dropped in its design, the more distant it is from the "species in Nature" being modeled, so the poorer and the more partial is it in its content. For instance, the genealogical species model is more reducing and less meaningful than the evolutionary one; there is "less" of the "species in Nature" in the former than in the latter.

It is clear that the ontological models are not the only possible. Epistemological models (concepts) figure along with them, which are construed with a minimum appeal to the objective component of cognitive situation. These include various types of operational concepts aimed at developing methods for identifying and describing some structural units by tradition called the species. But, from the conceptualism standpoint, such models and respective units they allow to recognize are biologically "empty" without reference to any and meaningful theory therefore cannot be related directly to the "species in Nature". It is possible to talk also about "subjective models" as manifestations of personal knowledge, i.e. of scientists' intuitive images about how the biota is structured at the species level.

It should be emphasized that degree of reduction of the ontological species model (concept) depends on degree of "meddling" of a subject (researcher) into the cognitive situation. As it was pointed out above, it is the subject that decides, which of the relations of the "species in Nature" with its "Umgebung" are to be omitted in order to make the "species in theory" meeting certain epistemological criteria, for example, to make it more operational. It is seen from this that the more reducing the ontological species model (concept) is due to its operationalization, the less of objective and more of subjective components is embedded in it. From this viewpoint, for example, definition of the species as a phylogroup is more "subjective" (in the sense just indicated) than its evolutionary definition. At best, such reducing models can be more appropriate, under conditions of operationalism, as "intersubjective" (in the sense of Popper), which does not indispensably implies they are more "objective".

There can be quite a lot of ways of reducing cognitively infinite Universe to particular ontological biota models and of further reducing the latter to some finite ontological species models. The potential number of such reducing models are just as many as informal theories of the biotic structure can be elaborated to infer essential criteria for construing the species models (presumably, they are not infinitely numerous). Any such finite "species in theory", as noted above, is necessarily a reducing partial representation of cognitively inexhaustible multidimensional "species in Nature". This means that certain natural phenomenon denoted by the "species" notion, in its most general understanding, may be represented by a number of partial ontological species models (concepts). This serves as a prerequisite for the "species pluralism" from the very beginning of construing the species problem at the onto-logical level.

In a more general and a more formal sense, each of the theories serving as reducing base for elaborating particular species models (concepts) can be considered as a "possible world" in sense of Kripke. Each of these worlds is defined by a variable (or a set of variables), which are treated as most significant for understanding and defining *the* species, be they genealogi-cal, ecological, ethological or any other possible consideration. This formalism might be of use from a semantic standpoint in considering definitions and naming different "kinds of species" (see 5.2). Besides, from a more practical viewpoint, it allows to distinguish, in some informal way, "good" and "bad" species, with the former being uniformly recognized in dif-ferent "possible worlds" defined by different variables [10].

In the analysis of objective component of the cognitive situation within the non-classical sci-entific paradigm, one of theoretically meaningful issues in the species problem becomes the determination of not competitive relations between the onto-epistemological species models (concepts) but the conditions of their mutual interpretability, i.e. of translation of statements of one concept into those of another with minimal loss of information. Obviously, the great-er is overlap of the areas in the general conceptual space corresponding to different species models (concepts), the more they are mutually interpretable. This standpoint makes cogni-tive situation of the species problem more clearly structured and allows a more accurate solving of practical tasks of comparison of particular species classifications based on differ-ent onto-epistemological models (concepts).

### 3.4. Species as "one of the many"

In the classical tradition, the species is considered *a priori* as a basic unit of the Natural System (see 2.2). This tradition is continued by the modern concept of biodiversity, ac-cording to which the species is the latter's basic unit [48]. But if the Natural System had a naturphilosophical status of the universal "law of Nature", in which the species took a unique place (see "Philosophy of Botany" of Linnaeus), the biodiversity is merely an epi-phenomenon of some fundamental property of the biota, namely of its structure. I be-lieve that, in modern biology, it is the biotic structure, and not some Natural System of naturphilosophy, that should be represented by certain informal model in the cognitive situation of the species problem. The implications of this substitution is that this struc-ture is not only multi-level, but also multifold, with the species can be seen as just "one of the many" units of this structure [23].

The currently dominating paradigm of biodiversity (or rather, of the biotic structure) im-plies that the latter is subdivided into two internested hierarchies, phylogenetic and ecologi-cal [49]. At the same time it is presumed that they are obviously not completely independent of each other but are, as a matter of fact, just mutually irreducible aspects of the single struc-tured biota.

Phylogenetic aspects of the biotic structure corresponds to the multi-level phylogenetic pattern in which the species is "one of the many" monophyletic groups of different levels of generality. This viewpoint was anticipated by those biologists of the 19th century who rejected fundamental status of the species as a unit of either classification or evolution (see 2.2), this idea is currently reflected in designation of the species, according to the phylogenetic species concept, as a *phylo*species or *clado*species or just as a phylogroup [50-52].

Ecological aspect of the biotic structure corresponds to the hierarchy of ecosystems, with its own basic structural units (elements). Within this general conception, it is possible to fix *eco*specis at some level of ecological hierarchy defined by its position in the niche structure of local communities [53-55]. However, there is another approach do describing community structure, which basic unit is the ecomorph, i.e. an array of organisms characterized by unity of ecological and morphological characters, irrespective of their phylogenetic history [56-57]. These ecomorphs may, for example, be age stages in organisms with "discrete" ontogeny (like larvae and imagoes in insects with complete metamorphosis), or gender groups performing different functions in the ecosystems (like mosquito's males and females), or occasionally castes in the social insects. In the terms of ecological structure, all these units are equivalent in the sense they take some comparable fixed positions in the hierarchy of ecosystems. In this perspective, the species in its "local" interpretation (as "non-dimensional species" of Mayr) is just "one of the many" of such ecomorphs. Indeed, it presumably does not matter for some waterfowl community, if respective ecological niches are occupied by different species of aquatic and terrestrial predatory insects or by larval and imago stages of the same dragonfly species.

The above consideration allows to emphasize that the species as a unit of the biotic structure is not an *a priori* given "basic" natural phenomenon, which is obligatory "the same" (in a sense) in all hierarchies of the biotic structure. It is just one of several manifestations (aspects) of that structure, so it is not *the* "species" but *a* "species unit", which is fixed somehow by a subject of the cognitive situation based on some ontological model (theory) of the biota. The latter model includes, as its part, indication of certain essential characteristics and parameters (structural, functional, temporal, etc.) that allow to fix certain units of the biotic structure (biodiversity), among which there might be the "species unit" in question. It is evident that various ontological models fitting certain research programs may presume various ways of fixation of the latter unit. In one case, it will be a phylospecies, in another — ecospecies, in the third — biospecies, etc. Taking into account the above ideas of the conceptual space, these units coincide to the extent that the parameters of the "species models" fixing them overlap in that space.

Such a theoretical (cognitive) determination of the ways of fixation of the "species units" of the biotic structure leads to a conclusion that the aforementioned "species pluralism" (see 3.1) is actually unavoidable. Moreover, its inevitable extension (hopefully asymptotic) can be assumed because of supposed progressive complication of the concepts of the biotic structure including causes and principles of its organization, functioning and evolution.

### 3.5. In what senses are the species "real"?

Within an intersection of the ontological and epistemological components of cognitive situation of the species problem (see 3.2), theoretical issues concerning the species "modes of being" are most important. One of these involves ontology of the "species units", which consideration is based on certain epistemic criteria of the species reality.

Approaches to solve this issue—or rather this problem, because it does not have any unique trivial solution—has being been discussed in a great amount of literature since the neo-Platonists (see 2.2). Previously, it most often was considered in the context of the classical scientific philosophical paradigm, according to which the species are either "real" in the sense that they exist objectively in the Nature (position of realism), or "unreal" being just outputs of some cognitive activity (position of nominalism).

Within the non-classical onto-epistemology, which important part is the contemporary conceptualism (see 3.1), diversity of the very "reality" is acknowledged; by this, I mean not the above S. Kripke's plural "possible worlds", but the "three worlds" in sense of K. Popper [58]. According to the latter, the "first world" corresponds to the objective reality, this is what exists "in fact" outside an observer. The "second world" corresponds to the subjective reality in consciousness (and unconsciousness) of a researcher, which is composed of subjective images reflecting what exists (or occasionally does not) "in fact". The "third world" (or a substantial part of it) corresponds to the theoretical reality, that is to the conceptual space in which the species problem is considered.

It is evident that those "three worlds" of Popper correspond to a degree to the three basic components of cognitive situation or, what is almost the same, to the "axes" of the conceptual space outlined above (see 3.1). From this it follows that the issue of the species reality as a part of the respective problem gains a particular emphasis; the question of whether the species is real or not should be raised with taking into account existence of those different realities. So this question becomes complete if only certain "cognitive axis" is indicated, as well. The "species in Nature" possess a reality which is used to be denoted as an objective. The "species in theory" is also "real", but its reality is different, it is that of a theoretical construct within the overall conceptual space. To a researcher, his/her own ideas of the species are part of his/her mental subjective reality, so it is also "real" in a peculiar manner. Thus, all these "species" existing in different Popperian "worlds", are obviously "real" in their own ways, though their realities are of essentially different ontology—and this is another aspect of the "species pluralism". With this perspective of considering species "realities", one of the key issues is to establish a correspondence between all of them.

In this regard, the "species in classification" deserves close attention. Classification can be considered as a model (representation) of some aspect of the structure of biological diversity, so it can be attributed with some reservations to the "third world" of Popper. But this is not a theoretical reality in its strict sense; rather, the "species in classification" is a judgment (hypothesis) about the "species in Nature" put forward on the basis of some data at hands within the scientific context provided by particular "species in theory". Thus, the "species in classification" is a kind of connecting link between all three "species realities" allowing to set a required correspondence between them.

## 3.6. Cognitive styles

The subjective component of cognitive situation is multidimensional and multilevel, like its other basic components. In referring to it in the non-classical theory of science, attention is most often paid to division of overall scientific community into research schools adhered to particular paradigms (research programs). This implies a particular theoretical interpretation of empirical data by members of this community according to a particular theoretical construct underlying respective paradigm (research program). This is, that is to say, an "apparent" non-personal manifestation of the subjective component. Relevance of this "paradigm effect" to the present issue is quite obvious; every sufficiently general species concept (biological, phenetic, phylogenetic, etc.) serves as a core for the formation of a particular paradigm (or is a part of respective research program). Therefore, this level of organization of the subjective component is considered in a lot of publications and so is hardly worth being discussed here any longer.

Much less attention is drawn to a lower level of the subjective component corresponding to the individual cognitive styles underlying researchers' personal (tacit) knowledge [59, 60]. These styles are responsible for forming an array of the Popperian "third worlds". Cognitive (thinking) styles are implied by researchers' way of perception of the world, they are diverse and multifaceted, can be ordered (in the simplest case) in pairs of opposites [61]. Examples include researchers' inclination for holistic or reduction vision of the whole biota and any of its structural elements, for intuitive or rational way of knowledge, etc. A pair of opposites "typological *vs.* population" thinking styles is known to be quite relevant to the species problem [62-63].

In the framework of contrasting classical and non-classical scientific paradigms (see 3.1), of special significance is the pair of "discrete *vs.* fuzzy" thinking, which corresponds evidently to the dichotomy of "discrete *vs.* fuzzy" logic [64]. The principal meaning of fuzzy thinking is that it frees a researcher from having to look for the sharp edges where they cannot in principle be drawn. "Splitting" phyletic lineages into fragments corresponding to the "vertical species" of paleontologists is an example of situations where such a thinking style is more than relevant. Another typical example is the interspecies hybridization; if it is not widespread in nature and not absorptive, it does not preclude recognition of the species status of respective units. In both these cases, the species are treated as "fuzzy" entities, contrary to the provisions of "xenotaxonomy" (in sense of [65]). This "fuzzy" term was suggested for a particular case of prokaryote species [66- 67], but it certainly deserves more wide treatment just outlined [10, 68]. Finally, this style allows to see not so dramatically the entire situation with the "species pluralism"; at least some of the ontological species models (concepts) are not exclusive but overlap and complement each others due to their having certain conceptual constructs in common, so, in a sense, these concepts are not "discrete" but "fuzzy".

## 4. Defining species: Conceptual pyramid

Any sufficiently advanced theoretical construct (theory, concept, etc.) is organized in a conceptual pyramid, which is caused by certain reason of logical nature.

According to the classical theory of definitions [69], each notion can be sufficiently strictly defined only within the above mentioned logical genus-species scheme (see 2.2). This means that (a) each particular notion must be related as a "logical species" to a more general notion as its "logical genus" and (b) within the latter, several "logical species" should be distinguished as the latter's partial notions, so that any each of them can be properly defined only with reference to its counterparts within the same "logical genus". Therefore, in order to define *the* species as a natural phenomenon, it is necessary to define, first, that natural phenomenon which notion can be considered as a "logical genus" for the biologically meaningful species notion and, second, those natural phenomena, which notions can serve, along with the species notion, as different "logical species" within the given "logical genus" properly defined.

Similar though less formal hierarchical scheme of definitions is implied by well-known Gödel's incompleteness theorem. Elaborated initially as purely mathematical, in its more general epistemological interpretation [70-71] it affirms that any theory (concept) cannot be exhaustively defined in the terms of the language of this theory (concept) itself. For such a definition to be properly construed, a kind of meta-language is required, which belongs to a theory (concept) of higher level of generality ("logical genus"), with respect to which the given notion is its partial interpretation ("logical species").

All the above has a direct bearing on the analysis of logical structure and content of the species problem. First of all, both argumentation schemes imply that species concepts should be arranged in a kind of "conceptual pyramid" of various levels of generality, with the most general concepts belonging to the "tip" of the pyramid and the least general ones being placed at its base. "Pyramidal" shape of the resulting structure is due to the fact that, at each level of generality, partial concepts are evidently more numerous than more general (inclusive) ones. Next, each species concept of lower generality level gains its substantiation only within the context provided by the concept of higher generality level. At last, what is quite important, such a "pyramidal" construction of the entire species problem means that within the species concept(s) proper, even of the highest generality level, the very notion of species cannot be well defined.

### 4.1. Pyramid(s) of the species concepts

There more than 20 species concepts are currently recognized [16, 17, 41, 72-74]; as it was pointed out (see 2.1), such a multiplicity is one of the core aspects of the species problem. Each of these concepts provides its own species definition (although not quite strict in most cases), based on a particular understanding of what are essential properties constituting the key parameter of the "specieshood".

Several classifications of the species concepts and definitions of different levels of generality were elaborated for ordering such a multiplicity of concepts. The latter are grouped in each of these classifications according to the parameters that are taken as the most important by respective authors for ordering the concepts. This appeared to be resulted in several hierarchical arrangements of the species concepts, with their amount reflecting number of the bases (ordering parameters), which can be fixed for classifying those concepts. This gives rise to a peculiar aspect of the species problem, now it is not diversity of the concepts proper, but of their classifications.

In one of the earlier versions of such conceptual pyramids, recognition of "primary" and "secondary" species concepts was proposed [40-41]. This implies that the primary concepts include more characteristics of the species than the secondary, so the former are more general and less in number while the latter are their partial interpretation and thus are more numerous. In a sense, this idea is similar to that of Gilmour [75] who suggested to recognize "general purpose" (primary) and "special purpose" (secondary) classifications. In the just mentioned Mayden's [41] classification, evolutionary species concept is referred to as the primary, because it actually is one of the most inclusive in its content. However, a systemic consideration of the species [9, 10, 76], though not explicitly formulated as a concept (see below), provides its even more general treatment, so it is the latter that can claim to be the primary, indeed, for this particular conceptual pyramid.

In another, more general approach to elaborating classifications of such kind, one of the most important grounds giving fundamentally different conceptual pyramids, I believe, might be consideration of the species in accordance to the ways they are considered within the conceptual space.

One such classification presumes distinguishing among concepts corresponding to either ontological or epistemological considerations of the species [77]. As it was mentioned above (see 3.2), they can be considered as different "projections" of the general species concept onto different "axes" of the conceptual space, so they may be considered as equivalent in this respect. The former are theoretically laden and give an idea of *what* is the species as a natural phenomenon (evolutionary, genealogical, reproductive, etc.). The latter are theory-neutral and indicate *how* to distinguish particular species whatever might be their theoretical foundations (operational taxonomic unit, minimal recognizable unit, etc.). However, from the conceptualism standpoint addressed to ontology (see 3.1), such a hierarchy cannot be considered as well established, because, in biology as a natural history science, formal operational concepts cannot function as sound scientific constructs outside the context given by biologically meaningful informal concepts. Attributing them an equal status (rank in the conceptual pyramid) yields a biased view of the entire species problem as it implies substitution of theoretical issues about meaningful species definitions by elaborating facilities for practical species identifications [10, 33, 39, 41, 74, 78]. Within the above hierarchy of the "primary" and "secondary" species concepts, operational ones are nothing more than "tertiary" ones belonging to the lowest level of the conceptual pyramid.

Close to the previous one by its meaning is a division of the species concepts reflecting their belonging to the "first" and the "third" worlds of Popper (see 3.4), which are the "species in

Nature" and the "species in theory" (or maybe the "species in classification"). Proponents of this division offer to use the term "species" to designate a unit of taxonomic classification, while natural units (populations) are to be denoted by some different terms [11, 79-81]; this idea goes back to Aristotle, see 2.3).

Another type of classification of the species concepts by general onto-epistemological criterion is a hotly debated interpretation of the species (in general sense) as a class, or as a cluster, or as a historical group, or as a individual (see 5.2). Such a classification by its content may be, with some reservations, considered as not actually biological but rather philosophical [39].

In the classification of species concepts elaborated on the basis of biologically meaningful criteria, a distinguish is made between diachronic and synchronic or, which is nearly the same, between historical and structural groups of concepts [82]. The former are evolutionary concepts, including the phylogenetic one, while the latter include, for example, typological and reproductive (genetic) concept. Recognition of structural and processual concepts [13] is close to this categorization; to them I would add a functional (ecological) group of concepts.

Some classification can be elaborated on the basis of what is taken as the principal parameter of the "specieshood" to be used for a theoretical species definition; this gives the following principal groups of the species concepts [17].

• the species as a similarity-based commonality unites such concepts as typological, phenetic, genetic, all presuming sharing particular traits by the species; also commonality of ontogenetic processes shared by conspecifics [83] and homeostatic property cluster concept [84] can be mentioned here;

• the species as a reproductive commonality summarizes generational and biological (in the narrow sense, i.e. "reproductive") concepts; fitting this category is also recognition concept [85-86], where emphasis is made not on the isolation, but on the integration, the latter gives the cohesion concept [87];

• the species as a historic commonality, these are phylogenetic, or genealogical concepts in both general and various partial interpretations;

• the species as a evolutionary commonality of both historical origin and peculiar "evolutionary role" of conspecifics;

• the species as a particular ecological commonality according to the ecospecies concept, or to the functional concept of Khlebosolov [88]. It is to be mentioned that biosystematics was the first to have developed a detailed hierarchy and nomenclature of ecologically treated "species units" in parallel to the taxonomic "Linnaean species" [53, 89];

• the species as a systemic unit [76] including its treatment as an element of the biota being a non-equilibrium system [9, 10, 23].

In discussion of the pyramid of the species concepts itself, one of the principal question is, whether it is possible to elaborate something like an "ideal" species concept, which would include in its definition all manifestations of the species units existing in the biological nature [10, 39, 41, 90-91]. The aforementioned evolutionary species was a suggested candidate for such a concept, as it is characterized by combination of evolutionary, genetic and occasionally ecological parameters [41, 92]. A more general definition of the species as a structural unit of the biota considered as evolving non-equilibrium system should also be mentioned in this respect. One of the promising ideas seemingly never discussed before can be an elaboration of a kind of general "framework concept" [93]; it provides a meaningful interpretation of the conceptual space and formulates biologically sound conditions, under which particular species concepts of different levels of generality can be inferred.

### 4.2. An "ultimate beginner" for the species concepts

Any of the conceptual pyramids of the species problem, in the ways of their construing considered in the previous section, remains closed on itself. However, in the terms of the above genus-species scheme supplemented with epistemologically interpreted incompleteness theorem (see 4), any kind of the "species pyramid" should be built into a concept (theory) of the next higher level of generality. The latter is designed to serve as a "logical genus" for any of "ideal" or "universal" species concepts as its partial "logical species". This provides a possibility to fix such a content-wise consideration context of the entire species problem, in which the most basic questions of the species theory (which is still absent) becomes meaningful; what is the species as a unit of the biotic structure, how it differs from other such units, why and how it emerged, and finally what (if any) is the species level of this organization.

Of the existing theories, which can serve as something like "superstructure" over the conceptual pyramid of the species problem, two have been most often being discussed for decades, evolutionary and ecological ones. In the context of the evolutionary theory, process of evolution is, rather metaphorically, represented in a form of (reduced to) branching phyletic lineages, which fragments are treated as (phylo)species. This theory sets the context for the phylogenetic species concepts. In the context of ecological theory, the (eco)species is treated as an element of the ecosystem structure; this serves as a justification for the ecological species concepts. As noted above (see 4.1), within the conceptual pyramid of the species problem proper, these two groups of concepts are thought to be generalized by the evolutionary species concept. But the latter itself remains without a more general justification. For such a justification, some higher-level ontological model (meta-model) is requested, which would treat the biota on a unified basis of both evolutionary and ecological standpoints.

Such a model would imply that the biota is a global evolving ecosystem. Within biology, a rather general theory of phylocenogenesis presumes such consideration, according to which phylogenetic development of the species units occurs within the ecosystems providing them with the diversity of ecomorphological units [94]. However, there is a more general ontological model (concept) treating the biota as a non-equilibrium system described in the terms of synergetics mentioned already in the previous section. From this perspective, any system of such kind is "doomed" to develop, and its development leads to its hierarchical structuring [95]. In the case of biota, its historical development, commonly referred to as the biological evolution, entails its structuring due to causal relationships that regulate flows of matter, energy, and information [96]. Thus, the biotic structure, with all its constituent elements (units), is an inevitable (axiomatic) consequence of historical development of the biota as a non-equilibrium system.

Meanwhile, according to this model, though presuming evolution of the biota as a whole, different categories of causes (proximate, initial, material, etc.), to the extent that they are independent and are not reducible to each other or to a single more general cause, act in a complementary manner and give rise to mutually irreducible and mutually complementary aspects (manifestations) of the overall biotic structure. Two such general aspects are being usually considered, the above mentioned ecological and phylogenetic, each with its own specific hierarchy; there are might be more of them, but these two are enough for the present issue. In each of them, their own structural elements (units) of different levels of generality are being patterned, which not only are not obliged to, but even cannot coincide, as they are generated by the discordant causes.

An important part of the structuring of the evolving biota is appearance of (quasi)discrete elements (units) of certain (not exactly fixed) levels of generality. One of these are of higher levels (such as local ecosystems or monophyla), others are of lower levels (such as eco-morphs or species). In the latter case, following the established tradition, at least some of these elements (units) can be uniformly designated as the species, though with explicit indication of the hierarchy they belong to (phylo-, eco-, etc.). At the same time, it is to be kept in one's mind that, in some approaches to describe these hierarchies, it is possible to do without the notion of species at all (see 3.3).

In this regard, again and inevitably, a fundamental question arises about what, if any, is exactly the "species in general" in its traditional meaning inherited from the classical science. To answer it within the above general causal model of the evolving biota, of primary importance becomes a task to elaborate a concept of some universal element (unit) of the biotic structure, with which it could be possible to associate actually "primary", or "ideal" species concept. The latter should probably include a reference to an area of intersection (or interaction) of general categories of causes, under which effect certain structural unit equally relevant to both (and other conceivable) hierarchies is emerged. Evolutionary species concept, not a once mentioned above, seems to fit this condition more than any other biologically meaningful concept. However, attempts to elaborate something more extensive in its content used to be resulted in so called "combinatorial" type of concept (in sense of [97]), with not definition proper, but with just a more or less long list of properties thought to be essential for the species (such as in [2]).

Such a "combinatorial" status of the general species concept seems to be due an effect of the so called "Hull principle" [10, 98-99], which means theoretical impossibility for the multidimensional "species in Nature" to be defied by a single exhaustive "formula". This principle, in its turn, is a consequence of (a) inverse relationship between strictness and richness of any natural science concept and/or notion and (b) the uncertainty relation between mutually irreducible species characteristics presumed by the principle of subsidiarity. With the "Hull principle" in effect, the above mentioned framework concept (see 4.1), and not a definition, might be a candidate for such a desired theoretical construct. It would allow to fix and to investigate certain pattern of structural organization of the biota at the level of generality attributed traditionally to the species.

In the cognitive situation given by the biota's ontological model just outlined above, any general definition of the species, whatever might it be, should be a final link in a downward cascade of definitions of higher levels of generality, forming their own conceptual meta-pyramid At the latter's tip, there appears such (or any other appropriate) biotic model as an "ultimate beginner" for the species concepts in general. At some lower level of this meta-pyramid, definitions of the causes of the biotic structure are fixed, then definitions of elements (units) of that structure go, and finally a definition of the species as one of these elements (units) of the biotic structure is formulated. Such a cascade of the inclusive definitions corresponds clearly to the sequential reduction of the initial basic ontological model to some particular species model (concept) (see 3.3).

It follows from the foregoing that, in a general biological theory relevant to the species issues, one of the principal notions should be not that of the species, but of a discrete element (unit) of the biotic structure. So, the "species problem" turns out to be the "biotic unit problem". As it was indicated above, such a unit (element) may be conventional species, phylogroup, ecomorph, age phase, etc., and the species in its current common understanding is just "one the many" of these elements. Accordingly, the tip of the conceptual pyramid of the species problem proper should be not any "ideal" species concept proper, but rather the general concept of the unit (element) of the biotic structure, a particular case of which is the species concept being sought.

It is evident that such a biotic model, whatever general might it be, should have its own meaningful foundation, which means that it itself should be built into a higher-level pyramid, in which the model in question becomes a "logical species" of some "logical genus". So the point is that an "ultimate beginner" at far higher level of generality is needed for substantiation of the very biotic model. This obviously extends the cognitive situation of the entire species problem beyond the biological issues.

Remaining within the framework of the above synergetic model, it might be reasonable, in order to substantiate a possibility to treat the biota as a particular kind of the non-equilibrium system, to look at some other versions of the latter to analyze how they are being structured and if there is something in common to all them that might somehow correspond to the species in its general biological understanding. System of scientific knowledge may serve as another instance of such kind of non-equilibrium systems, which development, according to evolutionary epistemology, can be liken to the biological evolution [100]. From this perspective, particular scientific ideas and concepts can be considered as particular "species" or some "species-like entities" that are born, live and extinct just like the biological species [101-102]. One cannot exclude that such an expanded way to consider the species problem would allow to formulate it more correctly for the biological science. In this context, of certain meaning could be an idea of the "ontological species" [11, 103] as a manifestation of the same type of organization of such systems, regardless of their particular natural, cognitive, or any other status.

## 5. Evolving specieshood

It was noted above that, within a given cognitive situation, designation of any natural phenomenon by a single notion implies that it is endowed with certain fundamental property that is preserved in all its appearances and thus distinguishes it from other natural phenomena of the same kind. In the classical terminology, such a property is routinely designated as the essence; as to the species in its most general sense, its essence was suggested to denote as the "specieshood" [9, 10, 99, 104]. By an initial assumption, it is the latter that makes the species what it is by its "nature", distinguishes it from other units of the biota's structure, and marks eventually the species level of organization of the living matter, i.e. defines the "species as a rank".

From this, it is evident that one of the key issues in the species problem is that about the "specieshood", namely, about that possible specific quality, which makes any species *the* species and distinguishes the latter from other units of the structured biota. The main part of this issue is, whether a fixed level can be found in the hierarchical structure of the biota that would correlate quite strongly with the "specieshood".

Addressing to the essence (essential characteristics) of the species as a natural phenomenon obviously involves the species problem in what is called the "modern essentialism". I do not intend to discuss here this very sophisticated matter; I would rather note only that, if the "species is Nature" is not supposed to be just an arithmetic sum of its constituent organisms, but is indeed a natural phenomenon endowed with some emergent properties, then it is quite normal to speak of its essence [105-106].

The main objections against essentialist interpretation of the species, within the biological consideration of the species problem (i.e. leaving aside philosophical arguments for and against essentialism), are as following: the species (a) evolve and (b) are organized in different ways. This contradicts an initial assumption of the classical essentialism, according to which essences should be permanent and universal for particular commonalities (such as "natural kinds"). The latter point corresponds evidently to the stationary world picture originated from the classical (Platonic) natural philosophy. However, within the contemporary global evolutionism, a fundamentally different interpretation of essentialism is rather admissible, allowing for a possibility of evolutionary changes of the essences themselves [107].

### 5.1. Evolution of *the* species

The above (see 4.2) synergetic model of the biota as an evolutionary non-equilibrium system may be taken as a background of a concept of the evolving specieshood. According to this model, life on Earth had historically originated and then was gradually developing; this is a kind of the "central dogma" of the whole modern evolutionism. This development implied gradual structuring of the biota, including before all formation of ecosystems and their complication by means of structuring the flows of matter, energy and information.

A part of this gradual structuring was as gradual formation and perfection of units of the biotic structure involved in the regulation of these main flows. Partly repeating (see 4.2), it should be emphasized that these flows are patterned by different categories of causality, and to the extent that the latter generate and arrange these flows more or less independently, structural units of the ecosystems are formed within each flow more or less independently from each other, as well. Using the current terminology, one can assume that structuring of the ecological component includes formation and specialization of ecomorphological units (ecomorphs), while structuring of the phylogenetic components includes formation and specialization (differentiation) of the phylogenetic units (let they be termed species).

This idealized model presumes that the "ecomorph way" of organizing the biota was being formed along the formation of ecosystems as a mode of structuring the flows of matter and energy. Respectively, the "species way" of organizing the biota was being formed with the formation of phylogeny as a mode of structuring the information flows. The both was being formed simultaneously but due to different causes. It follows from this consideration that such a dissociation of ecological (ecomorphs) and phylogenetic (species) ways of structuring the evolving biota lead to a well known discrepancy between units of ecological and phylogenetic patterns. On the one hand, this made it principally possible for the species units to become ecomorphologically differentiated, with emergence of ecologically different "morphs" (such as age phases) within them. On the other hand, different species evolved similar ecomorphological features to fit similar ecological niches. The above mentioned phylocenogenetic theory (see 4.2) allows to connect these ways of the biota structuring in a general model; this provides a meaningful theoretical background for a metaphorical interpretation of the species as "genealogical actors" playing particular roles in the "ecological theater" accordingly to certain environmentally and historically written "scripts" [39].

As it was noted above, this ontological model is more consistent with treatment of the species as a phyletic lineage. As to the "specieshood", it can be generally understood from this perspective as basically an ability of stable reproduction of species-specific epigenetic systems (in sense of [108]) in the course of their evolution. This reproduction is carried out through mechanisms that provide (a) certain closeness of the species gene pools, and (b) transfer of the genetic information with minimal distortion from generation to generation [109-110]. Cleavage of these gene pools (speciation) leading to the divergent phylogeny is, from the synergetic standpoint, a consequence of structuring the biota at the ecosystem level.

This model implies the following general picture of the evolution of both *the* "species" as a natural phenomenon and the "specieshood" as its essential characteristics. First, *the* species as a unit of the biotic structure was formed not immediately, but gradually with the evolution of biota. Second, the main direction of evolution of *the* species, as a biological phenomenon, was perfection of mechanisms for maintaining the integrity and stability of this unit at the epigenetic level. Finally, these mechanisms may be different in different groups of organisms, which expose different manifestations of the specieshood.

Specifying to a degree this evolutionary scenario, one can assume the following. At the beginning of historical formation of *the* species, there were loosely organized units of the prokaryotic diversity without effective mechanisms of epigenetic stability maintenance, so they cannot be strictly distinguished as ecomorphs or species proper [111]. At the end of this evolution, there are units with highly developed mechanisms of maintenance and transfer of relatively stable integrated epigenetic systems by means of bisexual reproduction. Thus, the peak of the specieshood evolution appears to be the biospecies in its "reproductive" understanding, i.e. that of Dobzhansky—Mayr.

## 5.2. So many kinds of species...

According to the above model, two main conclusions about ontology of the "species in Nature" can be drown.

First, a general framework for consideration of the species ontology in its general sense should be the process-structuralism treating the species as a "process-system" [112]. In a more particular version, the species such understood can be considered as a more or less tightly organized "historical group" [113]. The latter amendment allows to emphasize phylogenetic parameter as one of the key characters of the specieshood.

Second, this model can serve as one of the ways of ontological justification of *the* species being endowed objectively with different kinds of ontology. Indeed, both the "species in Nature" as a unit of the biotic structure and the specieshood as its essential characteristic change with the evolution of the biota. This results in that both degree and ways of integration of the species such understood may be different due to various natural history of the particular groups of organisms. So *the* species (even in its narrow phylogenetic meaning) appears to be a heterogeneous unit, and its heterogeneity is quite objective, though at least in part it does reflect different ways of looking at nature. In the traditional terms, this heterogeneity is referred to as different "kinds of species".

From the standpoint of ontology, the least integrated historical groups may correspond to "natural kinds" with an added historical dimension [104], this case is partly fits the category of "kumatoid" [114]. The most integrated groups may correspond to the ontological category of the individual or rather the "quasi-individual" [10, 39, 99, 113, 115-120].

Going back to the core of the species problem (see 2.1), it becomes more than clear that the ontological model just presented involves recognition of the "species pluralism" as an irremovable part of that problem. It should be acknowledged as a part of the objective reality, so it seems to be reasonable not to "fight" with it but to reflect it somehow in the thesaurus of the above "eidology" (see 1). This means, among other things, that recognition of heterogeneity of the "species in Nature" requires to fix different "kinds of species" terminologically to make the above thesaurus more adequate to that reality. An example with A. Dubois' "mayron", "simpson", "kyon", etc. [121] indicates that there is a big room for a "term-creativity". But do new terms actually provide any solution? [122].

A part of this issue should be terminological separation of different stages and forms of "being" of the species unit proper. A radical response to this question is a suggestion to call "species" only those units which meet the reproductive criterion and to treat any other forms of organismal diversity at this level of generality as simply "non-species" [8, 28, 109]. A more moderate and therefore more sensible would be to use existing apt terms such as "quasi-species", "para-species" and "eu-species" [10, 13, 109, 123-124] to refer to different stages and results of evolution of the specieshood.

Another part of the same issue is clarification of conditions of correct comparison of different species concept. For instance, routine direct contrasting phylogenetic (genealogical) and biological (reproductive, genetic) species concepts seems to be incorrect, because they are relevant not to the same but to clearly different aspects of the specieshood. Using the formalized terminology introduced above (see 3.2), they correspond to projection of the same notion of *the* species to different "sub-axes" of ontological axis of the conceptual space. Indeed, the phylogenetic species concept considers respective unite from the phenomenological point of view, fixing its place in the sequence of phylogenetic events. Unlike this, reproductive concept considers the species from the causal point of view, pointing to a specific mechanism that maintains integrity and isolation of the species. Therefore, they cannot directly compete in the same conceptual space. In order to eliminate this confusion, it seems reasonable to fix the term "biological" for the "species in Nature" of any living organisms in its general (mostly evolutionary) sense, and to use a special term for the species outlined by the Dobzhansky—Mayr's concept to refer correctly to its principal character; it might be *co*species, with its prefix borrowed from the *co*hesion concept of Tempelton [125].

For such a terminological fixation of various "kinds of species" to be sound, it is requested first of all to make it clear whether there actually is some fundamental unit in biotic structure, viz. the "species in general", which may be designated as a "logical generic" notion with respect to the "logical species" notions of the different "kinds of species". If supposedly there does not exist such a unit proved conclusively to be the same for the various aspects of the biotic structure (ecological, phylogenetic and occasionally any other), then perhaps it is unjustified to use a single rooted notion of the "species" (or "specion" of Duboi [121]), albeit with different prefixes.

To put this question a little bit more formally, it can be considered as a matter of semantics of the term "species". It seems to be clear that, for the latter to be really a rigid designator, as Ereshefsky [73] supposes, its denominator (referent) should be defined as strictly as possible. In this particular case, "strictly" means "narrowly"; accordingly to the terms adopted here, the subspace occupied by the species notion within the general conceptual space (see 3.2) should be restricted to a certain fixed meaning minimizing its different treatments. Otherwise, the term "species" will remain a non-rigid designator distinguishing different entities in each of possible worlds construed by either phylogenetic or ecological or ethological or else variables.

In such a case, the species notion should probably be restricted to the phylogenetic (genealogical, generational) understanding of the species. Accordingly, for the units recognized in the ecological hierarchy at the level of generality comparable to that of the species, it is possible to use such a term as "ecomorph" or any other proposed, say, in the framework of biosystematics (see 4.1).

As for the "species in classification" belonging to the "third world" of Popper (see 3.4), it makes sense to use the term "*taxo*species" to designate it. This allows to fix terminologically that single level of common structure of biota, which refers to different partial manifestations of a hypothetical "species in Nature" of the same, though also hypothetical, "species rank".

## 6. What if not the species?

One aspect of the overall species problem is the strong embeddedness of the species notion in the thesaurus of many fundamental and applied biological disciplines. This seems to prevent any actually radical solution of that problem presuming rejection of the species notion (as suggested in [126]), because it would lead to a substantial reorganization of the conceptual apparatus at the expense of that rejection. The reason is quite obvious; such a rejection entails necessarily rejection (or replacement) of other terms associated in one or another way with the species notion.

For example, replacing species by phylogroup should entail in an obvious way replacing of speciation by some other, such as phyliation [127]. Generally speaking, there is nothing critically wrong with such a change in case of strictly phylogenetic interpretation of biological evolution. However, it is not evident that other biological disciplines taking the latter in a more extended sense will enjoy abandon the concept of speciation in their descriptions of historical changes of the biological objects studied by them.

In ecology, as noted above (see 3.3), the species notion is not obligatory for description of the structure of local ecosystems, it is enough to deal with ecomorphs. However, in comparative analysis of different ecosystems, there is an evident need for some basic units of comparison that allow to relate soundly ecomorphs, recognized in each of the local ecosystems, to each other. It occurs that it is the species that fulfills currently such a function; for evolutionary ecologists, ecomorphs exist in the local ecosystems not by themselves but as manifestations of the local populations of widespread (different or same) species [128].

It seems to me that this particular aspect of the species problem is not just a consequence of conservatism of the conceptual apparatus of biology, but reflects one of the universally valid epistemological principles. According to the latter, in order to explore any differences between the objects, one must have some basis for comparison by which these objects can be considered as components of a single commonality (elements of the same set, tokens of the same natural kind, etc.) possessing some fundamental feature(s) in common. For many research tasks in biology, this basis means conspecificity, i.e. belonging of organisms, differing from each other in some way, to the same species as objectively existing natural unit possessing some unique particular manifestation of the specieshood. From this perspective, it is clear that, in order to get rid of the species notion in biology, it is necessary to introduce other basis for comparison, with substantiating such a replacement by reference to some biologically meaningful and sufficiently general theory.

## 7. In conclusion

Development of the species problem seems to be directed toward a better understanding of the following biologically meaningful questions: what is the species in its general (biological rather than formal) understanding, viz. if there is *the* species as a universal (all-embracing) unit of the biotic structure, or it has but a partial character; why and what are manifestations of this "species in general" and what are the causes of existence of both such "species in general" and its particular manifestations ("kinds of species").

It seems to me that a necessary condition for development of the species problem in such a way should be comprehension of its complexity, not allowing for any radical and simple (including purely empirical) solutions. This complexity of the problem in question reflects complexity of both the entire cognitive situation, in which this problem is explored, and that fragment of the ontological component, which corresponds to the general species concept.

One of manifestations of the species problem is arrangement of the species concepts and definitions of different levels of generality into a conceptual pyramid. Its "ultimate beginner" should be a kind of ontological model, in which a causally based conception of the species is inferred as one of the structural elements of the biota as an evolving non-equilibrium system.

The impetus for the further effective development of the species problem in the direction just pointed may be its consideration within the context of non-classical scientific paradigm. In particular, of great importance should be understanding of the cognitive situation as a conceptual space that is shaped by interaction of three components, viz. objective (ontological), epistemic and subjective [129]. Such a consideration provides eventually understanding of the species concept in its both general and partial senses as a particular cognitive construct. This will give a fresh look at the content of the entire multi-dimensional species problem, at its structure and key questions, as well as at relationships between different species concepts as "forms of being" of this problem.

## Author details

Igor Ya. Pavlinov

Zoological Museum, Moscow Lomonosov State University, Moscow, Russia

## References

[1] Zavadsky KM. [The doctrine of species]. Leningrad: Leningrad University Publ.; 1961. (in Russian)

[2] Zavadsky KM. [Species and speciation]. Leningrad: Nauka.; 1968. (in Russian)

[3] Skvortsov AK. [Principal stages of developments of the ideas of species]. Bulletin of Moscow Society of Naturalists 1967;72(5): 11-27. (in Russian)

[4] Skvortsov AK. [Problems in evolution and theoretical issues in taxonomy]. Moscow: KMK Sci. Pressl (2006). (in Russian)

[5] Stepankov NS. [Eidology: A lecture course program]. Krasnoyarsk: Krasnoyarsk State University Publ.; 2002. (in Russian)

[6] Dubois A. Phylogenetic hypotheses, taxa and nomina in zoology. Zootaxa 2008;1950: 51-86.

[7] Robson GC. The Species problem: An introduction to the study of evolutionary divergence in natural populations. Edinburgh: Oliver & Boyd; 1928.

[8] Dobzhansky T. A critique of the species concept in biology. Philosophy of Science 1935;2(3): 344-355.

[9] Pavlinov IYa. If there is the biological species, or what is the "harm" of taxonomy. Journal of General Biolofy 1992;53(5): 757-767. (in Russian, with English summary)

[10] Pavlinov IYa The species problem: Another look. In: Alimov AF, Stepanyanz SD. (ed.). Species and speciation: An analysis of new views and trends. Proceedings of Zoological Institute RAS. Add. 1.; 2009. p250-271. (in Russian, with English summary)

[11] Mahner M, Bunge M. Foundations of biophilosophy. Frankfurt: Springer Verlag; 1997.

[12] Ruse M. The species problem. In: Walter G, Lennox JG. (eds). Concepts, Theories and Rationalities in the Biological Sciences. Pittsburgh: University Pittsburgh Press; 1995. p172-193.

[13] Stamos DN The species problem. Biological species, ontology, and the metaphysics of biology. Oxford: Lexington Books; 2003.

[14] Hey J, Waples RS, Arnold ML, Butlin RK, Harrison RG. Understanding and confronting species uncertainty in biology and conservation. Trends in Ecology and Evoluiton 2003;18: 597-603.

[15] Coyne JA, Orr HA. Speciation. Massachusetts: Sinauer Associates Inc.; 2004.

[16] Wilkins JS. Species: A history of the idea. Berkeley: University California Press; 2010.

[17] Pavlinov IYa, Lyubarsky GYu. [Biological systematics: Evolution of ideas]. Moscow: KMK Sci. Press; 2011. (in Russian with English content)

[18] Zirkle C. Species before Darwin. Proceedings of the American Philosophical Society 1959;103(5): 636-644.

[19] Boethius. ["The Consolation of philosophy" and other treatises]. Moscow: Nauka; 1990. (in Russian)

[20] Volkova EV, Filyukov AI. [Philosophical issues in the species theory]. Minsk: Nauka & Teknhika Publ.; 1966. (in Russian)

[21] Panchen AL. Classification, evolution, and the nature of biology. Cambridge (UK): Cambridge University Press; 1992.

[22] Richards RA. The Species problem: A philosophical analysis. Cambridge (UK): Cambridge University Press; 2010.

[23] Pavlinov IYa. On the structure of biodiversity: some metaphysical essays. Schwartz J. (ed.). Focus on Biodiversity Research. New York: Nova Sci. Publ.; 2007. p101-114.

[24] Akutin AV. [The notion of "Nature" in Antiquity and in New Times]. Moscow: Nauka; 1988. (in Russian)

[25] Atran S. Origin of the species and genus concepts: An anthropological perspective. Journal of the History of Biology 1987;20(2): 195-279.

[26] Meyen SV, Shreider YuA. [Methodological issues in the theory of classification]. Voprosy Philosophii 1976;12: 67-79. (in Russian)

[27] Reig OA. The reality of biological species: A conceptualistic and a systematic approach. Studies in Logic Foundations of Mathematics 1982;104: 479-499.

[28] Mayr E The Growth of biological thought: Diversity, evolution, and inheritance. Cambridge (MA): Belknap Press; 1982.

[29] Bock WJ. Species: The concept, category and taxon. Journal of Zoological Systematics and Evolutionary Research 2004;42(1): 178-190.

[30] Rosenberg A. Why does the nature of species matter? Comments on Ghiselin and Mayr. Biology and Philosophy 1987;2(2): 192-197.

[31] Hey J. Genes categories and species. The evolutionary and cognitive cause of the species problem. New York: Oxford University Press; 2001.

[32] Hey J. The mind of the species problem. Trends in Ecology and Evoluiton; 2001;16(7): 326-329.

[33] de Queiroz K. Different species problems and their resolution. BioEssays 2005;27(12): 1263-1269.

[34] Loevtrup S. On species and other taxa. Cladistics 2008;3(2): 157-177.

[35] Ellis MW. The problem with the species problem. History and Philosophy of the Life Sciences 2011;33(3): 343-363.

[36] Morgun DV. [Epistemological Foundation of the Species Problem in Biology]. Moscow: Moscow State University Publ.; 2002. (in Russian)

[37] Il'in VV. [Philosophy of science]. Moscow: Nauka; 2003. (in Russian)

[38]  Breidbach O, Ghiselin M. Baroque classification: A missing chapter in the history of systematics. Annals of the History of Philosophy and Biology 2006;11: 1-30.

[39]  Hull DL. The ideal species concept - and why we can't got it. In: Claridge MF, Dawah AH, Wilson MR. (eds). Species. The units of biodiversity. London: Chapman & Hall; 1997. p357-380.

[40]  Mayr E. Species concepts and definitions. In: Mayr E. (ed.). The Species problem. A Symposium presented at the Atlanta Meeting of the American Association for the Advancement of Science, 28-29 Dec. 1955. Publ. 50. Washington (D.C.): Amer. Assoc. Advanc. Sci.; 1957. p1-22.

[41]  Mayden RL. A hierarchy of species concepts: The denouement in the saga of the species problem. In: Claridge MF, Dawah AH, Wilson MR. (eds). Species. The units of biodiversity. London: Chapman & Hall; 1997. p381-424.

[42]  Mayden RL. On biological species, species concepts and individuation in the natural world. Fish and Fisheries 2002;3: 171-196.

[43]  Sokal RR, Sneath RHA. Principles of Numercial Taxonomy. San Francisco: W.H. Freeman & Co; 1963.

[44]  Armand AD. [Two in one. The law of additivity]. Moscow: URSS; 2007. (in Russian)

[45]  McCray AT. Conceptualizing the world: Lessons from history. Journal of Biomedical Informatics 2006;39(3): 267-273.

[46]  Gärdenfors P. Conceptual spaces as a framework for knowledge representation. Mind and Matter 2004;2: 9-27.

[47]  Kim SH. Max Weber. 5.2 Ideal Type. In: Zalta EN. (ed.). The Stanford Encyclopedia of Philosophy; 2008. Available: http://plato.stanford.edu/entries/weber /#IdeTyp/.

[48]  Claridge MF, Dawah HA, Wilson MR (eds). Species. The units of biodiversity. London: Chapman & Hall; 1997.

[49]  Eldredge N, Salthe SN. Hierarchy and evolution. In: Dawkins R, Ridley M. (eds). Oxford surveys in evolutionary biology. Oxford: Oxford University Press; 1984. p184-208.

[50]  Eldredge N, Cracraft J. Phylogenetic patterns and the evolutionary process. New York: Columbia University Press; 1980.

[51]  de Pinna MCC. Species concepts and phylogenetics. Review of Fish Biology and Fisheries 1999;9(4): 353-373.

[52]  Mishler BD, Theriot EC. The phylogenetic species concept: Monophyly, apomorphy, and phylogenetic species concept. In: Wheeler QD, Meier R. (eds). Species concepts and phylogenetic theory: A debate. New York: Columbia Univ. Press; 2000. p44-54.

[53]  Turesson G. The species and the varieties as ecological units. Hereditas 1922;3(1): 100-113.

[54] Van Valen LM. Ecological species, multispecies, and oaks. Taxon 1976;25(2): 233-239.

[55] Andersson L. The driving force: Species concepts and ecology. Taxon 1990;39(3): 375-382.

[56] Krivolutsky DA. [Contemporary ideas of the animal life forms]. Ekologia 1971;3: 19-25. (in Russian)

[57] Chernov YuI. [Biological diversity: Its essence and problems]. Uspekhi Sovremennoy Boilogii. 1991;111(4): 499-507. (in Russian)

[58] Popper K. Three worlds. The Tanner lecture on human values, delivered at the University of Michigan, April 7, 1978; 1978. Available: http://tannerlectures.utah.edu/lectures/documents/popper80.pdf

[59] Riding RJ, Cheema I. Cognitive styles - An overview and integration. Educational Psychology 1991;11(3-4): 193-215.

[60] Kholodnaya MA. [Cognitive styles. On the nature of individual mind], 2nd ed. St-Petersbirg: Piter Publ.; 2004. (in Russian)

[61] Lyubarsky GYu. Classification of worldviews and taxonomic research. In: Pavlinov IYa. (ed.). Contemporary taxonomy: Methodological aspects. Moscow: Moscow State University Publ. 1986pp. 75-122. (in Russian)

[62] Mayr E. Darwin and the evolutionary theory in biology. evolution and anthropology: A centennial appraisal. Washington (D.C.): Anthropological Society of Washington; 1959. p409-412.

[63] Mayr E. Toward a new phylosophy of biology. Observations of an evolutionist. Cambridge (MA): Cambridge University Press; 1988.

[64] Kosko B. Fuzzy thinking: The new science of fuzzy logic. New York: Hyperion Books; 1994.

[65] Mccabe T. Studying species definitions for mutual nonexclusiveness. Zootaxa 2008;1939: 1-9.

[66] van Regenmortel MHV. Introduction to the species concept in virus taxonomy. In: van Regenmortel MHV, Fauquet CM, Bishop DHL, et al. (eds). Virus taxonomy. Classification and nomenclature of viruses. 7th Report on International Conference on the Taxonomy of Viruses. San Diego (CA): Academic Press; 2000. p3-16.

[67] Hanage WP, Fraser C, Spratt BG. Fuzzy species among recombinogenic bacteria. BMC Biology 2005;3: 6. Available: http://www.biomedcentral.com/1741-7007/3/6.

[68] González-Forero M. Removing ambiguity from the biological species concept. Journal of Theoretical Biology 2009;256(1): 76-80.

[69] Voishvillo EK. [The notion as a form of thinking: logical and gnoseological analysis]. Moscow: Moscow State University Publ.; 1989. (in Rissian)

[70] Antipenko LG. [Problem of the incompleteness theory and its gnoseological signifi-
cance]. Moscow: Nauka; 1986. (in Rissian)

[71] Perminov VYa. [Philosophy and foundations of mathematics]. Moscow: Progress-
Traditsia; 2001. (in Rissian)

[72] Mallet J. Species, concept of. In: Levin S. (ed.). Encyclopedia of biodiversity, vol. 5.
London: Academic Press; 2001. p427-440.

[73] Ereshefsky M. Foundational issues concerning taxa and taxon names. Systematic Biology 2007;56(2): 295-301.

[74] Wilkins JS. Philosophically speaking, how many species concepts are there? Zootaxa
2011;2765: 58-60.

[75] Gilmour JSL. Taxonomy and philosophy. In: Huxley J. (ed.). The new systematics.
Oxford: Oxford University Press; 1940.

[76] Malikov VG, Golenishchev FN. The systemic concept of formation and the problem
of species. In: Alimov AF, Stepanyanz SD. (eds). Species and speciation: An analysis
of new views and trends. Proceedings of Zoological Institute RAS. Add. 1; 2009. p
117-140. (in Russian, with English summary)

[77] Reif WE. Problematic issues in cladistics: 4. The species as a category. Neues Jahr-
buch für Geologie und Paläontologi 2004;233(1): 103-120.

[78] Hey J. On the failure of modern species concepts. Trends in Ecology and Evoluiton
2008;21(8): 447-450.

[79] Dupré J. On the impossibility of a monistic account of species. In: Wilson RA. (ed.).
Species: New interdisciplinary essays. Cambridge (MA): MIT Press; 1999. p3-21.

[80] Dupré J. In defence of classification. Studies in history and philosophy of science. Pt.
C: Studies in history and philosophy of biology and biomedical sciences 2001;32(2):
203-219.

[81] Rapini A. Classes or individuals? The paradox of systematics revisited. Studies in
history and philosophy of science. Pt. C: Studies in history and philosophy of biology
and biomedical sciences 2004;35(6): 675-695.

[82] Lee M, Wolsan M. Integration, individuality and species concepts. Biology and philosophy 2002;17(4): 651-660.

[83] Ho MW. Development, rational taxonomy and systematics. Biological forum
1992;85(2): 193-211.

[84] Boyd R. Homeostasis, species, and higher taxa. In: Wilson RA. (ed.). Species: New interdisciplinary essays. Cambridge (MA): MIT Press; 1999. p141-185.

[85] Paterson HEH. The recognition concept of species. In: Vrba ES. (ed.). Species and
Speciation. Transvaal Museum Monographs (Pretoria) 1985;4: 21-29.

[86]  Friedman VS. [Systems of the "friend or foe" identification and a renaissance of the biological species concept]. In: 21th Lyubishev readings: Contemporary problems of evolution. Ulyanovsk: Ulyanovsk State Pedagogical University Publ.; 2007. p201-215. (in Russian)

[87]  Templeton AR. The meaning of species and speciation: A genetic perspective. In: Otte D, Endler JA. (eds). Speciation and its consequences. Sunderland: Sinauer Assoc.; 1989. p3-27.

[88]  Khlebosolov EI. [The functional species concept in biology]. In: Ecology and evoluiton. Ryazan: Ryazan State Pedagogical University Publ.; 2003. p3-22. (in Russian)

[89]  Du Rietz GE. The fundamental units of biological taxonomy. Svensk Botanisk Tidskrift 1930;24(3): 333-428.

[90]  Ereshefsky M The poverty of the Linneaean hierarchy: A philosophical study of biological taxonomy. Cambridge (UK): Cambridge University Press; 2001.

[91]  Ereshefsky M. Species, taxonomy, and systematics. In: Matthen M, Stephens C. (eds). The handbook of philosophy of biology. Amsterdam: Elsevier; 2007. p403-427.

[92]  Hołyński RB. Philosophy of science from a taxonomist's perspective. Genus 2005;16(4): 469-502

[93]  Lyubarsky GYu. [A framework concept for the theory of biological diversity]. Zoologicheskie issledovania [Zoological Research] 2011;10: 5-44. (in Russian, with English summary)

[94]  Zherikhin VV. [Selected Works on palaeoecology and phylocenogenetics]. Moscow: KMK Sci. Press; 2003. (in Russian)

[95]  Barantsev RG. [Synergetics in contemporary natural science]. Moscow: URSS; 2003. (in Russian)

[96]  Brooks DR, Wiley EO. Evolution as entropy. Chicago: University Chicago Press; 1986.

[97]  Faegri K. Some fundamental problems of taxonomy and phylogenetics. Botanical Review 1937;3(8). 400-423.

[98]  Adams BJ. The species delimitation uncertainty principle. Journal of Nematology 2001;33(4): 153-160.

[99]  Pavlinov IYa. Etudes on metaphisics of contemporary taxonomy. In: Pavlinov IYa. (ed.). Linnaean miscellanea. Moscow: Moscow State University Publ.; 2007. p123-182.

[100]  Hull DL. Science as a process. Chicago: University Chicago Press; 1988.

[101]  Wilson DS. Species of thought: A comment on evolutionary epistemology. Biology and Philosophy 1990;5(1): 37-62.

[102]  Colin A, Bekoff M. Species of mind: The philosophy and biology of cognitive etholo-
       gy. Cambridge (MA): MIT Press; 1997.

[103]  Mahner M. What is a species? Journal for general philosophy of science 1993;24(1):
       103-126.

[104]  Griffiths PE. Squaring the circle: Natural kinds with historical essences. In: Wilson
       RA. (ed.). Species: New interdisciplinary essays. Cambridge (MA): MIT Press; 1999.
       p209-228.

[105]  Sober E. Evolution, population thinking, and essentialism. Philosophy of Science
       1980;47(3): 350-383.

[106]  Krasilov VA. Unresolved problems in the theory of evolution. Vladivostok: Far East
       Branch AS USSR Publ.; 1986.

[107]  Dumsday T. A new argument for intrinsic biological essentialism. The philosophical
       quarterly 2012;62(248): 486-504.

[108]  Shishkin MA. [Individual development and lessons from evolutionism]. Ontogenez
       2006;37(3): 179-198. (in Russian with English summary)

[109]  Dobzhansky T. Genetics and the origin of species. New York: Columbia University
       Press; 1937.

[110]  Dobzhansky T. Genetics of evolutionary process. New York: Columbia University
       Press; 1970.

[111]  Cohan FM. Towards a conceptual and operational union of bacterial systematics,
       ecology, and evolution. Philosophical transactions of the Royal Society ser. B 2006;
       361(1475): 1985-1996.

[112]  Griffiths PE. Darwinism, process structuralism, and natural kinds. Philosophy of sci-
       ence 63, Suppl. Proceedings of 1996 Biennial Meeting of Philosophy of Science Asso-
       ciation, Pt. I. Contributing Papers; 1996. pS1-S9.

[113]  Kluge AG. Species as historical individuals. Biology and Philosophy 1990;5(4).
       417-431.

[114]  Zuev VV. [Problem of reality in biological taxonomy]. Novosibirsk: Novosibirsk
       State University Publ.; 2002. (in Russian)

[115]  Ghiselin MT. Species, concepts, individuality and objectivity. Biology and philoso-
       phy 1987;2(1): 127-143.

[116]  Ghiselin MT. Metaphysics and the origin of species. New York.: State University of
       New York Press; 1997.

[117]  Hull DL. Are species really individuals? Systematic zoology 1976;25(1): 174-191.

[118]  Hull DL. A matter of individuality. Philosophy of science 1978;45(??): 335-360.

[119] Williams MB. Species are individuals: Theoretical foundations for the claim. Philosophy of Science 1985;52(4): 578-590.

[120] Pozdnyakov AA. On the individual nature of species. Journal of General Biology 1994;55(3): 389-397.

[121] Dubois A. Species and "strange species" in zoology: do we need a "unified concept of species"? Comptes rendus de l'Académie des sciences, Series IIA, Earth and planetary science, Paleol 2011;10(2-3): 77-94.

[122] McOuat G. From cutting nature at its joints to measuring it: New kinds and new kinds of people in biology. Studies in the history and philosophy of science 2001;32(4): 613-645,

[123] Eigen M. Viral quasispecies. Scientific American 1983;269: 42-49.

[124] Wilkins JS. The dimensions, modes and definitions of species and speciation. Biology and Philosophy 2007;22(2): 247-266.

[125] Templeton AR. Species and speciation. Geography, population structure, ecology, and gene trees. In: Howard DJ, Berlocher SH. (eds). Endless forms. Species and speciation. New York: Oxford University Press; 1998. p32-43.

[126] Kober G. Biology without species: A solution to the species problem. Dissertation. Udini; 2008. Available: http://udini.proquest.com/view/biology-without-species-a-solution-goid:89188033/

[127] Skarlato OA, Starobogatov YaI. [Phylogenetics and principles of elaboration of the natural system]. Proceedings of Zoological Institute AS USSR 1974;53: 30-46. (in Russian)

[128] Schwarz SS. [Ecological patterns on evolution]. Moscow: Nauka; 1980. (in Russian)

[129] Pavlinov IYa. [How it is possible to construct taxonomic theory]. Zoologicheskie issledovania [Zoological Research] 2011;10: 45-100. (in Russian with English summary)

# Conceptual Issues

# Chapter 2

# The Species Problem: A Conceptual Problem?

Richard A. Richards

Additional information is available at the end of the chapter

## 1. Introduction

There have long been multiple ways of conceiving *species* that divide up biodiversity in different and inconsistent ways. This "species problem" goes back at least to Aristotle, who used the Greek term *eidos* (translated as the Latin *species*) in at least three different ways. [1] Two thousand years later, Darwin confronted the species problem, listing some of the different ways species were conceived.

---

How various are the ideas, that enter into the minds of naturalists when speaking of species. With some, resemblance is the reigning idea & descent goes for little; with others descent is the infallible criterion; with others resemblance goes for almost nothing, & Creation is everything; with other sterility in crossed forms is an unfailing test, whilst with others it is regarded of no value. [2]

---

And one hundred years after Darwin, Ernst Mayr worried about this same problem in a book he edited, titled *The Species Problem*.

---

Few biological problems have remained as consistently challenging through the past two centuries as the species problem. Time after time attempts were made to cut the Gordian knot and declare the species problem solved either by asserting dogmatically that species did not exist or by defining, equally dogmatically, the precise characteristics of species. Alas, these pseudosolutions were obviously unsatisfactory. One might ask: "Why not simply ignore the species problem?" This also has been tried, but the consequences were confusion and chaos. The species is a biological phe-

nomenon that cannot be ignored. Whatever else the species might be, there is no question that it is one of the primary

levels of integration in the many branches of biology, as in systematics (including that of microorganisms), genetics,

and ecology, but also in physiology and in the study of behavior. Every living organism is a member of a species, and

the attributes of these organisms can often best be interpreted in terms of this relationship. [3]

---

More recently, the species problem seems to have gotten worse. In 1997, Richard Mayden identified at least twenty-two species concepts currently in use. [4]

This multiplicity of species concepts is a genuine problem in that different ways of conceiving species divide biodiversity in different and inconsistent ways, and no single species concept is adequate. What counts as a species under one concept may not count as a species under another. So whether a group of organisms counts as a species depends on which species concept is used. One researcher might, for instance, use morphological or genetic similarity to group into species, while another might use interbreeding, and yet another might appeal to history or phylogeny. In other words, one person might use a species concept based on morphological or genetic similarity, while another might use a concept based on interbreeding or phylogeny.

The consequences of using different species concepts are often striking. Species counts, one way of measuring biodiversity, depend on which concept is used. The replacement of other concepts with the *phylogenetic species concept*, for instance, has multiplied 15 amphibian species into 140. [5] A recent survey of taxonomic research [6] quantifies the effects of a shift to this particular species concept from other concepts, finding a 300% increase in fungus species, a 259% increase in lichen species, a 146% increase in plant species, a 137% increase among reptile species, an 88% increase among bird species, an 87% increase among mammals, and a 77% increase among arthropods. Running counter to this trend, there was a 50% decrease in mollusc species. Overall, there was an increase of 48.7% when a *phylogenetic species concept* replaced other concepts.

Given that this there is so far no consensus on species concepts, these differences in species counts suggest that the conventional grouping of organisms into species may be arbitrary and reflects only the subjective point of view assumed, as Joel Cracraft suggests (emphasis added):

---

The primary reason for being concerned about species definitions is that they frequently lead us to divide nature in

very different ways. If we accept the assumption of most systematists and evolutionists that species are real things in

nature, and if the sets of species specified by different concepts do not overlap, then it is reasonable to conclude that

real entities of the world are being confused. It becomes a fundamental scientific issue when one cannot even count the basic units of biological diversity. Individuating nature "correctly" is central to comparative biology and to teasing

apart pattern and process, cause and effect. Thus, time-honored questions in evolutionary biology--from describing

patterns of geographic variation and modes of speciation, to mapping character states or ecological change through

time, to biogeographic analysis and the genetics of speciation, or to virtually any comparison one might make--*will*

*depend for their answer on how a biologist looks at species.* [7]

This problem is magnified by the fact that which concept is used often depends on seemingly arbitrary facts, such as which organism is studied, as Cracraft explains:

There has been something of a historical relationship between an adopted species concept and the taxonomic group

being studied... Thus, for many decades now, ornithologists, mammalogists, and specialists from a few other disci-

plines have generally adopted a Biological Species Concept; most invertebrate zoologists, on the other hand, including

the vast majority of systematists, have largely been indifferent to the Biological Species Concept in their day-to-day

work and instead have tended to apply species status to patterns of discrete variation. Botanists have been somewhere

in the middle, although most have not used a Biological Species Concept. [7]

But even among those who study the same organisms, there is disagreement about which species concept is best. Those who are committed to the method of taxonomy sometimes known as "cladistics" tend to use different concepts than those who have adopted the more traditional "evolutionary systematics." And even those who regard themselves as cladists find little agreement. In a recent volume, five different cladistic species concepts were proposed and developed, seemingly without any consensus. [8]

This is clearly problematic for the understanding and preserving biodiversity, as Claridge, Dawah and Wilson recognize in their introduction to a recent collection of articles on species concepts:

The prolonged wrangle among scientists and philosophers over the nature of species has recently taken on added and

wider significance. The belated recognition of the importance of biological diversity to the survival of mankind and the

sustainable use of our natural resources makes it a matter of very general and urgent concern. Species are normally the

units of biodiversity and conservation... so it is important that we should know what we mean by them. One major

concern has been with estimating the total number of species of living organisms that currently inhabit the earth... In addition, many authors have attempted to determine the relative contributions of different groups of organisms to the

totality of living biodiversity... Unless we have some agreed criteria for species such discussions are of only limited

value. [9]

---

Moreover, if the application of endangered species legislation is affected by species counts, then the consequences of the species problem spreads beyond biology and into public policy. [1] There are clearly costs if the adoption of a particular species concept results in increased species counts. The authors of the survey quoted above, have estimated the costs of the proliferation of species taxa, based on the fact that the adoption of the *phylogenetic species concept* results in increased species counts that reduce the geographic range of species, and that in turn make more species protected.

---

Any increase in the number of endangered species requires a corresponding increase in resources and money devoted

toward conserving those species. For example, it has been estimated that the complete recovery of any of the species

listed by the U.S. Endangered Species Act will require about $2.76 million... Thus, recovering all species listed current-

ly would cost around $4.6 billion. With widespread adoption of the PSC [*phylogenetic species concept*], this already for-

midable amount could increase to $7.6 billion, or the entire annual budget for the administering agency (U.S. Fisheries

and Wildlife Services) for the next 120 years. [6]

---

These additional costs might be justified *if* the increased species counts represented an objective improvement in the measure of biodiversity. But the additional costs are hard to justify if they are merely a consequence of some arbitrary choice of species concepts.

There are theoretical concerns here as well. If species are the fundamental units of evolution and classification, as is typically assumed, surely we need a satisfactory, unambiguous way to determine what counts as the fundamental units in these ways. [1] We need to have a good idea, for instance, about what counts as a species in order to identify and study speciation events. After all, only if a new species has been formed can there be speciation. And as long as species are the fundamental, basal units of classification, as is usually assumed, we need to know unambiguously what counts as a species to generate an unambiguous classification.

We might make progress on this long-standing species problem by thinking about scientific problems in general. Some scientific problems are empirical in the sense that they are solved by the addition of new empirical data or information. For example, we might solve a problem of disease by the observation of some bacterial or viral pathogen. As is well known, this is happening with a variety of cancers. On the other hand, some scientific problems are largely conceptual in the sense that they are solved not so much by the addition of new empirical information, but through some conceptual innovation, change or clarification. For example, problems related to planetary motion were solved by Johannes Kepler through the use of a new orbital concept based on elliptical rather than circular motion. And around the turn of the twentieth century, Wilhelm Johannsen coined the terms 'gene,' 'genotype,' and 'phenotype' to introduce new and useful concepts to the many problems in the study of heredity. [10] Sometimes old concepts get modified, as we see in relativistic physics with its new ways of conceiving *mass, space* and *time*. In each of these latter cases, progress was made through thinking about the concepts used, *not just* through the addition of new empirical information.

There is a general insight to be gained in thinking about scientific problems in this way. From at least Plato and Aristotle on, it has been recognized that knowledge of the world is based on the application of language, ideas or concepts to the world. Consequently, successful inquiry depends in part on getting our language, ideas or concepts right. This can be relatively straightforward, as in Kepler's application of the ellipse to planetary motion, or in the invention of the concepts of *quark, atom, electron, element, compound, gene, protein, homology, enzyme, genotype, population, species, ecosystem, neuron,* etc. Or less straightforwardly, progress may be found in the relation between concepts. How is the idea of an *electron* related to the idea of an *atom*? And how is the concept of a *species* related to that of a *population*? It may also be that getting concepts right involves something less concrete and more abstract. Scientific progress might be predicated partly on getting clear about *scientific law, evidence, explanation, theory, testing, observation* and so on. And at an even more abstract level, scientific progress might result from thinking more clearly about the nature of various concepts, and how they are related. For instance, how is *observation* related to *evidence* and *theory*? And what is the relation between *scientific law* and *scientific explanation*?

So is the species problem empirical, conceptual or both? If empirical it will be solved by more empirical data or information. Present trends suggest that the problem is not exclusively empirical. The last century has made great progress in the empirical investigation of biodiversity and evolution, but the species problem seems to instead be getting worse! We now have more jointly inconsistent and individually inadequate concepts than ever. It is my contention here that the species problem is at least partly conceptual, and it is solved at an abstract level: the nature and relation of various species concepts. The solution is not merely a matter of introducing a new species concept, or modifying an existing concept. Rather it is to be found in an understanding how the various species concepts are related within a framework, how each concept works individually, and how this all has resulted in the species problem.

I shall argue that the species problem is solved first, by understanding the division of labor within the conceptual framework. Some species concepts are theoretical and are concerned with the nature of species things. Others are operational, telling us how to identify and individuate species things. Here I follow the lead of Richard Mayden and Kevin de Queiroz, but go a step further and argue that these operational concepts are better conceived as *correspondence rules*. The second component of the solution is based on an understanding of the structure of theoretical species concepts. I will argue that the primary, theoretical species concept has a structure - a definitional core and a descriptive periphery - and once we see how this conceptual structure works, we can understand why there has been an enduring species problem and how to solve it. Finally, I will conclude with a few thoughts about scientific concepts and how they get used within the social, "demic" structure of science. Part of understanding the species problem is to be found in how researchers and theorists in different fields, with different interests, engage the species concept.

## 2. The conceptual framework

The recent history of the species problem is not promising. Along with the increase in our understanding of biodiversity and the evolutionary processes that produced it has come a proliferation of species concepts. Richard Mayden [4] has identified and individuated over twenty species concepts currently in use. Some species concepts he identifies are based on similarity. The *morphological species concept* asserts that "species are the smallest groups that are consistently and persistently distinct, and distinguishable by ordinary means." The *phenetic species concept* is based on overall similarity and phenetic clustering. Some species concepts are based on molecular similarity, such as the *genotypic cluster concept* and the *genealogical concordance concept*. Other concepts are based on evolutionary processes. The *biological species concept*, advocated by Ernst Mayr, and Hugh Paterson's *recognition species concept* are based on sexual reproduction. But since not all organisms reproduce sexually, the *agamospecies concept* was proposed to serve as an umbrella concept for all taxa that are uniparental and asexual. Some process concepts are based on ecology, such as the *ecological species concept*, which identifies species with unique adaptive zones. Historical species concepts treat species as historical entities, extended in time. Here we find the *evolutionary species concept*; the *successional species concept*; the *paleospecies concept*; and the *chronospecies concept*, that each conceives of species as segments of a changing lineage. The *cladistic species concept*, the *composite species concept*, the *internodal species concept*, and the *phylogenetic species concept* are all based on the idea that speciation events can serve to demarcate the beginnings and endings of species lineages.

The details of each of these species concepts are not important for purposes here. What is important is that first, with increased empirical understanding, species concepts seem to be proliferating, second, these concepts are inconsistent, carving nature in different and inconsistent ways; third, no single concept is adequate, applying across biodiversity. The *biological species concept*, for instance, applies only to sexually reproducing organisms, and therefore cannot be used to group asexual organisms. Fourth, the species problem does not seem to be solved by additional empirical information. This suggests that the problem is not exclusively empirical, and requires a conceptual solution.

Mayden recognizes this. After outlining all these species concepts, he argues that there are really two main kinds of species concepts: *primary theoretical concepts* tell us what kinds of things species taxa are; *secondary operational concepts* tell us how to identify and individuate species taxa. This approach is hierarchical because the operational concepts depend on the theoretical concepts. Operational concepts do not tell us what species are, but *given a particular theoretical concept*, how to identify and individuate them. [4] Operational and theoretical concepts are therefore not competing but supplementary ways of thinking about species. It is therefore possible to use different operational concepts, without dividing biodiversity up in inconsistent ways – if a single theoretical concept is used.

This hierarchical thinking about species may have the potential to solve the species problem, but only if there is a single, adequate theoretical concept. Mayden argues that there is such a concept, based on the fundamental idea of a lineage: the *evolutionary species concept (ESC)*. Mayden [4] gives three statements of this concept. The first from G. G. Simpson asserts that a species is "a lineage (an ancestral-descendent sequence of populations) evolving separately from others and with its own unitary evolutionary role and tendencies." The second statement, from Edward Wiley, identifies species as "a single lineage of ancestor-descendent populations which maintains its identity from other such lineages and which has its own evolutionary tendencies and historical fate." The third formulation, from Wiley and Mayden, is that a species is "an entity composed of organisms which maintains its identity from other such entities through time and over space, and which has its own independent evolutionary fate and historical tendencies." According to Mayden, the *ESC* is theoretically significant and universal. It can apply across biodiversity. [4]

The *Evolutionary Species Concept* is not obviously operational. One cannot just observe lineages of the relevant kind in nature. The *ESC* therefore requires other operational, species concepts:

---

While the ESC is the most appropriate primary concept, it requires bridging concepts permitting us to recognize entities compatible with its intentions. To implement fully the ESC we must supplement it with more operational, accessory notions of biological diversity – secondary concepts. Secondary concepts include most of the other species concepts. While these concepts are varied in their operational nature, they are demonstrably less applicable than the ESC because of their dictatorial restrictions on the types of diversity that can be recognized, or even evolve. [4]

---

Secondary operational concepts are those that can be readily applied to biodiversity, and are indicative of species lineages. Species concepts based on morphological or genetic similarity, for instance, can help identify lineages, since organisms within a single lineage will generally share some morphological and genetic traits. Concepts based on processes such as reproductive isolation and cohesion, mate recognition systems and ecological niches, can also be used to identify lineages since these are processes that operate in the formation and persistence of lineages.

Kevin de Queiroz has proposed a similarly hierarchical way to think about species concepts. According to de Queiroz, there are the *species concepts* proper that give the necessary properties of species and provide theoretical definitions. Then there are *species criteria* that give contingent properties and are "standards for judging whether an entity qualifies as a species." [11] Many of the species concepts in use are to be understood as species criteria rather than species concepts proper.

---

The species criteria adopted by contemporary biologists are diverse and exhibit complex relationships to one another (i.e. they are not necessarily mutually exclusive). Some of the better-known criteria are: potential inter-breeding or its converse, intrinsic reproductive isolation... common fertilization or specific mate recognition systems... occupation of a unique niche or adaptive zone... potential for phenotypic cohesion... monophyly as evidenced by fixed apomorphies... or the exclusivity of genic coalescence... qualitative... or quantitative... Because the entities satisfying these various criteria do not exhibit exact correspondence, authors who adopt different species criteria also recognize different species taxa. [11]

---

Like Mayden, de Queiroz argues that there is single, primary species concept that is adequate – applying across biodiversity. This is, according to de Queiroz, the *general lineage concept*:

---

Species are segments of population-level lineages. This definition describes a very general conceptualization of the species category in that it explains the basic nature of species without specifying either the causal processes responsible for their existence or the operational criteria used to recognize them in practice. It is this deliberate agnosticism with regard to causal processes and operational criteria that allows the concepts of species just described to encompass virtually all modern views on species, and for this reason, I have called it the general lineage concept of species. [11]

---

In a later paper, de Queiroz describes this general theoretical concept in terms of a "metapopulation lineage," which he describes as "sets of connected subpopulations, maximally inclusive populations." [12]

Mayden and de Queiroz are largely right about the conceptual framework and the potential solution to the species problem. There may be multiple, seemingly inconsistent ways of thinking about species, but these ways of thinking are not all equivalent. The *biological species concept* and the *evolutionary species concept*, for instance, are not competing ways to think about species. Rather they a complementary. The biological species concept, based on interbreeding, is valuable insofar as it indentifies the kind of lineages required by the evolutionary species concept. Nor need each concept be individually adequate, applying across biodiversity. The biological species concept, based on reproductive cohesion and isolation, need only apply to sexually reproducing organisms. The important insight here is that some ways of thinking are substantive (evolutionary species concept) and tell us what species things are, and some ways are operational (biological species concept), telling us how to identify and individuate species taxa. But it may be misleading to think about all these ways of thinking as "concepts."

This division of conceptual labor echoes a debate early in the twentieth century about how to define scientific concepts in physics, such as *length* and *mass*. The physicist P. W. Bridgman proposed that we should define these concepts in terms of the *operations* we use to measure them. *Mass*, for instance, would be defined in terms of the ways of measuring mass - the resistance to acceleration, or the operation of gravity. Bridgman's proposal that operations give definitions became known as "operationalism," and came to be seen as an answer to the philosophical problem of how to connect theoretical laws, that contain only non-observational terms, to observation. How can we connect laws about unobservable particles, for instance, to the empirical regularities we observe in nature? The philosopher of science Rudolf Carnap explained this problem:

---

Our theoretical laws deal exclusively with the behavior of molecules, which cannot be seen. How, therefore, can we deduce from such laws a law about observable properties such as the pressure or temperature of a gas or properties of sound waves that pass through the gas? The theoretical laws contain only theoretical terms. What we seek are empirical laws containing observable terms. Obviously, such laws cannot be derived without having something else given in addition to the theoretical laws.... That something else that must be given is this: a set of rules connecting the theoretical terms with the observable terms. [13]

---

Carnap called these rules connecting theoretical and observable terms "correspondence rules." What is significant in Carnap's proposal is that these *correspondence rules* connecting theoretical concepts to observation are not really concepts in the usual theoretical sense. This is clear in Carnap's rejection of the view (disagreeing with Bridgman) that operational rules can provide definitions: "There is a temptation at times to think that the set of rules provides a means for defining theoretical terms, whereas just the opposite is really true". [13] Nor can correspondence rules function *as* definitions: "What we call these rules is, of course, only a terminological question; we should be cautious and not speak of them as definitions. They are not definitions in any strict sense." [13] Rather, the definitions give operational guidance, telling us what operations are relevant. What is important here is that those concepts that function operationally are different from those that function theoretically. In effect, they tell us how to observe a thing, not what sort of a thing it is.

We can apply Carnap's insight here to the species problem. As argued by Mayden and de Queiroz, some species concepts are theoretical. They tell us how to conceive species. They define species taxa and constitute the species category. But some species concepts are operational. They tell us how to identify and individuate the groups that are properly species *giv-*

*en a particular theoretical concept.* But these so-called operational concepts are really *rules* that help us to determine if a group of organisms satisfies the demands of the theoretical concept. The biological species concept is really a rule about how to identify and individuate the sexually reproducing lineages that constitutes species taxa. By using this terminology, referring to operational concepts as "correspondence rules," Carnap thought we could avoid the general confusion of definitions and operations. He also thought this solved a problem in science, the tendency of philosophers to ask scientists for definitions of scientific concepts in familiar, non-theoretical terms.

They want the physicist to tell them just what he means by "electricity", "magnetism", "gravity", "a molecule". If the

physicist explains them in theoretical terms, the philosopher may be disappointed. "That is not what I meant at all", he

will say. "I want you to tell me, in ordinary language, what those terms mean." [13]

The problem here is that the scientist is being asked for something he or she cannot give – a definition in something other than theoretical terms. Each of these concepts has satisfactory definitions, but they are in terms of the theoretical framework. That is the proper source for definitions – telling us how to interpret these concepts – not the operations to measure or identify the things that satisfy them. Carnap concluded:

The answer is that a physicist can describe the behavior of an electron only by stating theoretical laws, and these laws

contain only theoretical terms. They described the field produced by an electron, the reaction of an electron to a field,

and so on…. There is no way that a theoretical concept can be defined in terms of observables. We must, therefore,

resign ourselves to the fact that definitions of the kind that can be supplied for observable terms cannot be formulated

for theoretical terms. [13]

There are three things to note here about Carnap's analysis. First is his emphasis on the role of theoretical frameworks in the interpretation of scientific concepts. Theoretical terms are to be understood in terms of the overarching theory. For a species concept the overarching theory is evolutionary theory. Second is the proposal that we think about operations as rules rather than concepts. What we might call operational *concepts*, are really *rules* for connecting theoretical concepts to observation. Third, there soon came to be a general consensus against operationalism in physics. That consensus continues today. There are obvious reasons for the rejection of operationalism. It seemed to lead to the multiplication of concepts of such basic things as *length.* [14] The operations to measure length at the scale of stars and galaxies are different than the operations to measure length at the scale of a football field or at the atomic level. The problem is that if we base concepts on operations, there seems to be a new concept for each distinct operation. With operationalism comes the proliferation of concepts like *length*. Similarly we should expect the proliferation of concepts like *mass, density* and *temperature, etc.* If there is more than one way to measure each, there will be multiple concepts. It is no wonder that operationalism was rejected in physics. It should similarly be rejected in evolutionary biology and with respect to the species problem.

If we adopt Carnap's approach, distinguishing theoretical definitions from operational "correspondence rules," and apply this approach to the species problem, the species problem largely dissolves. So, for instance, given a particular theoretical concept (either Mayden's *ESC* or de Queiroz's general lineage concept) some kinds of similarities will be indicative of the relevant kind of population lineage, and will therefore help identify and individuate species taxa. For those population lineages of sexually reproducing organisms, reproductive cohesion, reproductive isolation and mate recognition systems will be relevant to the identification and individuation of species taxa. And for non-sexually reproducing lineages, there will other correspondence rules. If so, then the species problem is largely a consequence of not discriminating between species concepts proper and the correspondence rules that help apply species concepts to the world.

Implicit in this division of conceptual labor are two distinct sets of evaluative criteria. Theoretical concepts best serve the needs of evolutionary theory and biosystematics if they are universal – apply across biodiversity. This is in effect, a unification requirement. A single concept will ideally *unify* phenomena – the apparently discrete groupings of organisms that we see across biodiversity and processes that produce these groupings. On the other hand, operational concepts, or better "correspondence rules," do not need to unify. Rather as different processes operate across biodiversity there is instead a proliferation of rules. We need different rules, for instance, for sexual and asexual organisms. The more we find out about evolutionary processes, the more rules we will discover to identify and individuate species things. Rather than unification, with correspondence rules/operational concepts there is proliferation. The more the merrier!

There are good reasons to think that Mayden and de Queiroz have the broad outlines of a primary theoretical concept right as well - even though there may be differences in each of the three formulations Mayden provides of the ESC and de Queiroz's general lineage concept. A primary theoretical concept must be theoretically significant and consistent with evolutionary theory. At the most basic level, the theory of evolution tells us that there is change over time. Darwin thought that this involved the origin of new species through divergent change, whereby mere varieties become species. [15] This principle of divergence then explained the branching evolutionary tree diagram that in turn served to illustrate his approach to classification.

---

I request the reader turn to the diagram illustrating the action, as formerly explained, of these several principles; and

he will see that the inevitable result is that the modified descendants proceeding from one progenitor become broken

up into groups subordinate to groups... So that we here have many species descended from a single progenitor grouped into genera; and the genera are included in, or subordinate to, sub-families, families and orders, all united into one

class. [15]

---

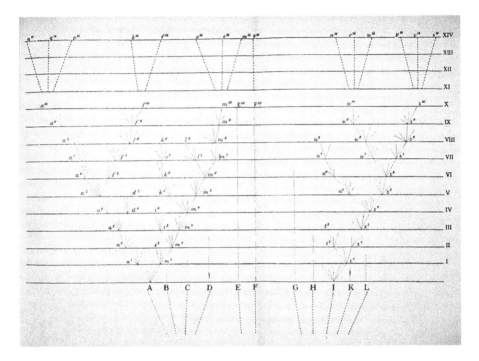

**Figure 1.**

What is important here is that this tree (figure 1) [16] emphasized the temporal, historical dimension of evolution, and the branching associated with speciation. It tells us that species have beginnings in speciation events. They have duration. They change. And they have endings. Since Darwin, this historical component has become further entrenched in evolutionary thinking about species.

This is not to say, of course, that species taxa are *just* historical entities. Evolutionary theory tells us that they exist as well, as groups of organisms at particular times, groups that share similarities, sometimes interbreed, occupy ecological niches, vary geographically, form gene pools, and have a variety of social structures. This way of thinking about species has been developed and refined most notably by the thinkers of the Modern Synthesis, such as Mayr,

Dobzhansky, and Simpson. This *population* dimension, along with the historical, suggests that there are two ways to think about species taxa. We can think about them over time, as historical, *diachronic* entities that originate, change and go extinct. Or we can think about them at a single time, as *synchronic* groups of organisms that are connected or given some sort of structure by some biological process. If so, then evolutionary theory tells us that whatever else species taxa are, they have two dimensions – diachronic and synchronic, and an adequate theoretical species concept must reflect that fact.

Evolutionary theory also tells us that species are the things that evolve. First, they have beginnings and endings in speciation and extinction events. Accordingly, each species taxon also has its own distinctive fate, in its trajectory of change or stasis and ultimate extinction. But species taxa also have some sort of cohesion, whether through reproduction, social interaction, gene transfer or the operation of natural selection. But to be universal, a theoretical concept must be indeterminate about which processes produce these general features. If there is a solution to the species problem, as I think there is, it will surely be based on something like what Mayden and de Queiroz propose – a primary theoretical species concept that treats species taxa as segments of populations lineages with cohesion and distinctive fates. And the more researchers find out about the processes that segment these population lineages and that produce cohesion, and that preserve or produce morphological, behavioral and molecular similarities, the more correspondence rules they will have at hand to identify and individuate species taxa. Since these correspondence rules are subservient to the primary theoretical species concept, *when understood correctly* they cannot ultimately divide biodiversity in inconsistent ways. It is only when they are taken to be independent of a primary theoretical concept that they can conflict.

This is not to say, however, that the nature and application of the correspondence rules is obvious and unproblematic. It is not always obvious which correspondence rules are appropriate in particular instances. That will often depend on empirical facts about the relevant organisms and processes in question - facts that may or may not be known. Nor is the nature of the primary theoretical concept unproblematic. The lineage and population concepts both require clarification. It is not always clear what kinds of cohesion are relevant and operate in the various groups of organisms. But more pertinent to purposes here, this division of conceptual labor is only half of the solution to the species problem. The other half is found at a lower level, the level of the individual theoretical species concept, and how it functions.

## 3. Conceptual structure

Mayden and de Queiroz suggest that even with the use of multiple operational concepts/ species criteria there is general agreement that species are segments of population lineages. This is what evolutionary theory requires. There is good reason to agree with them. Ironically though, an historical conception of species as lineages predates evolution in the ideas of John Ray and Linnaeus. [1] And Darwin noted that an historical way of thinking about species was largely accepted by his contemporaries.

With species in a state of nature, every naturalist has in fact brought descent into his classification; for he includes in

his lowest grade, or that of a species, the two sexes; and how enormously these sometimes differ in the most important

characters, is known to every naturalist: scarcely a single fact can be predicated in common of the males and hermaph-

rodites of certain cirripedes, when adult, and yet no one dreams of separating them. The naturalist includes as one

species the several larval stages of the same individual, however much they may differ from each other and from the

adult... He includes monsters; he includes varieties, not solely because they closely resemble the parent-form, but be-

cause they are descended from it... [15]

But as Darwin's evolutionary tree diagram in the Origin shows, this historical thinking about species is also central to evolutionary theory. The idea here is that even with the use of other criteria for grouping into species, and identifying and individuating species taxa, there has been guidance from the basic conception that species are lineages. In Darwin's tree diagram, species are the branches of the tree. If so, a systematist might use morphological or molecular similarity to identify and individuate species, but in ways that are constrained by a population lineage conception of species. This requires that the systematist ignore irrelevant morphological traits based on sexual dimorphism and developmental stages. If so, then there is an implicit hierarchy here, ust as Mayden and de Queiroz have argued.

There are puzzles about actual usage that remain. When naturalists, evolutionists and systematists actually *use* the term *species*, they don't always seem to mean "segments of population lineages with cohesion and distinctive fates." Rather they seem often to have other things in mind, as Darwin recognized in his own time:

[H]ow various are the ideas, that enter into the minds of naturalists when speaking of species. With some, resemblance

is the reigning idea & descent goes for little; with others descent is the infallible criterion; with others resemblance goes

for almost nothing, & Creation is everything; with other sterility in crossed forms is an unfailing test, whilst with oth-

ers it is regarded of no value. [2]

This is still the case. One person might apply the term *species* to a frog population on the basis of a distinctive morphology and behavior, without consciously thinking or claiming that the morphological or behavioral similarities are subservient to some other theoretical concept. Another might take the term to mean interbreeding and reproductive isolation as applied to a population of birds. A geneticist might take the term to mean something related to genotypic similarity. The point is that even if there is a primary theoretical concept *availa-ble* to guide thinking about species, not always does this theoretical concept get manifest in the actual usage and meaning of the term *species*. A molecular systematist might mean one thing, a naturalist might mean another, and a geneticist might mean yet something else when using the term *species*. How do we account for this variability in usage, given a single primary theoretical concept? The division of conceptual labor doesn't seem to adequately answer this question. It just tells us that there are different ways *available* to think about species, some theoretical some operational. To answer this question we need to look more closely at how concepts get structured and actually used.

Much modern thinking about concepts begins with a framework laid out by Gottlob Frege, in a classic German paper of 1892, and its English translation, "On Sense and Reference." [17] Here Frege addressed the question of how language can represent things in the world. He argued that linguistic entities such as concept terms function in propositions in two ways: first, through a "nominatum," what the term *refers* to (what it *designates* or *denotes*); second, through the "sense," or *meaning* of the term. According to Frege, the sense of a term is grasped by anyone who knows the language, and is to be identified with the description that would be associated with the term in that language. The sense or meaning of the term *water* for instance would be identified with an associated description of *water*. The meaning of a term must be distinguished from what it refers to, or denotes, because, according to Frege, co-referential terms (terms that refer to the same thing) often have different meanings. Two terms that referred to the planet Venus, for instance, have different meanings based on the descriptions that designate different times of appearance in the sky: "The nominata of 'evening star' and 'morning star' are the same, but not their senses." [17] Meaning is therefore more fine-grained than reference, in that two terms can refer to the same thing, yet still have different meanings. (As we shall see, the meaning of 'species' is more fine-grained than its reference.)

If meaning is to be associated with some descriptive content – a description that gives conditions for the application of the concept, then to understand the meaning of a term we need to know the descriptive content. One standard, "classical" approach conceives the description in terms of a definition with a particular definitional structure, a set of singly necessary and jointly sufficient conditions for falling under the concept. The meaning of the concept term is then this set of necessary and sufficient conditions. [18] This does not rule out non-definitional descriptive content though. Alongside the definitional core is a set of conditions that are associated with the term, but in an "accidental" way.

The term *water*, for instance, has a theoretically provided definitional core based on its particular composition of two hydrogen atoms and one oxygen atom. But it also has a descriptive periphery: its density, freezing point, where it is found, its taste and appearance, recreational potential etc. On the classical approach, the term *species* similarly has a definitional core based on a set of singly necessary and jointly sufficient conditions, and a descriptive periphery. As argued here, the definitional core of *species* is constituted by the conditions that species are segments of a population lineage with cohesion and a distinct fate. These are singly necessary conditions – each one is required, and together they are sufficient for being a species taxon.

There are, however, other ways to think about definitional structure. One limitation of the classical approach is that it implies that falling under a concept is all or nothing. Either the necessary and sufficient conditions are satisfied or they are not. But it seems possible for this to be a matter of degree. The "cluster" approach asserts that something can fall under a concept to varying degrees depending on how many conditions are met, and how typical or characteristic the particular conditions satisfied are. [19] This way of thinking about concepts as probabilistic clusters of conditions has lead some to advocate a "prototype" or "exemplar" approach, where some instance of the concept that instantiates the core set of conditions comes to represent it as an exemplar or ideal instance. [19] Here there are then degrees of concept application. Something can more or less fall under a particular concept depending on how many and which conditions are satisfied, or how close the analogy is with the exemplar. Definitional structure on the cluster approach then, is a conceptual core that has greater definitional weight than other conditions, without thereby constituting a set of singly necessary and jointly sufficient condition. The definition of a term would then be some weighted cluster or other of the descriptive properties or conditions associated with the concept.

On this approach, *water* would still have a conceptual core, its molecular composition, but that may not be strictly necessary or sufficient. It might be that we would require some additional conditions. To count as *water*, there must some cluster of other conditions met. Perhaps we might require that it be made of up a certain range of proportion of "light" (with protium hydrogen atoms) versus "heavy water" (with deuterium hydrogen atoms). Or perhaps we might require some set of conditions related to functioning – safe for humans to drink etc.

But neither of these theories of meaning is fully adequate. Neither can answer questions about what determines the inclusion of conditions in the definition, or about how these conditions are related. They only designate the structure of concepts. What then determines the descriptive content and makes it cohere? Recently, an approach known as the "theory theory," has provided an answer to these questions. The idea is that the definitional structure of concepts is filled out and made coherent by some *theory*, scientific or otherwise, that contains the relevant concept. Chemical theory, for instance, gives the definitional conditions of the concept of *water* (whether the structure is classical or cluster) based on its molecular composition of two hydrogen atoms and one oxygen atom. Other descriptive conditions provided include freezing point, density, appearance and so on. These conditions *cohere* because the concept of water is given its meaning – definitional and descriptive content – by a chemical theory that identifies which attributes or conditions are important and how they are related. 'Found in beer and wine,' for instance, is an attribute of water that is unimportant according to chemical theory and is therefore not included in the definitional content.

So given the *theory theory*, which of the competing conceptual models is correct? I suspect that they are both applicable, depending on the concept. In the use of everyday non-technical concepts, such as 'vegetable,' the cluster model may be better. What counts as a *vegetable*, for instance, depends on a variety of factors, nutritive functioning, tradition, menu organization, etc. In a typical "meat and three" restaurant of the southern United States, for instance, french fries and peach cobbler sometimes count as vegetables for purposes of ordering from the menu. But in science, the conceptual structure may typically be more tightly specified and the classical model may better. What counts as a *quark, electron, element, compound, gene, population, reproduction, neuron,* etc. is in normal situations tightly specified by a well-defined set of definitional conditions, and the descriptive periphery does not play a significant role in determining whether something falls under a concept or not. (This may not be true in cases where the concept is still contested.) Similarly, for *species,* there might be a tightly specified structure and what counts as a species is limited to a well-defined conceptual core. More specifically, what makes something a species is determined by whether it is a segment of a population lineage with cohesion and distinctive fate, and nothing else is ultimately determinative. This is clearly more consistent with the classical approach.

This is not to say that there is no vagueness in the application of classical concepts. The condition themselves may be vague. In the case of species, what counts as a population may be borderline vague in the way that *town* and *city* are vague. There is no well-defined boundary between the two even if there are clearly significant differences in terms of size. Instead there is a range of population values that are borderline and could go either way. Similarly with *species,* there may be borderline vagueness in terms of *population.* There are no well-defined boundaries that demarcate non-populations and populations. There may also be vagueness (or perhaps ambiguity if there are several well-defined alternatives) in the other conditions. There might be different kinds of lineages, different kinds of cohesion, and different ways to think about evolutionary fates or trajectories.

This vagueness goes hand-in-hand with a *referential indeterminacy.* If the definitional conditions are vague or ambiguous, in the ways just outlined, then we may not know precisely how to apply the concept to the world. There may be groups of organisms that might be populations but not clearly and unambiguously so. Or there may be within some populations some level of cohesion, but not clearly and unambiguously enough cohesion to count as a species. In these cases, there is referential indeterminacy. It is not precisely clear how to apply the term *species.* [1] That is not to say, however, that the application of *species* is not restricted in some way. There will be a *reference potential,* limits within which the concept might be applied. Some groups of organisms clearly do not count as populations in the right way, and are therefore outside the reference potential. [1] In short, even though the classical conceptual structure may be unambiguous, the application of the concepts may still be problematic.

It may also be that a concept is not yet settled on theoretical grounds, in that there is some dispute about which definitional conditions are correct. This may be because there is some disagreement about the theoretical significance of certain conditions. After Darwin's *Origin,* for instance, speciation processes came to be important theoretically in thinking about species in a way that they were not before. Later theorists, especially those associated with the Modern Synthesis, developed a framework for thinking about speciation. In the terms William Whewell used, a species concept may become "explicated" – developed, refined and clarified as it gets applied to the world. [20, 1]

There are some important implications to this analysis of species concepts. First, there is an abstract, objective meaning constituted by a descriptive content associated with the term *species*, that is independent of any particular use of the term. This content is structured into a definitional core, determined in large part by the overarching theory of evolution, and a descriptive periphery, established at least in part by the contingent, empirical facts about those things that satisfy the theoretical definition. Included here are facts about morphological, behavioral, molecular similarities, processes operating in speciation and cohesion in both sexual and asexual organisms.

Second, this descriptive content is available in part or whole, to anyone who uses the term *species*, and has knowledge of evolutionary theory and the relevant empirical facts. Depending on context and interests though, focus may fall on either the definitional core or various parts of the descriptive periphery. Like the person for whom *water* means 'stuff to drink, bathe with, or swim in,' one could focus only on specific parts of the periphery. But he or she need not be thought of as denying that water is a compound of hydrogen and oxygen – whether or not he or she knows the molecular composition of water. Similarly, a person could concentrate on limited parts of the descriptive content of the term *species*, depending on theoretical or practical interests. This person could focus on reproductive isolation, genetic similarity or ecological functioning, without denying a theoretical definition of species that identifies species as segments of population lineages with cohesion and a distinctive fate – whether or not he or she is aware of that particular theoretical definition. So even if the subjective meaning of the terms *species* may vary in the actual usage of different researchers, there is an objective meaning of the term that is independent of the interests and backgrounds of those who use the term.

## 4. The demic structure of science

There is yet another factor relevant to a full understanding of the species problem. In the practice of science, scientists do not interact with *all* other scientists. Theoretical physicists, for instance, typically interact little with biologists. And even those within these disciplines scientists don't interact equally. Rather science gets practiced mostly within smaller groups. Vertebrate systematists, for instance, interact mostly just with other vertebrate systematists. And even within this group there are subgroups based on other factors such as the particular vertebrates studied, and whether molecules, morphology or behavior is the focus. Similarly geneticists who work on very specific problems are most likely to interact. There is then a hierarchy within the practice of science. Those within a particular discipline interact more than they do with those outside the discipline. And those within subdisciplines interact more. At the lowest level there are small groupings where the interaction is greatest. Following David Hull [21, 1], we can think of these small groups as *demes* - groups of interacting scientists that share distinctive subject matter, problems, methods and values.

Each of the demes may need to engage the species concept in various ways, depending on their distinctive interests, problems, methods and values. And most important for purposes here, each deme may focus on various parts of the descriptive content of the species concept, and ignore other parts. So a geneticist may not need to worry about the morphological similarity typical of species, or the historical dimensions of species in engaging the species concept. And an ecologist may not need to worry so much about genetic similarity. De Queiroz recognizes these differences in interests:

The existence of diverse species concepts is not altogether unexpected, because concepts are based on properties that are of the greatest interest to subgroups of biologists. For example, biologists who study hybrid zones tend to emphasize reproductive barriers, whereas systematists tend to emphasize diagnosability and monophyly, and ecologists tend to emphasize niche differences. Paleontologists and museum taxonomists tend to emphasize morphological differences, and population geneticists and molecular systematists tend to emphasize genetic ones. [12]

We need not follow de Queiroz there though, in thinking of these as different concepts. Rather these are just different emphases on the descriptive content of the theoretical species concept. Moreover, researchers need not focus on just one part of the descriptive content. In behavioral genetics, both genes and behavior are obviously important. And for evolutionary theorists all aspects of species may be relevant.

What is important here is first that particular interests may guide how the members of each deme thinks about species. Second, this does not entail that across demes researchers are using different theoretical concepts. The primary theoretical concept is still available to all. And most importantly, the primary concept constrains the usage of the term *species*. A geneticist may, for instance, think about species in terms of genes, but not in ways that are inconsistent with the fact that species have two dimensions – populational and historical, as segments of population lineages. The bottom line is that researchers may focus on different parts of the descriptive content that constitute the objective meaning of the term *species*. Which part they focus on may be determined by contingent, pragmatic factors that are unique to their particular deme.

Not all of these uses of the species term across demes are equally authoritative though. There is a linguistic division of labor. Since evolutionary theory plays an important role in determining the definitional core of the term *species*, those who work most directly on evolutionary theory have some linguistic authority over the term *species*. Just as theoretical particle physicists have linguistic authority over terms like *quark* and tell us what quarks really are through a theoretical definition, evolutionary theorists have linguistic authority over the term *species*, and tell us what they really are through a theoretical definition. So what constitutes the definitional core of a term is determined by those with linguistic authority. What this means is that some uses of the term *species* are parasitic on other more constitutive uses. And just as anyone who uses the term *quark* should know and respect the authoritative meaning established by theoretical physicists, anyone who uses the term *species* should know and respect the authoritative meaning of that term as determined by those with linguistic authority.

The species problem has been in part a consequence of the neglect of two facts: first, there is a social hierarchy in science that governs interaction, ultimately into demes; and second, there is a division of linguistic labor that arises out of this hierarchy. Those who work in these demes do not always recognize or respect this division of linguistic labor, and sometimes treat their own usage as authoritative. If so, then it would *seem* that there really are different concepts in use. The invertebrate systematist's concept is seemingly not the vertebrate systematist's, which is not the ecologist's, and which is in turn not the evolutionary theorist's. Here the use of apparently inconsistent concepts is an illusion, generated by a misunderstanding of the structure and content of the theoretical species concept, and a neglect of the division of linguistic labor. By understanding all this, we can, in effect, "dissolve" the species problem

## 5. Conclusion

Some conceptual problems are relatively easy to solve. We propose or invent a new concept that works better. Or we modify a current concept to better serve theoretical purposes. Both kinds of solutions are central to the practice and progress of science. While these solutions are not easy in the sense that the solutions are always or even ever obvious, they are easy in that they are straightforward and uncomplicated. The species problem is not easy in this way though. Its solution requires a sophisticated understanding of how scientific concepts work, are structured and get content. It also requires an understanding of how they work within the social structure of science. This complexity explains the long-endurance of the species problem. In part, the understanding of how concepts work was lacking. Only recently do we have the theoretical framework to understand such conceptual problems. So, just as we need evolutionary theory to understand what species are, we need a satisfactory conceptual theory to understand complex conceptual problems like the species problem.

There are, however, worries still lurking. What if there are theoretically important differences between the various segments of population lineages that we are identifying as species? Perhaps there are crucial differences between vertebrates, invertebrates, fungi and bacteria such that they should not all be regarded as forming the same kinds of species. What if, on our best theoretical understanding, there really do seem to be different kinds of species things? Is there really then, a single, fully adequate species concept? Or might there be multiple, irreducible concepts? If so, then the species problem returns, and not just as an illusion.

Marc Ereshefsky argues for just this kind of possibility. He accepts the basic idea that species are genealogical - historical lineages, but denies that they are all the same kinds of lineages. First he begins by noting there are three main ways of thinking about species - in terms of interbreeding, ecology and monophyly. Then he argues that these are different kinds of lineages produced by different evolutionary forces.

---

The positive argument for species pluralism is simply this: according to contemporary biology, each of the three approaches to species highlights a real set of divisions in the organic world... All of the organisms on this planet belong to a single genealogical tree. The forces of evolution segment that tree into a number of different types of lineages, often causing the same organisms to belong to more than one type of lineage. The evolutionary forces at work here include interbreeding, selection, genetic homeostasis, common descent, and developmental canalization... The resultant lineages include lineages that form interbreeding units, lineages that form ecological units, and lineages that form monophyletic taxa. [22]

---

These different kinds of lineage concepts apply in different ways to biodiversity. Some organisms, for instance, may not form ecological lineages. Consequently, that lineage concept would therefore not apply.

It is not initially obvious how to respond to Ereshefsky's pluralism. He considers and then rejects the suggestion that there is an additional parameter that can unite these three different kinds of lineages under one conception. [22] But at some level he seems to be already thinking of them under one conception. To even think of them *as three kinds of lineages* seems to assume that there is an overarching, more general way of thinking about species based on the idea of a lineage. It seems that if the theoretical concept is general enough, then surely it can be universal.

More worrisome perhaps, what if the species concept itself is ultimately unnecessary and misguided, the way the outdated ideas of *phlogiston* and *vital force* are? From my perspective as a philosopher, this anti-realist worry is abstract and not given much force by either evolutionary theory or what we know about the world. Nonetheless it cannot by dismissed on purely philosophical grounds. The answer, I believe, will be found ultimately in the practice of science. Is the theoretical species concept discussed here ultimately necessary for the practice of the biological sciences? It seems to me that it is, but the future might prove otherwise.

## Author details

Richard A. Richards

University of Alabama, Tuscaloosa, Alabama, USA

## References

[1]  Richards RA. The Species Problem: A Philosophical Analysis. Cambridge, U.K.: Cambridge University Press; 2010.

[2] Stauffer RC. *Charles Darwin's Natural Selection,* Cambridge U.K.: Cambridge University Press; 1975.

[3] Mayr E. Species Concepts and Definitions. In: Mayr E. (ed.) The Species Problem. Washington DC, U.S.A.; American Assoc. for the Advancement of Science; 1957.

[4] Mayden, RL. A Hierarchy of Species Concepts: the Denouement in the Saga of the Species Problem. In: Claridge MF., Dawah HA , Wilson MR., eds., Species: the Units of Biodiversity, London U.K.: Chapman and Hall; 1997.

[5] MacLaurin J, and Sterelny K. What is Biodiversity? Chicago, Il.: University of Chicago Press; 2008.

[6] Agapow P. and Bininda-Edmunds ORP. Crandall KA., Gittleman JL. Mace GM. Marshall JC. and Purvis A. The Impact of Species Concept on Biodiversity Studies. The Quarterly Review of Biology 2004; 79(2) 161-179.

[7] Cracraft J. Species Concepts in Theoretical and Applied Biology: A Systematic Debate with Consequences. In: Wheeler QD, and Meier R (eds.) Species Concepts and Phylogenetic Theory, New York, N.Y.: Columbia University Press; 2000.

[8] Wheeler QD and Meier R. Species Concepts and Phylogenetic Theory, New York, N.Y.: Columbia University Press; 2000.

[9] Claridge MF. Dawah HA. and Wilson MR. Practical Approaches to Species Concepts for Living Organisms. In: Claridge MF. Dawah HA. and Wilson MR (eds.) Species: the Units of Biodiversity. London U.K.: Chapman and Hall; 1997.

[10] Roll-Hansen N. Sources of Wilhelm Johannsen's Genotype Theory. Journal of the History of Biology 2009;42 457-493.

[11] de Queiroz K. The General Lineage Concept of Species and the Defining Properties of the Species Category. In: Wilson RA. (ed.) Species: New Interdisciplinary Essays. Cambridge Ma.: MIT Press; 1999.

[12] de Queiroz, K. Ernst Mayr and the Modern Concept of Species, Proceedings of the National Academy of Sciences. 2005;102(1) 6600-6607.

[13] Carnap R. An Introduction to the Philosophy of Science, N.Y., N.Y.: Basic Books; 1966.

[14] Chang H. Operationalism. *The Stanford Encyclopedia of Philosophy. Fall 2009.* URL = <http://plato.stanford.edu/archives/fall2009/entries/operationalism/>.

[15] Darwin C. On the Origin of Species, A Facsimile of the First Edition, Cambridge, Mass.: Harvard University Press; 1964.

[16] Reproduced with permission from John van Wyhe ed. The Complete Work of Charles Darwin Online. (http://darwin-online.org.uk/)

[17] Frege G. Sense and Reference. The Philosophical Review 1948;57(3) 209-230.

[18] Margolis E. and Laurence S. Concepts, The Stanford Encyclopedia of Philosophy. Spring 2007. URL = http://plato.stanford.edu/archives/spr2007/entries/concepts/

[19] Murphy GL. and Medin DL. The Role of Theories in Conceptual Coherence. Psychological Review 1985;92(3) 289-316.

[20] Whewell W. Selected Writings on the History of Science, Chicago, Il.: University of Chicago Press; 1984.

[21] Hull DL. Science and Selection: Essays on Biological Evolution and the Philosophy of Science, Cambridge U.K.: Cambridge University Press; 2001.

[22] Ereshefsky M. The Poverty of the Linnaean Hierarchy: A Philosophical Study of biological Taxonomy, Cambridge, U.K.: Cambridge University Press; 2001.

# Chapter 3

# Defining 'Species,' 'Biodiversity,' and 'Conservation' by Their Transitive Relation

Kirk Fitzhugh

Additional information is available at the end of the chapter

## 1. Introduction

---

*"...it follows that we should not regard the organism or the individual (not to speak of the species) as the ultimate element of the*

*biological system. Rather it should be the organism or the individual at a particular point of time, or even better, during a certain,*

*theoretically infinitely small, period of its life. We will call this element of all biological systematics... the character-bearing*

*semaphoront."* —W. Hennig [1: p6, emphasis original]

---

Definitions offer meanings of words by way of associations with other terms [2, 3]. Accordingly, a definition relates a concept to a name [4]. Like any field of science, precision of communication in biology is dependent upon the use of terms, while actions taken as consequences of the acceptance of those terms are constrained by definitions. The importance of the meanings of words is especially critical when actions in one subfield of biology are dependent on terms developed in other subfields. A case in point involves the terms *species, biodiversity,* and *conservation.* With a voluminous literature regarding what species are supposed to be, the topic has suffered from an overemphasis on 'species concepts' that are largely detached from more inclusive biological systematics principles (e.g. [5]) required to answer the question 'What is a species?' While upwards of 26 species 'concepts' are recognized [6–8], there is inconsistency among these in that they either refer to causal or acausal constructs. Causal characterizations typically present species as entities involved in past events, whereas acausal accounts refer to differentially shared features of organisms. Wilkins [8: p58, emphasis original] refers to causal-based concepts as indicating *"what* species *are,"* and acausal concepts as "how we *identify* species." The implication of either perspective leans toward species being entities that can be perceived by way of their properties, yet too often species and the organisms to which they refer are conflated (cf. [9]).

Settling the matter of what species are and the definition of the term requires a perspective that goes beyond the traditional, inordinately narrow consideration of just species. In acknowledging that species are taxa, we first must consider the formal definition of *taxon*. By extension, the definition of taxon must be consistent with the goal of biological systematics, and that goal must be accordant with that of scientific inquiry. The perspective regarding species, as with all taxa, must be reflective of scientific practice in not only systematics but also other fields in biology. For instance, we too often neglect to acknowledge that because evolutionary biology is foundational to systematics, taxa are not class constructs, but rather the products of discreet inferential actions intended to impart causal understanding, e.g. the inferences of phylogenetic hypotheses—as cladograms—from observations of differentially distributed organismal characters. Systematics deals with systematization, not classification [10–14]. As will be shown in this chapter, the scope of taxa extends to a number of types of biological phenomena involving organisms that are not routinely recognized by our formal nomenclatural systems, i.e. the *International Codes of Nomenclature* (e.g. [15]) or the *PhyloCode* [16]. The challenge is to place species into the more inclusive context of taxa. In doing so, we have an opportunity to not only formally define the term *species* such that it is concordant with the definition of *taxon*, but also acknowledge that the one term species has taken on a responsibility beyond its means in service of scientific inquiry.

In parallel with the significance and difficulties associated with the terms *species* and *taxon*, biologists have yet to settle the matter of defining *biodiversity*. Consider the following characterizations: "the collection of genomes, species, and ecosystems occurring in a geographically defined region" [17; see also 18–21; but see 22 for alternative opinions]; "the number of species observed or estimated to occur in an area (species richness)... This results from widespread recognition of the significance of the species as a biological unit..." [23: p220; see also 24–27]; "variation at all levels of biological organization" [28: p3; see also 29]; or "the variability among living organisms from all sources including… terrestrial, marine and other aquatic ecosystems and the ecological complexes of which they are part; this includes diversity within species, between species and of ecosystems" [30: p89; 31: p13, 109]. While biodiversity encompasses notions of spatial and temporal variation, the currency of that variation remains unsettled, ranging from the properties of organisms to taxa, but with species being the most popular subject [9, 32]. The definitional ambiguities regarding species and biodiversity carry with them potential negative epistemic and operational implications in fields as diverse as systematics, evolutionary biology, ecology, and conservation.

The vagaries of defining biodiversity stem in large part from the lack of consensus among systematists as to the definition of species, much less the more inclusive term taxon (e.g. [7, 33–36]). Thus, justifying actions under the guise of conservation will remain ambiguous per the largely arbitrary focus on the one taxon, species. As will be shown in this chapter, settling the matter of how to characterize biodiversity and conservation requires acknowledging the transitive relations between taxa, biodiversity, and conservation: {(taxa, biodiversity), (biodiversity, conservation), (taxa, conservation)}. These relations manifest themselves as follows:

*biodiversity* must be determined in the context of *taxa*; *biodiversity* determines the limits on *conservation*; *taxa* determine the realization of *conservation*. Regarding species, these relations have the added benefit of emphasizing the relevance of formal definitions of biodiversity and conservation to a formal definition of species that is consistent with biological systematics practice. These definitions follow from the fact that they reflect the inferential products of our observations of organisms. There is the added consequence of recognizing that the one term species cannot accommodate the variety of causal events to which it has been associated and also be cogently defined. The time has come to begin giving serious consideration to a more scientifically and operationally reasonable parsing of 'species' with new sets of terms that accurately convey a variety of non-comparable causal events.

Pursuant to the transitive relations outlined above, this chapter will pursue three interrelated issues. The first is to present a solution regarding the role of taxa when speaking of systematics, biodiversity, and conservation. This entails identifying the objects that cause our perceptual beliefs, i.e. individual organisms, and the inferential relations of objects/beliefs to the variety of explanatory hypotheses referred to as taxa. It is then a straightforward matter to show that species, as a taxon, is but one of the classes of hypotheses used in the pursuit of causal understanding in systematics. This offers an advance toward defining the term species as developed by Fitzhugh [12–14, 37] and recognizing the inherent limits of that definition and the need for additional terms.

If the objects from which our perceptual beliefs are derived in the biological sciences are individual organisms (more properly 'semaphoronts' *sensu* [1]; cf. quote above **Introduction**), not species or other taxa, then we need to assess whether or not genes, species, ecosystems, etc., can be regarded as 'units' or 'entities' to which biodiversity and conservation refer. The second goal of this chapter is to present a formal, operational definition of biodiversity that is not only consistent with conservation, but also the nature of taxa as explanatory hypotheses. From a systematics perspective, biodiversity cannot be construed in terms of genes, specific/phylogenetic taxa, or ecosystems, but rather in the context of past, proximate tokogenetic (reproductive) events related to groups of organisms that exist in the present. The restriction imposed is one directly related to the objects and events that conservation attempts to conserve: circumscribed tokogenetic events into the future – *not* species, *not* taxa. When referring to biodiversity, one can argue that population- or intraspecific-level variation/polymorphism, inclusive of smaller-scale heterogeneity (e.g. 'distinct population segments' [38]; 'evolutionarily significant units' [39–42]; 'designatable units' [43]) can be considered along with specific- and supraspecific-level (i.e. phylogenetic) hypotheses. The only roles population-, intraspecific-, specific-, or phylogenetic-level taxa have to play in this regard is in the capacity of surrogates for denoting inferred, temporally proximate systems of past tokogeny. Taxa, including species, are inferential products functioning to provide causal understanding of what is perceived of observed organisms, thus taxa are not objects to which biodiversity can directly apply. The subjects to which biodiversity refer are, instead, individual organisms that exist in the present – not species, genomes (which, while a functional term, refer to parts of organisms), or ecosystems (the interactions of organisms with their surroundings as well as other organisms). The reference to organisms in the context of biodiversity is operational only in the sense of maintaining geographically circumscribed sets of tokogenetic events into the future, i.e. conservation.

In speaking of maintaining future tokogeny in reference to biodiversity, the third goal of this chapter will be a formal, operational definition of conservation. This will entail the view that conservation is an activity focused on future tokogeny among or between individuals, not the maintenance of taxa, including species. It is individual organisms and their potential to undergo future tokogeny that is the subject of conservation. The definition of conservation in terms of individuals and potential tokogeny imposes limits not only on the definition of biodiversity but also definitions of taxon and species. As conservation pertains to the maintenance of tokogenetic events among groups of individuals into the future, the subject of biodiversity is also those groups of individuals. But in the case of biodiversity, taxa provide the epistemic, as well as causal basis for delimiting those groups, upon which conservation has a basis for implementation.

## 2. The nature of taxa

As indicated in the **Introduction**, definitions of taxon, biodiversity, and conservation are interdependent by way of their transitive relations. For instance, a key element in discussions of biodiversity and conservation has been species [9, 32, 44–46], and sometimes by extension, intra- and supraspecific taxa. Addressing the matter of defining biodiversity and conservation cannot solely rely on presentation of a definition of species. The more general question, 'what are taxa?', must first be addressed. Answering this question has significant consequences for characterizations of biodiversity and conservation if taxa are to play any role, per their transitive relations.

Definitions of taxon have taken two general approaches: (1) taxa are class constructs: e.g. "a taxonomic group of any rank" [47: p9; 48–59]; or (2) taxa are quasi-explanatory accounts, e.g. "A taxon or taxonomic group… can mean only a group of organisms related genetically (or so related to the best of our knowledge).... It is a taxonomic group or assemblage of plants or animals, having certain characteristics in common which we take as evidence of genetic relationships, and possessed of some degree of objective reality" [60: p38; 15, 61, 62]. This class versus explanation distinction is often ambiguous. It is not uncommon to see definitions of taxon as a class construct yet also referred to in an evolutionary context, such that taxa must be interpreted as explanatory vehicles. For instance, Kardong [63: p330] states, "A taxon is simply a named group of organisms. A taxon may be a natural taxon, one that accurately depicts a group that exists in nature resulting from evolutionary events. Or a taxon may be an artificial taxon, one that does not correspond to an actual unit of evolution." The causal connotations associated with (2) have also taken on the form of considering taxa as natural kinds, or 'homeostatic property cluster kinds' [64–71].

In contrast to characterizations of taxa, discussions of species *per se* tend to consider three alternatives. Species are (1) class constructs [48, 53, 72], (2) entities with the ontological status of individuals [5, 35, 36, 70, 73–90], or (3) natural kinds [64–71, 88, 91–95]. A fourth option is the view that species are 'segments of lineages' [16, 45, 94]. Discussions of individuality and natural kinds, whether in regard to species or taxa as a whole, have degraded into little more than rhetorical devices, e.g. [71, 88, 95, 96]. Considering species in the broader context of taxa, arguments have been provided by Fitzhugh [10–14, 37, 97, 98] outlining that neither the class nor individuality thesis is appropriate to the issue (see [34] for extensive critiques), and the lineage and natural kinds concepts are incomplete characterizations relative to the treatment of taxa as explanatory hypotheses. Taxa are not mere classes, given that the goal of science is not classification but rather systematization, i.e. the placement of objects into some theoretical framework for the purpose of acquiring causal understanding [55, 99–101; *contra* 93]. Taxa, including species, cannot be conceived as individuals for the fact that there are no discernible properties by which such 'entities' could be discovered, recognized, or described [102, 103]. While the lineage and natural kinds concepts approach the necessary explanatory tone, neither has been sufficiently developed in terms of their inferential relations to the nature of other taxa and the intersections with the objects that are our interest, i.e. organisms. What was proposed by Fitzhugh [12, 13, 37; cf. 34 for a similar, independently derived perspective] is that as our perceptions are caused by individual organisms, we naturally apply a variety of explanatory hypotheses to those perceptions, and at least some of those hypotheses are formally represented by names controlled by international bodies governing nomenclatural actions (e.g. [15]; cf. [104, 105] for examples of implementation in this context). Several classes of these hypotheses are shown in Figure 1, derived from Hennig [1: fig.6; see also 12: fig.1]. Note that what are observed in the present are individual organisms (semaphoronts). Such perceptions are possible because of the particular properties instantiated by those objects. An observation statement is itself an explanatory account of one=s sense perceptions, where the causes of those perceptions are the existence of objects [106]. Beyond our perceptual hypotheses are a variety of other causal accounts that are routinely invoked to answer specifiable questions regarding the properties of organisms. The relations between causal questions that might prompt explanatory hypotheses of the types shown in Figure 1 and how they are communicated by way of informal and formal names are presented in Table 1.

What is apparent is that species *sensu lato*, much less all taxa, are not the fundamental objects or entities of which we speak when referring to observations (*contra* e.g. [9]). Indeed, as outlined later, questions like "Are species real?" can be only answered by a very qualified 'yes.' Species are real to the extent that they, as well as all taxa, represent specifiable sets of past causal events involving individual organisms, not for the fact that species/taxa are entities (*fide* [9, 35, 107]). As well, the nature of inferences to species cannot entail the notion that the relation between organismal characters and species is one of 'diagnosis,' or that diagnoses are a matter of hypothesis testing (e.g. [108]).

The inferential derivation of all other taxa becomes even more apparent when we acknowledge that the goal of biological systematics is (or should be, cf. [10, 12–14, 98, 109]) consistent with that of scientific inquiry [55, 110–118]: to acquire ever-increasing descriptive and, more importantly, causal understanding. Indeed, the view that systematics is not a process of classification *qua* discerning and arranging classes via acausal 'relationships' (e.g. [57]) or discovering metaphysical individuals (e.g. [85]), but instead an endeavor to pursue causal understanding clearly extends from Darwin's [119: chapter XIII] own views. Ghiselin [61: p87] correctly notes that,

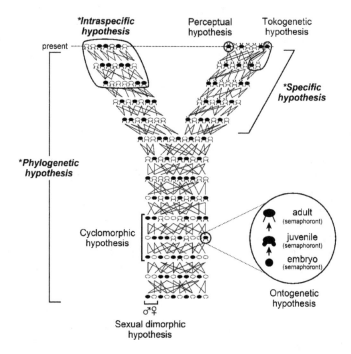

**Figure 1.** Relations between perceptual, ontogenetic, tokogenetic, intraspecific, specific (cf. Table 1: 'species₁,' 'species₂,' 'species₄') and phylogenetic hypotheses (adapted in part from [1: fig.6; 12: fig.1; 37: fig.1; 98: fig.1]). Those classes of taxa that might be referred to in the context of biodiversity and conservation are marked with an asterisk (*), i.e. intraspecific, specific, and phylogenetic.

"Systematics has scientific value as explanation, not as mere description. The purpose of classification is not the accurate pigeonholing or identification of enzymes or dried specimens, but the assertion of meaningful propositions about laws of nature and particular events."

If taxa are explanatory hypotheses directed at our observations of organisms, then there are two interesting consequences for biodiversity. The first is that species are not the exclusive units or entities to be considered in the determination or communication of biodiversity or conservation. It has been the pervasive conflation of individual organisms with species that has led to the misconception that species are discernible things and the principle objects of interest. The second consequence is that supraspecific hypotheses are not epistemically equivalent to specific hypotheses, thus negating the notion of taxonomic surrogacy [26, cf. references therein promoting its use; also referred to as 'taxonomic sufficiency,' e.g. 120–123; not to be confused with 'surrogacy' as applied in conservation, cf. 32]. These are distinct classes of hypotheses inferred from different sets of theories [12–14]. Similarly, supraspecific hypotheses assigned the same rank are not epistemically equivalent relative to hypotheses of other ranks [12, 26, 120; *contra* 124]. Viewed from the perspective of cladograms, there are no objective criteria for selectively assigning some taxa formal, ranked names to the exclusion of remaining hypotheses implied by those cladograms. Lacking such equivalence is the strongest argument against the use of taxonomic surrogacy, at least with regard to supraspecific taxa serving as representations of species.

| SEQ CHAPTER \h \r 1 Causal questions: | SEQ CHAPTER \h \r 1 Relations: | SEQ CHAPTER \h \r 1 Represented by: |
|---|---|---|
| SEQ CHAPTER \h \r 1"Why do I have these sense perceptions?" | SEQ CHAPTER \h \r 1*Perceptual hypothesis* – An individual exists. | SEQ CHAPTER \h \r 1Observation statement |
| SEQ CHAPTER \h \r 1"Why does this individual have character $X$ at time $t_2$ in contrast to $Y$ at $t_1$?" | SEQ CHAPTER \h \r 1*Ontogenetic hypothesis* – This individual has character $X$ at time $t_2$ because it is part of the ontogenetic trajectory. | SEQ CHAPTER \h \r 1Semaphoront names, e.g. "embryo," "larva," "adult" |
| SEQ CHAPTER \h \r 1"Why are these individuals observed at this location in contrast to some other location?" "Why does this individual, or individuals, have character $X$ in contrast to $Y$?" | SEQ CHAPTER \h \r 1*Tokogenetic hypothesis* – Individuals are at this location because they are products of past tokogenetic events among other individuals. This individual, or individuals, has character $X$ because the genetic capacity to exhibit the character was passed on from their parent(s). | SEQ CHAPTER \h \r 1Families, demes, populations, communities, etc. |
| SEQ CHAPTER \h \r 1"Why do individuals to which species hypothesis *x-us* refers have either character $X$ or $Y$ in contrast to only character $X$ observed among other individuals?" | SEQ CHAPTER \h \r 1*"Intraspecific" hypothesis* – The reproductively isolated population is polymorphic because character $Y$ originated in the population, such that observed individuals with $X$ and $Y$ are products of past tokogenetic events among individuals with those characters. | SEQ CHAPTER \h \r 1Polymorphism |
| SEQ CHAPTER \h \r 1"Why do these individuals have character $X$ in contrast to character $Y$ observed among individuals to which other species hypotheses have been applied?" | SEQ CHAPTER \h \r 1*Species$_1$ hypothesis* – character origin, with subsequent fixation via tokogeny by sexual reproductive events. Individuals have character $X$ because it originated among individuals with character $Y$, and $X$ eventually became fixed throughout the population, such that individuals observed in the present are products of past | SEQ CHAPTER \h \r 1Species *sensu lato* names |

| SEQ CHAPTER \h \r 1 Causal questions: | SEQ CHAPTER \h \r 1 Relations: | SEQ CHAPTER \h \r 1 Represented by: |
|---|---|---|
| | sexual-based tokogenetic events involving individuals with that character. SEQ CHAPTER \h \r 1 *Species$_2$ hypothesis – simultaneous character origin/fixation via tokogeny by sexual reproductive events, i.e. hybridization, polyploidy.* Individuals have character *X* because it immediately originated as a consequence of hybridization, such that individuals observed in the present are products of past sexual-based tokogenetic events involving hybrid individuals with that character. SEQ CHAPTER \h \r 1 *Species$_3$ hypothesis – simultaneous character origin/fixation, with subsequent tokogeny by asexual, apomictic/ parthenogenetic, or self-fertilizing hermaphroditic reproductive events.* Individuals have character *X* because it originated in an individual, such that individuals observed in the present are products of subsequent past asexual-, apomictic- / parthenogenetic-, or self-fertilizing-based tokogenetic events derived from that original individual. SEQ CHAPTER \h \r 1 *Species$_4$ hypothesis – character origin, with subsequent fixation via tokogeny by alternations of sexual and asexual reproductive events.* Individuals have character *X* because it originated among individuals with character *Y*, and *X* eventually became fixed throughout the population, such that individuals observed in the present are products of past alternating sexual and asexual tokogenetic events involving individuals with that character. SEQ CHAPTER \h \r 1 *Species$_5$ hypothesis – immediate character origin/fixation via horizontal genetic exchange.* Individuals have character *X* because it immediately originated in an individual from a horizontal genetic exchange event, such that individuals | |

| SEQ CHAPTER \h \r 1 Causal questions: | SEQ CHAPTER \h \r 1 Relations: | SEQ CHAPTER \h \r 1 Represented by: |
|---|---|---|
| | observed in the present are products of subsequent past asexual-based tokogenetic events derived from that original individual. | |
| SEQ CHAPTER \h \r 1"Why do these Individuals, to which species hypotheses *a-us* and *b-us* refer, have character *X* in contrast to character *Y*?" (applicable to gonochoristic or cross-fertilizing hermaphroditic organisms) | SEQ CHAPTER \h \r 1*Phylogenetic hypothesis* Individuals have character *X* because this character originated within a population with character *Y*, and *X* eventually became fixed throughout the population, and there was subsequent splitting of that population into two or more populations. | SEQ CHAPTER \h \r 1Supraspecific names |

**Table 1.** Comparisons of hypotheses commonly encountered in biological systematics (modified from [12]). Note that within what has traditionally been considered the scope of species it is necessary to distinguish five different sets of causal events that account for shared features. See Figure 1 for graphic representations of each hypothesis (only 'species$_i$' hypotheses shown).

## 3. Inferences of taxa

The previous section outlined the link between biological systematics hypotheses, colloquially known as taxa, and what exists in the form of observed organisms with properties explained by those hypotheses (Table 1, Fig. 1). As the principle goal in all fields of science is the acquisition of causal understanding, the inferential relations in systematics that span our observation statements and explanatory hypotheses can be summarized as follows:

[A]

a.  organisms, as objects, are perceived as a matter of the properties they instantiate;

b.  as consequences of perceptions in (a), represented by our causal questions, the focal point of biology is individual organisms (semaphoronts), or parts or remnants of organisms (Table 1);

c.  there are a variety of explanatory hypotheses that serve as answers to questions in (b) regarding perceptions of organisms; the following classes of hypotheses are common in systematics and other biological fields (Table 1; Fig. 1):

  i.     ontogenetic (via semaphoronts)

  ii.    tokogenetic

  iii.   intraspecific/polymorphic

  iv.    specific (species) *sensu lato* (cf. Table 1)

  v.     phylogenetic;

**d.** the goal of inferring the hypotheses in **(c)** is to provide at least preliminary explanatory accounts of differentially distributed properties observed among semaphoronts.

It is from these inferential relations that implications can be identified for considerations of intraspecific/polymorphic, specific, and phylogenetic hypotheses for biodiversity and conservation. Notice that each class of explanatory hypothesis in **[A](c)** has as its basis particular properties of organisms (Table 1). Each type of hypothesis serves to address a subset of the totality of properties one perceives, and as such, hypotheses inferred from different causal theories are not epistemically equivalent [12, 13]. These conditions present consequences for the view that taxa, no matter the rank, are the relevant subjects in the definitions or discussions of biodiversity and conservation.

The inferential links between our perceptions of organisms and conclusions in the form of taxa [cf. **[A](c)**] have been analyzed by Fitzhugh [10, 12–14, 37, 97, 98, 109, 125, 126], thus only briefly described here. The process of reasoning from observed effects, in this case the properties of organisms, to explanatory hypotheses (taxa) is known as abduction or abductive inference [127–154; cf. 10–13, 37, 97, 98, 109, 125, 126 for considerations of abduction in relation to biological systematics and evolutionary biology]. Abduction can be schematically represented as:

**[B]**

- auxiliary theory(ies)/hypotheses

- theory(ies) relevant to perceived effects

- perceived effects

- explanatory hypothesis.

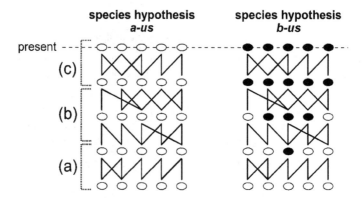

**Figure 2.** Diagrammatic representations of two specific hypotheses (cf. Table 1: 'species₁,' 'species₂,' 'species₄'), indicating (a) origins and (b) fixation of the respective properties of gray and black body walls, with (c) subsequent tokogeny resulting individuals observed in the present.

While a form of non-deductive inference, abduction is distinct from induction sensu stricto, which entails the process of hypothesis testing, contra [155]. Deduction provides for stipulating potential test evidence.

Consider the example in Figure 2. Specimens are observed with the unexpected or surprising properties of either gray or black bodies. It is the nature of these observations that prompts the implicit or explicit causal questions of why these properties are present among individuals in contrast to what has been previously observed (and expected). Specific hypotheses $u$-$us$ and $b$-$us^1$ provide respective explanatory accounts of the observations. The theory used for the purposes of inferring these hypotheses would have the form described by Fitzhugh [12, 13; see also 37]:

**[C]**

If property $Y$ originates by mechanisms $a, b, c... n$ among gonochoristic [*partim*] or cross-fertilizing hermaphroditic individuals of a reproductively isolated population with character $X$, and $Y$ subsequently becomes fixed throughout the population during tokogeny by mechanisms $d, e, f... n$, then individuals observed in the present will exhibit $Y$.

The basis for giving the label 'species' to these hypotheses is contingent on the same theory being applied to particular features, which is consistent with the formal definition of species offered by Fitzhugh [13: p207]:

**[D]**

An explanatory account of the occurrences of the same character(s) among gonochoristic or cross-fertilizing hermaphroditic individuals by way of character origin and subsequent fixation during tokogeny.

Given the theory in **[C]** and definition in **[D]**, Fitzhugh [13] pointed out that specific-level hypotheses cannot be applied to organisms with obligate asexual, parthenogenetic, or self-fertilizing hermaphroditic reproductive strategies, suggesting that causal accountings applicable to such individuals would be akin to phylogenetic hypotheses (cf. **[G]**, **[H]**); not to be confused with the 'phylogenetic species concept' [6, 156, 157; cf. 158]. The situation is, however, more complex as is apparent in Table 1. At least five sets of general causal conditions are subsumed under species *sensu lato*—'species$_1$' through 'species$_5$'. Contrary to Fitzhugh's [13] suggestion, it would be more accurate to regard 'species$_3$' (asexual, apomictic/parthenogenetic, self-fertilizing hermaphroditic hypotheses) and 'species$_5$' (horizontal genetic exchange hypotheses; [159–163]) distinct from phylogenetic hypotheses since the causal events entailed by the latter involve character origin/fixation as part of tokogeny that is to some extent sexual-based interbreeding, with subsequent population splitting (cf. **[H]**). The remaining classes of hypotheses, 'species$_1$,' 'species$_2$,' and 'species$_4$,' refer to causal events among individuals to which phylogenetic hypotheses *also* apply. While the definition of species by Fitzhugh ([13];

---

1 I have intentionally avoided using the binomial arrangement of genus and specific epithet to denote specific hypotheses. As supraspecific (= phylogenetic) hypotheses are the products of inferential acts that are separate from inferences of specific hypotheses (Table 1), it can be argued that there should be no requirement that formal recognition of a specific hypothesis be made in conjunction with any phylogenetic (= genus ranked) hypothesis [12, 14, 104, 105].

cf. [D]) satisfies the requirement of limiting specific hypotheses to obligate sexual-based interbreeding ('species$_1$,' 'species$_2$'), this definition can be extended to organisms with more complex life histories involving combinations of asexual and sexual reproductive events ('species$_4$').

Two alternative strategies might be considered for handling the five connotations of species in Table 1. Either approach might appear unorthodox, but what we save by holding to tradition comes at the expense of precisely conveying hypotheses and accurately representing biodiversity and conservation. What is apparent is that effective communication of a multitude of causal events represented in systematics hypotheses cannot be accomplished with the one term, *species*. The more radical approach would be to limit specific hypotheses to 'species$_1$,' via the theory in [C], while referring to 'species$_{2-5}$,' as inferential products of four additional theories:

[E]

a.  *species theory* ('species$_1$'): See [C]; phylogenetic hypotheses, *sensu* [H], are also applicable to individuals to which such specific hypotheses apply;

b.  *interspecific hybrid theory* ('species$_2$'): if property $Y$ simultaneously originates and is fixed by hybridization, e.g. polyploidy, among gonochoristic or cross-fertilizing hermaphroditic individuals to which respective 'species$_1$' hypotheses refer, such that subsequent tokogeny is limited to individuals with $Y$, then individuals observed in the present will exhibit $Y$; phylogenetic hypotheses, *sensu* [H], are also applicable to individuals to which such specific hypotheses apply;

c.  *asexual-autogamic theory* ('species$_3$'): if character $X$ exists among individuals with obligate reproduction that is asexual, apomictic/parthenogenetic, or self-fertilizing, and character $Y$ originates by mechanisms $a, b, c... n$, during tokogeny, then individuals observed in the present exhibiting $X$ and $Y$ are respective tokogenetic products of individuals with those characters [13: p210]; phylogenetic hypotheses are not applicable to individuals to which such specific hypotheses apply;

d.  *asexual-sexual theory* ('species$_4$'): if property $Y$ originates by mechanisms $a, b, c... n$ among individuals with $X$ during one of the alternative phases of asexual or sexual tokogenetic events, and $Y$ subsequently becomes fixed throughout the population during tokogeny by mechanisms $d, e, f... n$, then individuals observed in the present will exhibit $Y$; phylogenetic hypotheses, *sensu* [H], are also applicable to individuals to which such specific hypotheses apply;

e.  *horizontal genetic exchange theory* ('species$_5$'): if character $X$ exists among individuals and character $Y$ subsequently occurs by horizontal genetic exchange mechanisms $d, e, f... n$ with other individuals, then individuals observed in the present exhibiting $X$ and $Y$ are respective tokogenetic products of individuals with those characters.

A more conservative approach entails using the specific theory and definition in [C] and [D], respectively, to broadly accommodate 'species$_1$,' 'species$_2$,' and 'species$_4$.' 'Species$_3$' and 'species$_5$' remain classes of explanatory accounts among organisms that cannot be coordinately accommodated by either specific *sensu stricto* (cf. [D]) or phylogenetic (cf. [G], [H]) hypotheses and both types of hypotheses would require new, separate formal names distinct from the rank of species. Acknowledging theories [E](c) and (e) for 'species$_3$' and 'species$_5$,' respectively, makes apparent the importance of the suggestions offered here: the applications of these theories are not necessarily mutually exclusive for obligate asexually reproducing individuals [80, 159, 163–166]. The solution to conveying both classes of events must be by way of taxon names distinct from the one term 'species.' No doubt exceptions to the theories in [E] can be pointed to. Rather than intended as an exhaustive list, the objective here is to recognize that protocols need to be pursued for accommodating a variety of classes of events to be communicated in a manner more effective than under the one heading, 'species.'

Applying the specific theory in [C] or [E](a) to the observations in Figure 2, the inference can be summarized as follows:

**[F]**

**Theory:** If property $Y$ originates by mechanisms $a$, $b$, $c$... $n$ among gonochoristic or crossfertilizing hermaphroditic individuals of a reproductively isolated population with character $X$, and $Y$ subsequently becomes fixed throughout the population during tokogeny by mechanisms $d$, $e$, $f$... $n$, then individuals observed in the present will exhibit $Y$.

**Observed effects:** Individuals have gray bodies, in contrast to white or black.

**Explanatory hypothesis (species taxon *a-us*, cf. Fig. 2):** The gray character arose in a reproductively isolated population of white individuals, and the gray condition subsequently became fixed throughout the population during tokogeny, leading to gray individuals observed in the present.

While this inference pertains to 'species$_1$' hypotheses, there are several aspects that can be extended to all of the classes of hypotheses in Table 1 that are inferred from theories in [E], and will also figure prominently in the definitions of biodiversity and conservation presented later. Note that individuals observed in the present are products of proximate instances of tokogeny (Fig. 2c). Regardless of whether these tokogenetic events occurred subsequent to the origin (Fig. 2a) and fixation (Fig. 2b) of the character to which the hypothesis refers, or were contemporaneous with either of those events, observed individuals are products of tokogeny. What will be shown later regarding proximate tokogeny is that specific *sensu lato* as well as supraspecific taxa (phylogenetic hypotheses) are only tangentially relevant to biodiversity and conservation. Specific- and phylogenetic-level taxa serve as surrogates, referring to separate, hypothesized proximate systems of tokogeny (Fig. 2c) that are used to denote biodiversity. But in the case of conservation, taxa are largely irrelevant, as the scope of the act of conservation is potential processes of tokogeny into the future, not the maintenance of intraspecific, specific, or supraspecific taxa (Fig. 1, Table 1). Taxa cannot be conserved because they are not spatiotemporally restricted things – they are explanatory constructs only relevant to observed organisms.

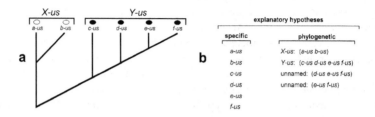

**Figure 3.** a) Cladogram schematically summarizing hypothesized causal events. (b) List of all hypotheses indicated in the cladogram in (a).

Consider next the example in Figure 3a. Presented as a cladogram, phylogenetic hypothesis *X-us* implies that the property gray originated and became fixed among individuals as a result of an unspecified causal event(s), subsequent to which was a population splitting event, the specifics of which are also not indicated in the diagram. *Y-us*, on the other hand, entails that the character black originated and became fixed, subsequent to which were a series of population splitting events, to which other unspecified hypotheses refer. Unlike the previous example of inferences of specific hypotheses *a-us* and *b-us* (cf. **[F]**, Fig. 2), phylogenetic hypotheses *X-us* and *Y-us* are derived from a phylogenetic theory; somewhat different from the theory used to infer *a-us* and *b-us* (cf. Table 1: 'species₁,' 'species₂,' 'species₄,' **[E](a), (b), (d)**; [10, 12–14, 37, 98, 126]):

**[G]** [2]

If character X exists among individuals of a reproductively isolated, gonochoristic or cross-fertilizing hermaphroditic population and character Y originates by mechanisms a, b, c... n, and becomes fixed within the population by mechanisms d, e, f... n [= ancestral species hypothesis], followed by event or events g, h, i... n, wherein the population is divided into two or more reproductively isolated populations, then individuals to which descendant species hypotheses refer would exhibit Y.

The abductive inference of hypothesis Y-us, for instance, would have the form:

**[H]**

**Theory:** If character *X* exists among individuals of a reproductively isolated, gonochoristic or cross-fertilizing hermaphroditic population and character *Y* originates by mechanisms *a, b, c... n*, and becomes fixed within the population by mechanisms *d, e, f... n* [= ancestral species hypothesis], followed by event or events *g, h, i... n*, wherein the population is divided into two or more reproductively isolated populations, then individuals to which descendant species hypotheses refer would exhibit *Y*.

---

2 While this 'phylogenetic theory' offers a common cause accounting, it actually refers to two classes of common cause events: character origin/fixation (applied to each set of shared characters) and population splitting. The implementation of the theory in **[H]** obviates Sober's [167: p157] view that phylogenetic inference is a two-step process, i.e. "The first problem is… one infers a tree;… one uses an inferred tree to solve a further problem [of ancestral character transformation]."

**Observed effects:** Individuals to which species hypotheses *c-us*, *d-us*, *e-us*, and *f-us* refer have black bodies, in contrast to white or gray.

**Explanatory hypothesis (supraspecific taxon *Y-us*, cf. Fig. 3a):** The black character among individuals to which species hypotheses *c-us*, *d-us*, *e-us*, and *f-us* refer arose in a reproductively isolated population of white individuals, and the black condition subsequently became fixed throughout the population during tokogeny, followed by three events of population splitting to produce isolated populations to which descendant species hypotheses refer (*c-us*, *d-us*, *e-us*, *f-us*).

Note that the cladogram in Figure 3a refers to two classes of hypotheses – specific (*sensu stricto*; **[F]**) and phylogenetic. Of these, eight (six specific, two phylogenetic) are formally named, while two phylogenetic hypotheses are unnamed (Fig. 3b), i.e. (*d-us*, *e-us*, *f-us*) and (*e-us*, *f-us*). Acknowledging the presence of these latter, unnamed hypotheses highlights the erroneous view that supraspecific hypotheses can serve as 'taxonomic surrogates' for specific hypotheses. Supraspecific taxa invariably subsume any number of less general phylogenetic hypotheses, such that attempts to treat formally named hypotheses as surrogates for specific hypotheses is not only arbitrary at the level of phylogenetic hypothesis selection, but also arbitrary with regard to only focusing on the one class of systematics hypothesis called species (cf. Fig. 1, Table 1).

The inference in **[H]** would typically be said, albeit erroneously, to be a 'parsimony' approach (cf. [167–169]), in contrast to implementations of either maximum likelihood or Bayesianism (cf. [170– 172]). Parsimony and likelihood cannot, however, be treated as separate concepts in abduction [10, 14]. So-called maximum likelihood methods require the assumption that one's observation statements regarding shared characters are inconsistent with the causal questions to be asked in relation to explaining those observations, given that a cladogram subsumes a series of incomplete explanatory hypotheses that take 'branch length' into consideration. By extension, likelihood methods utilize theories that are not necessarily common cause theories (cf. **[G]**) because rates of change are involved. Strictly speaking, using a rate-based theory applies to tokogenetic-, not phylogenetic-level phenomena. The consequence is that the scope of such inferences moves from being phylogenetic to either specific, polymorphic, or intra-specific/tokogenetic [10, 13, 14]. As noted by Cleland ([173: p572, emphasis original; see also 174–177]),

"The scientifically most fruitful common cause explanations appeal to *last* (proximate) common causes. A last common cause represents the causal juncture at which the items in the collection cease to share a more recent common cause. Because they maximize causal unity last common cause explanations have greater explanatory power than other common cause explanations."

Bayesianism, on the other hand, addresses changes in hypothesis belief relative to test evidence (*contra* e.g. [178–180]). Bayes' Theorem has no role to play in abductive inference for the fact that character data, as the basis for inferring phylogenetic hypotheses, cannot in turn be used to test those hypotheses [10, 14, 126].

## 4. Defining biodiversity and conservation

Most of the definitions of biodiversity cited in the **Introduction** make reference to three qualities: genomes, species, and ecosystems. Noss [22: p356] included an additional dimension, claiming that "Biodiversity is not simply the number of genes, species, ecosystems, or any other group of things [*sic*] in a defined area... Ecologists usually define 'diversity' in a way that takes into consideration the relative frequency or abundance of each species or other entity, in addition to the number of entities in the collection." But as taxa are explanatory hypotheses rather than objects, Noss commits the common mistake (cf. [9]) of conflating species with individual organisms. As shown in Figures 1–2, **[A]**, **[F]**, and Table 1, individuals (represented/communicated by observation statements) and species are separate subjects produced from different inferential actions. Noss [22: p356, fig.1] further states there are "three primary attributes of ecosystems: composition, structure, and function," and within each are interdependent, internested relationships (Fig. 4). Interdependence, much less hierarchical organization is at best dubious. For instance, 'genes,' if one assumes this to mean a discrete sequence of nucleotides, are not dependent upon 'genetic structure' and 'genetic processes.' The latter two are dependent on the presence of genes, as properties of organisms, not vice versa. Given what was discussed earlier regarding individual organisms as the focal point of observations in biology, we need only consider the compositional aspects in Figure 4, where the hierarchy includes genes, populations, and species. A key question to ask is whether or not any of these actually match what we perceive. Species *sensu lato* (Table 1, **[E]**) are explanatory hypotheses, not individuals, entities, or things. While a sequence of nucleotides can be indirectly discerned, they are only referred to as genes because of the functional role they play in causal processes during the life (ontogeny) of an organism. Sequences, as opposed to functional units *qua* genes, are properties of organisms. A population is a group of individuals, and speaking of that group is by way of one's perceptions of individual organisms, and perhaps by extension past causal relations (Fig. 1, Table 1). There are no applicable emergent properties of a population one might perceive that are not manifestations of the component organisms of that population or their collective actions. The biological components that can be perceived in the more inclusive concepts shown in Figure 4, communities/ecosystems and landscape types, are individual organisms. Biodiversity could not have as its focus genes, species, or populations. The pertinent functional subjects are individual organisms. If we are to characterize a concept of biodiversity, it must be in terms of the tangible aspects of organisms, i.e. their intrinsic properties, that provide the basis for making relational statements about spatially or temporally different areas as consequences of the range of inferences in **[A](c)** (Table 1, Fig. 1). By their very nature, specific-level hypotheses – indeed all taxa – do not represent the totality of what is perceived of organisms. This is apparent from what is shown in Table 1, Figures 1–2, and **[A](c)**. Different classes of biological systematics hypotheses provide answers to causal questions addressing any number of different classes of properties among organisms. To limit a conception of biodiversity to species [*sic*] would not only be to deny the explanatory relevance of other hypotheses used to account for other organismal properties, but also arbitrary. Circumventing this problem requires orienting focus away from species, to individual organisms and the properties they instantiate.

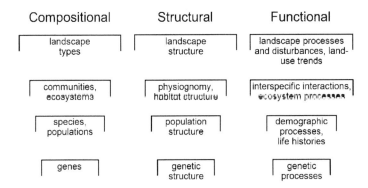

**Figure 4.** The three principle attributes of biodiversity according to Noss [22: fig.1], compositional, structural, and functional, and the hierarchical arrangements of components within each.

In contrast to the three-part hierarchies of Noss [22], Sarkar [32] suggested that among the enti-ties [*sic*] considered in biodiversity and conservation, two hierarchies are used: spatial and taxonomic. The spatial hierarchy entails molecules, cell organelles, cells, individuals, popula-tions, communities, and ecosystems. The taxonomic hierarchy contains alleles, linkage groups, genotypes, subspecies, species, genera, families, orders, classes, phyla, and kingdoms. While Sarkar's spatial hierarchy conveys part-whole relations, the taxonomic hierarchy is a mix of part-*cum*-function and explanatory hypotheses. As noted in the previous paragraph, such a 'hi-erarchy' fails for the fact that it does not contain entities to which biodiversity and conservation can refer.

If there is no epistemic basis for restricting a definition of biodiversity to genes, species *sensu lato*, or other taxa, and the objects of our observation statements are individual organisms, to what does biodiversity refer? Sarkar [32] is correct in pointing out that the answer is contingent upon identifying what is being conserved. If conservation is the act of conserving something, and that something is conveyed by the term biodiversity, then clearly we are not conserving specific or supraspecific taxa *sensu* **[F]** and **[H]** (Figs. 1–3, Table 1), or any taxon for that matter. Taxa are explanatory accounts, as inferential reactions to our observations of organisms, providing understanding of relevant features by way of past causal processes. Taxa are not tangible qualities existing in the present to which conservation efforts can be applied into the future. But it can be argued that there is a relation between conservation and taxa that allows for a cogent and operational definition of biodiversity – exemplifying the transitive relations between taxa, biodiversity, and conservation, and solidifying the status of species, regardless of connotation, as explanatory hypotheses. Recall the inferences of specific hypotheses in **[F]** (Fig. 2). It was noted that while such hypotheses have causal components that include character origin and fixation, the hypotheses also imply that individuals are products of proximate events of tokogeny (Fig. 2c). Per a given specific hypothesis, those proximate tokogenetic events would be separate from other such sets of tokogeny outlined by other specific hypotheses. Similarly, since the phylogenetic inference in [H] (Fig. 3a) entails separately-inferred specific hypotheses (cf. [F], Fig. 2), the 'terminal branches' of the cladogram imply the same sets of proximate tokogenetic events. In point of fact, as exemplified in Figure 1 (Table 1), proximate tokogenetic events are components of not only specific and phylogenetic hypotheses but also intraspecific/polymorphic, and cyclomorphic hypotheses.

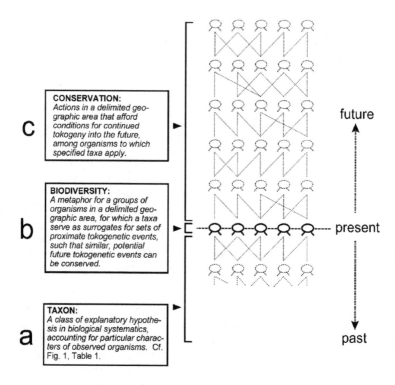

**Figure 5.** Schematic representation of relations between observed organisms, taxa, biodiversity, and conservation. Formal definitions of biodiversity and conservation are provided. Note that explanatory hypotheses referred to as taxa will be of different forms depending on the questions to which those hypotheses serve as answers (cf. Table 1, Figure 1). Yet, biodiversity and conservation focus on future tokogenetic events, regardless of the taxon used (cf. Figs. 3–5). Taxa serve as surrogates to denote the geographic scope of individuals to which biodiversity refers, and conservation is to be applied. Hypotheses regarding obligate asexual, apomictic/parthenogenetic, or self-fertilizing hermaphroditic tokogeny, or horizontal genetic transfer (cf. Table 1: 'species$_3$,' 'species$_5$') can be conveyed in the same type of diagram

Identifying temporally proximate tokogenetic events as components of most (ontogenetic excluded) classes of hypotheses in systematics provides not only the basis for defining biodiversity, but furnishes the relevant conceptual link to conservation – again illustrating the transitive relations between taxa, biodiversity, and conservation. Defining biodiversity can be accomplished by illustrating the relations between taxa (*qua* sets of proximate tokogenetic events), organisms existing in the present, and conservation. These relations are schematized in Figure 5. Given our observations of organisms in the present, we abductively infer explanatory hypotheses referred to as taxa. Based on what was described earlier, a formal definition of taxon would be (cf. [12, 13]):

[I]

**Taxon:** A class of explanatory hypotheses in biological systematics, causally accounting for particular characters of observed organisms.

Contrary to what was suggested by Wilkins ([90: p146; see also [89]), taxa are not "phenomena that call for a theoretical explanation." Taxa *are* the explanations. The phenomena to which they refer are organisms with differentially shared characters. Regarding intraspecific/ polymorphic, specific *sensu lato*, and phylogenetic taxa, all of these hypotheses entail proximate tokogeny (Figs. 1, 5a). But taxa, inclusive of hypothesized proximate tokogenetic events, are not the actual objects of concern. Biodiversity cannot be characterized in terms of taxa *simpliciter*. Rather, biodiversity is best conceived as a metaphor, relating a circumscribed, contemporaneous group of organisms (Fig. 5b) to taxa, insofar as taxa serve as surrogates for proximate tokogeny (e.g. Fig. 2c). A formal definition of biodiversity would then be:

[J]

**Biodiversity:** A metaphor for groups of organisms in a delimited geographic region, for which taxa serve as surrogates for sets of proximate tokogenetic events, such that similar, potential future tokogenetic events can be conserved.

The definition allows for making rational decisions about what taxa serve as surrogates. Depending on the spatial scope of organisms under consideration, biodiversity cannot be strictly equated with just species or phylogenetic hypotheses (cf. 'phylogenetic diversity,' [107, 181–185]). That one speaks of biodiversity is first contingent upon observations of organisms by way of the properties they instantiate, followed by abductive inferences to hypotheses that account for certain of those properties, in part as matters of past tokogeny. Associated with reference to proximate tokogeny, itself a taxon (Fig. 1, Table 1), the definition of biodiversity in [J] has its greatest strength in providing an operational relation to conservation. The most tangible aspect of biodiversity as defined here is that it is either prompted by considerations of conservation or is the unambiguous concept to which conservation is directed, which is consistent with the perspective offered by Sarkar [32]. In terms of proceeding from the present – with observations of organisms in given areas – into the future, what are to be conserved are conditions offering the greatest potential for tokogenetic events like those referred to or implied by taxa (Fig. 5c). Regarding biodiversity as sets of tokogenetic systems is then consistent with this definition of conservation:

[K]

**Conservation:** Actions in a delimited geographic region that afford conditions for continued tokogeny into the future, among organisms to which specified taxa apply.

The intent of conservation is to enable future events of tokogeny among individuals, not to conserve taxa. To reiterate, taxa are inferential reactions to observations of individuals, indicating relevant past causal events accounting for particular organismal properties. It is epistemically meaningless to say conservation seeks to maintain explanatory hypotheses or taxa, especially since those hypotheses only have relevance to effects observed in the present, as individual organisms, not potential causal events among future individuals. But this error is the case when one suggests, for instance, that a habitat should be maintained as part of an effort to conserve particular species. Beyond correctly characterizing the nature and role of taxa, the utility of the definitions in [J] and [K] is that they are fully operational in relation to those taxa that would serve as relevant surrogates for biodiversity (Fig. 1, Table 1). Generalized examples of application, derived from Figure 5, are provided next.

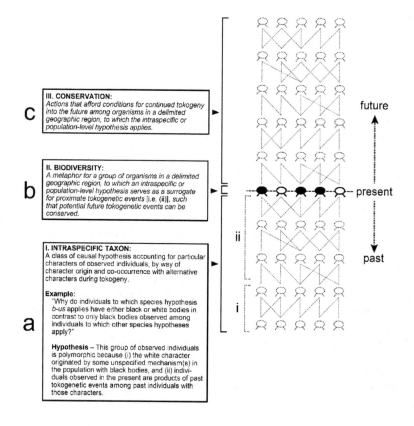

**Figure 6.** Example of the relations between observed individuals to which an intraspecific hypothesis is applied (cf. Figs. 1–2, Table 1), and biodiversity and conservation. See Figure 5 and text for additional explanation.

Speaking of intraspecific/polymorphic variation within a population implies past causal events of character origin [Fig. 6a(i)] and incomplete fixation [Fig. 6a(ii)] as accounting for observed conditions. Biodiversity within the present population is a matter of referring to variation or polymorphism as results of recent tokogeny (Fig. 6b). By extension, conservation involves actions that allow for continued tokogeny that might ensure the properties of individuals to which biodiversity refers (Fig. 6c). The consideration of biodiversity in terms of specific *sensu stricto* taxa as surrogates is essentially identical (Fig. 7). Biodiversity in relation to supraspecific taxa (Fig. 8), i.e. phylogenetic hypotheses, is similar to the two previous examples, once again indicating that biodiversity and conservation pertain to tokogeny, not species *sensu stricto* (i.e. 'species$_1$,' 'species$_2$,' 'species$_4$') or supraspecific taxa. Referring to supraspecific taxa (Fig. 8a) in the context of biodiversity necessitates considering specific hypotheses (Fig. 8b), since it is the latter that entail the relevant proximate sets of tokogenetic systems. Conservation then pertains to maintaining the continuity of those sets of tokogenetic systems into the future (Fig. 8c).

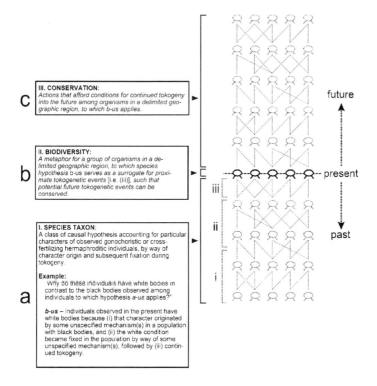

**Figure 7.** Example of the relations between observed individuals to which a species hypothesis is applied (cf. Figs. 1–2, Table 1), and biodiversity and conservation. See Figure 5 and text for additional explanation.

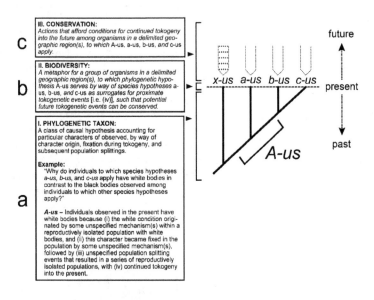

**Figure 8.** Example of the relations between observed individuals to which a phylogenetic hypothesis is applied (cf. Figs. 1–2, Table 1), and biodiversity and conservation. See text for additional explanation.

## 5. Discussion

The crux of the analysis presented in this chapter is that defining *species* (including *taxa*), *biodiversity*, and *conservation* are best approached from the perspective of the transitive relations between the concepts to which each of these terms refers. Recognizing these relations follows from the pursuit of causal understanding that is not only the goal of science in general but also the subfield known as systematics. But instead of limiting the question to "What are species?", it has been shown that the more relevant question is, "What are taxa?" If species are taxa, then species have a standing no more and no less epistemically important than what are accorded other taxa. The consequences of answering the two questions can be summarized as follows: (1) at best, *species* [Table 1: 'species$_1$,' 'species$_2$,' 'species$_4$'; **[E](a), (b), (d)]**] is a class of systematics hypotheses applied to those organisms among which to some extent there is obligate genetic exchange by way of tokogenetic-based sexual reproductive events; the matter of other causal events to which species have been referred, i.e. 'species$_3$' and 'species$_5$' (Table 1, **[E](c), (e)]**), require separate consideration; (2) *biodiversity* is a rhetorical device indicating relations between observed organisms and hypothesized sets of past sexual or asexual proximate tokogeny, not limited to species *sensu lato*, inferred from characters of organisms (cf. **[J]**); and (3) *conservation* is a reaction to the metaphor of *biodiversity*, in the form of actions that promote tokogeny into the future (cf. **[K]**).

In the context of taxa as explanatory hypotheses, as reactions to our observations of organisms, the scope of species in biology is in need of revision. Consider the insightful treatment by Wilkins [8][3]. Among the 26 'species concepts' identified by Wilkins [6], Wilkins [8: p58, emphasis original] suggests there are "seven 'basic' species concepts:"

---

"...*agamospecies* (asexuals), *biospecies* (reproductively isolated sexual species), *ecospecies* (ecological niche occupiers),

*evolutionary species* (evolving lineages), *genetic species* (common gene pool), *morphospecies* (species defined by their form,

or phenotypes), and *taxonomic species* (whatever a taxonomist calls a species)."

---

Wilkins segregates morphospecies and taxonomic species as focusing on "how we *identify* species," whereas remaining concepts refer to "*what* species *are.*" It is among these latter 'basic' concepts that Wilkins [8: p59] narrows the distinction to "ecospecies" and "biospecies." But he regards ecospecies and biospecies as referring not to a concept of species but rather causes of species. In other words, they are explanatory accounts. Wilkins [8: p59] (see also [89]) raises the important point that "Theory-based [species] concepts presume the universal applicability of that theory outside the groups on which it was formulated.... Discovery techniques [*sic*] that are based on explanatory concepts are hostage to empirical fortune." To that end, he suggests that species be defined as "those groups of organisms that resemble their parents." As a necessity borne out of the need to subsume asexual- and sexual-based tokogeny and horizontal genetic exchange under the one term species, this definition is reduced to being largely equivalent to tokogeny (cf. Fig. 1, Table 1). Notice also that Wilkins' definition is imbued with causal content. As all taxa refer to explanatory accounts, tokogeny already serves that role in a capacity that is more restricted relative to species. But to say "explanatory concepts are hostage to empirical fortune," cannot be regarded as a negative consequence. At best, it is nothing more than acknowledging that conclusions of all abductive inferences, from obser-vation statements to taxa, are never theory neutral. To think that systematics should be divorced from considerations of explanation is contrary to the very objective of scientific inquiry. Rather than attempting to reduce the definition of species to that of tokogeny, we need to face the issue that the one term species has been given the unrealistic task of encompassing a spectrum of causal conditions intended to account for select properties of organisms ([E], Table 1). Since it is our observations of properties of organisms that compel our inferential actions leading to the explanatory conclusions we call taxa, the challenge with respect to species is straightforward. We need to limit the scope of specific hypotheses to the explanatory realm of character origin/fixation within reproductively isolated populations of organisms with obligate, albeit not exclusive, sexual reproduction (i.e. [C]; Table 1: 'species$_1$,' 'species$_2$,' 'species$_4$'), and recognize other classes of hypotheses-as-taxa when speaking of organisms that are strictly asexual, apomictic/parthenogenetic, self-fertilizing hermaphroditic, or engage in horizontal genetic exchange ([E], Table 1: 'species$_3$,' 'species$_5$'). With respect to obligate asexual,

---

3 It should be noted that Wilkins (pers. comm.) does not subscribe to the view that species are explanatory hypotheses.

apomictic/parthenogenetic, or self-fertilizing hermaphroditic organisms, or events of horizontal genetic exchange, neither specific *sensu stricto* nor phylogenetic hypotheses apply in the capacity required for sexually-reproducing organisms.

The concerns expressed above regarding parsing the various classes of hypotheses we call taxa brings to light significant consequences for considerations of biodiversity and conservation. That biodiversity is a metaphor is a consequence of our reliance on intraspecific/polymorphic, specific, and phylogenetic hypotheses serving as surrogates for hypotheses of proximate tokogeny (Fig. 5). It is the continuation of tokogenetic events, asexual or sexual, that conservation seeks to preserve. The inherent constraints imposed by the nature of taxa as explanatory hypotheses determine these characterizations. The most notable outcome is that the relevance of species to biodiversity and conservation is both diminished and placed in proper perspective with all other taxa. The definition of biodiversity in [J] is thus at odds with the more standard view that biodiversity is a hierarchy of 'things,' e.g. genes, species/taxa, ecosystems, in need of conservation. For instance, Margules et al. [19: p310] state that,

"The concept of biodiversity encompasses the entire biological hierarchy from molecules to ecosystems. It includes entities [sic] recognisable at each level (genes, taxa, communities, etc.) and the interactions between them (nutrient and energy cycling, predation, competition, mutation, and adaptation, etc.). These entities are heterogeneous, meaning that all members at each level can be distinguished from one another; they form a hierarchy of nested individuals... The complete description of each level requires the inclusion of all members. The number of viable entities at all levels is phenomenally large and in practice unknown. Yet sustaining this variety, unknown and unmeasured, the variety of life on earth, is the goal of biodiversity conservation. To achieve this goal it will be necessary to retain the complex hierarchical biological organization that sustains characters within taxa, taxa within communities or assemblages, and assemblages within ecosystems."

Biodiversity purportedly encompasses a hierarchy of individuals, e.g. nucleotides, organisms, taxa, communities, ecosystems, etc, and conservation is the attempt to preserve some semblance of the variety of individuals at each level. The assumption that the individuality thesis is ontologically appropriate to taxa, communities, and ecosystems cannot be defended. Consider the variety of hypotheses in Figure 1 (see also Table 1, Figs. 2–3, 5–8). What are graphically depicted is past causal events relative to specimens observed in the present. The totality of events entailed by any of these hypotheses does not connote individuals, things, or 'historical entities.' The illustrated events involved past individuals *qua* organisms, but the events themselves are not individuals [12–14, 37, 126]. None of these hypotheses present emergent properties that would allow one the opportunity to perceive more inclusive individuals beyond the specimens upon which the hypotheses have been inferred. The inferential path from observation statements regarding organisms, which is contingent upon our perceptions of the properties of those organisms, does not lead to observation statements regarding taxa, communities, or ecosystems as individuals or entities (cf. [F], [H]). This chapter has echoed what was recognized by Fitzhugh [10, 12–14, 37, 97, 98, 126], that taxa are consequences of our inferential actions, prompted by perceptions of individual organisms.

With taxa acknowledged as representations of our explanatory hypotheses, it becomes immediately apparent that we are not speaking of a hierarchy of ever-inclusive individuals [*contra* 32, 186–191]. What is critical to correctly speaking of biodiversity is that we identify the subjects to which we are referring (Figs. 5–8). The present analysis has provided arguments intended to reorient the focus of biodiversity and conservation away from genes, taxa, communities, or ecosystems as ontological individuals, to the subjects that are perceived by biologists, i.e. organisms, and the relations of organisms to past proximate tokogeny (biodiversity) and potential future tokogeny (conservation). The subjects of biodiversity and conservation can be conceived as intersections between a multitude of fields of research, including ecology, systematics, developmental biology, paleontology, biogeography, and population genetics. But in each instance, taxa are not the ontologically or biologically relevant units of interest. Our interest is in individual organisms per our inferential reactions to the properties they instantiate, e.g. **[F]**, **[H]**. Some of those reactions are communicated under the rubric of taxa, and conveyed in the context of biodiversity. Via biodiversity, our selective endeavors to ensure tokogeny among organisms are the acts of conservation.

Recall the quote from Noss [22: p356] given in **Defining Biodiversity and Conservation**, that biodiversity is not typically considered only in terms of the number of groups of things, "the number of genes, species, ecosystems, or any other group of things [*sic*] in a defined area," but also "the relative frequency or abundance of each species or other entity, in addition to the number of entities in the collection." It was noted that Noss' characterization conflates species with the individuals to whom those hypotheses refer. The relative abundance of hypothesized sets of tokogenetic systems between given areas, indicated by taxa as surrogates, is not the same as the number of individuals in those areas. The correct relation is between abundance of individuals in an area relative to particular systematics hypotheses. But that relation would be one predicated on biodiversity, not part of it. There is the alternative view that species richness is an index of biodiversity. Maclaurin and Sterelny [46: p173] suggest that "Species richness, supplemented in various ways, is a good multipurpose measure of biodiversity, because many processes affect richness...and it is causally relevant to many outputs." The problem is that species richness is limited to relations with specific hypotheses, and do not subsume explanations of observed properties of individuals denoted by other classes of taxa. There is no epistemic basis for restricting biodiversity to species, just as conservation is not the act of maintaining species or other taxa. Rather, biodiversity encompasses our hypotheses of past causal events that give us at least tentative understanding of what we observe in the present, with conservation a concerted effort to maintain opportunities for specified organisms, not species or other taxa, to continue to engage in tokogenetic events into the future. And just as specific *sensu stricto* (Table 1: 'species$_1$,' 'species$_2$,' 'species$_4$') and phylogenetic hypotheses (cf. **[H]**) applied to sexually reproducing organisms serve as proxies for tokogeny in both the context of biodiversity and conservation, the taxa employed to account for properties among obligate asexual, apomictic/parthenogenetic, or self-fertilizing hermaphroditic organisms (Table 1: 'species$_3$') and horizontal genetic exchange (Table 1: 'species$_5$') also connote surrogates for tokogeny when speaking of the biodiversity or conservation of those organisms.

The misconception that species are the units to which biodiversity refers and conservation seek to maintain also extends to phrases such as 'species extinction' or 'endangered species.' Extinction is not the transition from existence to non-existence of species or taxa, but rather the cessation of reproductive events among a group of organisms. What is endangered is not a species, but rather the opportunities for tokogeny conveyed in intraspecific/polymorphic, specific *sensu lato*, or phylogenetic hypotheses, among others (cf. Table 1, Fig. 1). Orienting focus away from the ontologically specious view that taxa are spatio-temporally restricted things or individuals, to the relevant representations of our select, biased explanatory accounts of what we observe of organisms provides the most effective route to ensuring the integration of biological disciplines that go into 'biodiversity studies' and conservation efforts.

A final observation is in order. In conjunction with establishing that taxa are explanatory hypotheses for particular features of organisms, and species are taxa, this chapter set out to define the terms *species/taxa, biodiversity*, and *conservation* by acknowledging their transitive relations, {(taxa, biodiversity), (biodiversity, conservation), (taxa, conservation)}. Pursuant to these definitions, it is evident that these relations are not representative of our actions in various fields of biological study. We observe individual organisms, and infer classes of hypotheses referred to as taxa. On the basis of particular taxa, we attempt to conserve organisms and their opportunities for tokogeny into the future. Notice that it is taxa, *not biodiversity*, that serves as the conceptual link. What is apparent is that *biodiversity* offers no tangible contributions to our characterizations of organisms, references to taxa, or implementations of conservation. At best *biodiversity* is redundant relative to *taxa*, at worst gratuitous for *conservation* (cf. Figs. 5–8). The transitive relations that in fact exist are, {(individuals, taxa), (taxa, conservation), (individuals, conservation)}. These revised relations manifest themselves as follows: *taxa* must be determined in the context of *individuals*; *taxa* establish the limits on *conservation*; *individuals* determine the realization of *conservation*. *Biodiversity* is an unnecessary concept—a contrivance of unbridled reification, along the same lines of excess that have befallen *species*.

## Acknowledgments

I am grateful for discussions and suggestions offered by Brent Karner, John Long, Brian V. Brown, and Mark Thompson.

## Author details

Kirk Fitzhugh

Address all correspondence to: kfitzhug@nhm.org

Research & Collections Branch, Natural History Museum of Los Angeles County, Los Angeles, CA, USA

# References

[1]  Hennig W. Phylogenetic Systematics. Urbana: University of Illinois Press; 1966.

[2]  Hanson NR. Perception and Discovery: an Introduction to Scientific Inquiry. San Francisco: Freeman, Cooper & Company; 1969.

[3]  Psillos S. Philosophy of Science A–Z. Edinburgh: Edinburgh University Press; 2007.

[4]  Hempel CG. Fundamentals of Concept Formation in Empirical Science. Chicago: The University of Chicago Press; 1952.

[5]  Agapow P-M, Bininda-Emonds ORP, Crandall KA, Gittleman JL, Mace GM, Marshall JC, Purvis A. The Impact of Species Concept on Biodiversity Studies. Quarterly Review of Biology 2004; 79(2): 161–179.

[6]  Wilkins JS. Defining Species: a Sourcebook from Antiquity to Today. New York: Peter Land; 2009.

[7]  Wilkins JS. Species: a History of the Idea. Berkeley: University of California Press; 2009.

[8]  Wilkins JS. Philosophically Speaking, How Many Species Concepts are There? Zootaxa 2011; 2765: 58-60.

[9]  Wheeler QD, Knapp S, Stevenson DW, Stevenson J, Blum SD, Boom BM, Borisy GG, Buizer JL, De Carvalho MR, Cibrian A, Donoghue MJ, Doyle V, Gerson EM, Graham CH, Graves P, Graves SJ, Guralnick RP, Hamilton AL, Hanken J, Law W, Lipscomb DL, Lovejoy TE, Miller H, Miller JS, Naeem S, Novacek MJ, Page, LM, Platnick NI, Porter-Morgan H, Raven PH, Solis MA, Valdecasas AG, Van Der Leeuw S, Vasco A, Vermeulen N, Vogel J, Walls, RL, Wilson EO, Woolley JB. Mapping the Biosphere: Exploring Species to Understand the Origin, Organization and Sustainability of Biodiversity. Systematics and Biodiversity 2012; 10(1): 1–20.

[10]  Fitzhugh K. The Abduction of Phylogenetic Hypotheses. Zootaxa 2006; 1145: 1–110.

[11]  Fitzhugh K. The Philosophical Basis of Character Coding for the Inference of Phylogenetic Hypotheses. Zoologica Scripta 2006; 35(3): 261–286.

[12]  Fitzhugh K. Abductive Inference: Implications for 'Linnean' and 'Phylogenetic' Approaches for Representing Biological Systematization. Evolutionary Biology 2008; 35(1): 52–82.

[13]  Fitzhugh K. Species as Explanatory Hypotheses: Refinements and Implications. Acta Biotheoretica 2009; 57(1–2): 201–248.

[14]  Fitzhugh K. The Limits of Understanding in Biological Systematics. Zootaxa 2012; 3435: 40–67.

[15] International Commission on Zoological Nomenclature. International Code of Zoological Nomenclature. London: The International Trust for Zoological Nomenclature; 1999.

[16] Cantino PD, de Queiroz K. PhyloCode: International Code of Phylogenetic Nomenclature, Version 4c. The International Society for Phylogenetic Nomenclature; 2010. http://www.ohio.edu/phylocode/ (accessed 27 March 2012).

[17] Committee on Biological Diversity in Marine Systems. Understanding Marine Biodiversity: a Research Agenda for the Nation. Washington DC: National Academy Press; 1995.

[18] Solbrig O. The IUBS-SCOPE-UNESCO Program of Research in Biodiversity. In: Solbrig OT. (ed.) From Genes to Ecosystems: A Research Agenda for Biodiversity. Cambridge: The International Union of Biological Sciences (IUBS); 1991. p5–14.

[19] Margules CR, Pressey RL, Williams PH. Representing Biodiversity: Data and Procedures for Identifying Priority Areas for Conservation. Journal of Biosciences 2002; 27(4): 309–326.

[20] Koricheva J, Siipi H. The Phenomenon of Biodiversity. In: Oksanen M, Pietarinen J. (eds.) Philosophy and Biodiversity. New York: Cambridge University Press; 2004. p27–53.

[21] Rawles K. Biological Diversity and Conservation Policy. In: Oksanen M, Pietarinen J. (eds.) Philosophy and Biodiversity. New York: Cambridge University Press; 2004. p199–216.

[22] Noss RF. Indicators for Monitoring Biodiversity: a Hierarchical Approach. Conservation Biology 1990; 4(4): 355–364.

[23] Gaston KJ. Global Patterns in Biodiversity. Nature 2000; 405(11 May): 220–227.

[24] Agapow P-M. Species: Demarcation and Diversity. In: Purvis A, Gittleman JL, Brooks T. (eds.) Phylogeny and Conservation. New York: Cambridge University Press; 2005. p57–75.

[25] Ricotta C. A Semantic Taxonomy for Diversity Measures. Acta Biotheoretica 2007; 55(1): 23–33.

[26] Bertrand Y, Pleijel F, Rouse G. Taxonomic Surrogacy in Biodiversity Assessments, and the Meaning of Linnean Ranks. Systematics and Biodiversity 2006; 4(2): 149–159.

[27] Barraclough TG, Fontaneto D, Herniou EA, Ricci C. The Evolutionary Nature of Diversification in Sexuals and Asexuals. In: Butlin RK, Bridle JR, Schluter D. (eds). Speciation and Patterns of Diversity. New York: Cambridge University Press; 2009. p29–45.

[28] Gaston KJ, Spicer JI. Biodiversity: an Introduction. Malden: Blackwell Publishing; 2004.

[29] Reyers B, Polasky S, Tallis H, Mooney HA, Larigauderie A. Finding Common Ground for Biodiversity and Ecosystems Services. Bioscience 2012; 62(5): 503-507.

[30] Secretariat of the Convention on Biological Diversity. Handbook of the Convention on Biological Diversity Including its Cartagena Protocol on Biosafety, 3rd Edition. Montreal: Friesen; 2005.

[31] Expert Panel on Biodiversity Science. Canadian Taxonomy: Exploring Biodiversity, Creating Opportunity. Ottawa: The Council of Canadian Academies; 2010. http://www.scienceadvice.ca/uploads/eng/assessments%20and%20publications%20and%20news%20releases/biodiversity/biodiversity_report_final_e.pdf (accessed 5 April 2012).

[32] Sarkar S. Defining "Biodiversity"; Assessing Biodiversity. The Monist 2002; 85(1): 131–155.

[33] Mayden RL. A Hierarchy of Species Concepts: the Denouement in the Saga of the Species Problem. In: Claridge MF, Dawah HA, Wilson MR. (eds.) Species: The Units of Biodiversity. New York: Chapman & Hall; 1997. p381–424.

[34] Stamos DN. The Species Problem: Biological Species, Ontology, and the Metaphysics of Biology. New York: Lexington Books; 2003.

[35] Claridge MF. Species are Real Biological Entities. In: Ayala FJ, Arp R. (eds.) Contemporary Debates in Philosophy of Biology. Chichester: Wiley-Blackwell; 2010. p91–09.

[36] Richards RA. The Species Problem: a Philosophical Analysis. New York: Cambridge University Press; 2010.

[37] Fitzhugh K. The Inferential Basis of Species Hypotheses: the Solution to Defining the Term 'Species.' Marine Ecology 2005; 26(3–4): 155–165.

[38] Anonymous. 2002 Endangered Species Act of 1973 [amended 2002]. Public Law 93-205, 87 Stat. 884, 16 U.S.C. 1531–1544; 2002. http://www.nmfs.noaa.gov/pr/pdfs/laws/esa.pdf (accessed 5 April 2012).

[39] Ryder, OA. Species Conservation and Systematics: the Dilemma of Subspecies. Trends in Ecology and Evolution 1986; 1(1): 9–10.

[40] Moritz C. Defining "Evolutionarily Significant Units" for Conservation. Trends in Ecology and Evolution 1994; 9(10): 373–375.

[41] Moritz C. Strategies to Protect Biological Diversity and the Evolutionary Processes that Sustain It. Systematic Biology 2002; 51(2): 238–254.

[42] Fraser DJ, Bernatchez L. Adaptive Evolutionary Conservation: Towards a Unified Concept for Defining Conservation Units. Molecular Ecology 2001; 10(12): 2741–2752.

[43] Green DM. Designatable Units for Status Assessment of Endangered Species. Conservation Biology 2005; 19(6): 1813–1820.

[44] Wilson EO. Introductory Essay: Systematics and the Future of Biology. In: Hey J, Fitch WM, Ayala FJ. (eds.) Systematics and the Origin of Species: on Ernst Mayr's 100th Anniversary. Washington DC: The National Academies Press; 2005. p1–7.

[45] de Queiroz K. Ernst Mayr and the Modern Concept of Species. In: Hey J, Fitch WM, Ayala FJ. (eds.) Systematics and the Origin of Species: on Ernst Mayr's 100th Anniversary. Washington DC: The National Academies Press; 2005. p243–263.

[46] Maclaurin J, Sterelny K. What is Biodiversity? Chicago: University of Chicago Press; 2008.

[47] Camp WH, Rickett HW, Weatherby CA. Proposed Changes in the International Rules of Botanical Nomenclature. Brittonia 1949; 7(1): 1-51.

[48] Gregg JR. The Language of Taxonomy: an Application of Symbolic Logic to the Study of Classificatory Systems. New York: Columbia University Press; 1954.

[49] Simpson GG. Principles of Animal Taxonomy. New York: Columbia University Press; 1961.

[50] Blackwelder RE. Taxonomy: a Text and Reference Book. New York: John Wiley & Sons; 1967.

[51] Mayr E. Principles of Systematic Zoology. New York: McGraw-Hill Book Company; 1969.

[52] Mayr E. The Growth of Biological Thought: Diversity, Evolution, and Inheritance. Cambridge: Harvard University Press; 1982.

[53] Jardin N, Sibson R. Mathematical Taxonomy. New York: John Wiley & Sons, Ltd.; 1971.

[54] Abbott LA, Bisby FA, Rogers DJ. Taxonomic Analysis in Biology: Computers, Models, and Databases. New York: Columbia University Press; 1985.

[55] Mahner M, Bunge M. Foundations of Biophilosophy. New York: Springer; 1997.

[56] Futuyma DJ. Evolution. Sunderland: Sinauer Associates, Inc.; 2005.

[57] Williams DM, Ebach MC. Foundations of Systematics and Biogeography. New York: Springer; 2008.

[58] Schuh RT, Brower AVZ. Biological Systematics: Principles and Applications. Ithaca: Cornell University Press; 2009.

[59] Wiley EO, Lieberman BS. Phylogenetics: Theory and Practice of Phylogenetic Systematics. Hoboken: John Wiley & Sons, Inc.; 2011.

[60] Rickett HW. So What is a Taxon? Taxon 1958; 7(2): 37–38.

[61] Ghiselin M. The Triumph of the Darwinian Method. Berkeley: University of California Press; 1969.

[62] Wägele J-W. Foundations of Phylogenetic Systematics. München: Verlag Dr. Friedrich Pfeil; 2005.

[63] Kardong KV. An Introduction to Biological Evolution. New York: McGraw-Hill; 2008.

[64] Boyd R. Homeostasis, Species, and Higher Taxa. In: Wilson RA. (ed.) Species: New Interdisciplinary Essays. Cambridge: MIT Press; 1999. p141–185.

[65] Keller RA, Boyd RN, Wheeler QD. The Illogical Basis of Phylogenetic Nomenclature. The Botanical Review 2003; 69(1): 93–110.

[66] Franz NM. Outline of an Explanatory Account of Cladistic Practice. Biology and Philosophy 2005; 20(2–3): 489–515.

[67] Rieppel O. Monophyly, Paraphyly, and Natural Kinds. Biology and Philosophy 2005; 20(2): 465–487.

[68] Rieppel O. Modules, Kinds, and Homology. Journal of Experimental Zoology, Part B 2005; 304(1): 18–27.

[69] Rieppel O. The PhyloCode: a Critical Discussion of its Theoretical Foundation. Cladistics 2006; 22(2): 186–197.

[70] Rieppel O. Species: Kinds of Individuals or Individuals of a Kind. Cladistics 2007; 23(4): 373–384.

[71] Assis, LCS, Brigandt I. Homology: Homeostatic Property Cluster Kinds in Systematics and Evolution. Evolutionary Biology 2009; 36(2): 248–255.

[72] Sneath PHA, Sokal RR. Numerical Taxonomy: the Principles and Practice of Numerical Classification. San Francisco: W.H. Freeman and Company; 1973.

[73] Ghiselin MT. A Radical Solution to the Species Problem. Systematic Zoology 1974; 23(4): 536–544.

[74] Ghiselin MT. Ostensive Definitions of the Names of Species and Clades. Biology and Philosophy 1995; 10(2): 219–222.

[75] Ghiselin MT. Metaphysics and the Origin of Species. Albany: SUNY Press; 1997.

[76] Hull DL. Are Species Really Individuals? Systematic Zoology 1976; 25(2) 174–191.

[77] Hull DL. A Matter of Individuality. Philosophy of Science 1978; 45(3): 335–360.

[78] Wiley EO. Phylogenetics: the Theory and Practice of Phylogenetic Systematics. New York: John Wiley & Sons; 1981.

[79] Wiley EO. Kinds, Individuals, and Theories. In: Ruse M. (ed.) What the Philosophy of Biology Is. Dordrecht: Kluwer; 1989. p289–300.

[80] Cohan FM. Bacterial Species and Speciation. Systematic Biology 2001; 50(4): 513–524.

[81]  Brogaard B. Species as Individuals. Biology and Philosophy 2004; 19(2): 223–242.

[82]  Härlin M, Pleijel F. Phylogenetic Nomenclature is Compatible with Diverse Philo-
      sophical Perspectives. Zoologica Scripta 2004; 33(6): 587–591.

[83]  Colless DH. Taxa, Individuals, Clusters and a Few Other Things. Biology and Philos-
      ophy 2006; 21(3): 353–367.

[84]  Ereshefsky M. Species and Linnaean Hierarchy. In: Wilson RA. (ed.) Species: New In-
      terdisciplinary Essays. Cambridge: MIT Press; 1999. p285–305.

[85]  Ereshefsky M. The Poverty of the Linnaean Hierarchy: a Philosophical Study of Bio-
      logical Taxonomy. New York: Cambridge University Press; 2001.

[86]  Ereshefsky M. Foundational Issues Concerning Taxa and Taxon Names. Systematic
      Biology 2007; 56(2): 295–301.

[87]  Ereshefsky M, Matthen M. Taxonomy, Polymorphism, and History: an Introduction
      to Population Structure Theory. Philosophy of Science 2005; 72(1): 1–21.

[88]  Rieppel O. Monophyly and the Two Hierarchies. In: Williams DM, Knapp S. (eds.)
      Cladistics: the Branching of a Paradigm. Berkeley: University of California Press;
      2010. p147–167.

[89]  Wilkins JS. How to be a Chaste Species Pluralist-Realist: the Origins of Species
      Modes and the Synapomorphic Species Concept. Biology and Philosophy 2003; 18(5):
      621–638.

[90]  Wilkins JS. What is a Species? Essences and Generation. Theory in Biosciences 2010;
      129(2–3): 141–148.

[91]  Griffiths PE. Squaring the Circle: Natural Kinds with Historical Essences. In: Wilson
      RA. (ed.) Species: New Interdisciplinary Essays. Cambridge: MIT Press; 1999. p209–
      228.

[92]  Wilson RA. Realism, Essence, and Kind: Resuscitating Species Essentialism? In: Wil-
      son RA. (ed.) Species: New Interdisciplinary Essays. Cambridge: MIT Press; 1999.
      p187–207.

[93]  Reydon TAC. Classifying Life, Reconstructing History and Teaching Diversity: Philo-
      sophical Issues in the Teaching of Biological Systematics and Biodiversity. Science &
      Education 2011; doi: 10.1007/s11191-011-9366-z.

[94]  de Queiroz K. The General Lineage Concept of Species and the Defining Properties
      of the Species Category. In: Wilson RA. (ed.) Species: New Interdisciplinary Essays.
      Cambridge: MIT Press; 1999. p49–89.

[95]  Assis LCS. Individuals, Kinds, Phylogeny and Taxonomy. Cladistics 2011; 27(1): 1–3.

[96]  Assis LCS. Species, Reality and Evidence: a Reply to Reydon. Cladistics 2011; 27(1):
      6–8.

[97] Fitzhugh K. Les Bases Philosophiques de l'Inférence Phylogénétique: Une vue d'Ensemble. Biosystema 2005; 24: 83–105.

[98] Fitzhugh K. The 'Requirement of Total Evidence' and its Role in Phylogenetic Systematics. Biology and Philosophy 2006; 21(3): 309–351.

[99] Griffiths GCD. On the Foundations of Biological Systematics. Acta Biotheoretica 1974; 23(3–4). 85–131.

[100] O=Hara RJ. Systematic Generalization, Historical Fate, and The Species Problem. Systematic Biology 1993; 42(3): 231–246.

[101] Bunge M. Philosophy of Science, Volume 1, From Problem to Theory. New Brunswick: Transaction Publishers; 1998.

[102] Strawson PF. Individuals: An Essay in Descriptive Metaphysics. London: Methuen; 1959.

[103] Gracia JJE. Individuality: an Essay on the Foundations of Metaphysics. Albany: State University of New York Press; 1988.

[104] Nogueira JMM, Fitzhugh K, Rossi MCS. A New Genus and New Species of Fan Worms (Polychaeta: Sabellidae) from Atlantic and Pacific Oceans—the Formal Treatment of Taxon Names as Explanatory Hypotheses. Zootaxa 2010; 2603: 1–52.

[105] Fitzhugh K. Revised Systematics of Fabricia oregonica Banse, 1956 (Polychaeta: Sabellidae: Fabriciinae): an Example of the Need for a Uninomial Nomenclatural System. Zootaxa 2010; 2647: 35–50.

[106] Haack S, Kolenda K. Two Fallibilists in Search of the Truth. Proceedings of the Aristotelian Society, Supplementary Volume 1977; 51: 63–104.

[107] Mishler BD. Species are not Uniquely Real Biological Entities. In: Ayala FJ, Arp R. (eds.) Contemporary Debates in Philosophy of Biology. Chichester: Wiley-Blackwell; 2010. p110–122.

[108] Hey J. On the Arbitrary Identification of Real Species. In: Butlin RK, Bridle JR, Schluter, D. (eds). Speciation and Patterns of Diversity. New York: Cambridge University Press; 2009. p15–28.

[109] Fitzhugh K. Fact, Theory, Test and Evolution. Zoologica Scripta 2008; 37(1): 109–113.

[110] Hempel CG. Aspects of Scientific Explanation and Other Essays in the Philosophy of Science. New York: The Free Press; 1965.

[111] Rescher N. Scientific Explanation. New York: The Free Press; 1970.

[112] Popper KR. Objective Knowledge: an Evolutionary Approach. New York: Oxford University Press; 1983.

[113] Popper KR. Realism and the Aim of Science. New York: Routledge; 1992.

[114]  Salmon WC. Scientific Explanation and the Causal Structure of the World. Princeton: Princeton University Press; 1984.

[115]  Van Fraassen BC. The Scientific Image. Oxford: Clarendon Press; 1990.

[116]  Strahler AN. Understanding Science: an Introduction to Concepts and Issues. Buffalo: Prometheus Books; 1992.

[117]  Hausman DM. Causal Asymmetries. New York: Cambridge University Press; 1998.

[118]  de Regt, HW, Leonelli, S, Eigner K., editors. Scientific Understanding: Philosophical Perspectives. Pittsburgh: University of Pittsburgh Press; 2009.

[119]  Darwin C. On the Origin of Species by Means of Natural Selection, or the Preservation of Favoured Races in the Struggle for Life. London: John Murray; 1859.

[120]  Maurer D. The Dark Side of Taxonomic Sufficiency (TS). Marine Pollution Bulletin 2000; 40(2): 98–101.

[121]  Terlizzi A, Bevilacqua S, Fraschetti S, Boero F. Taxonomic Sufficiency and the Increasing Insufficiency of Taxonomic Expertise. Marine Pollution Bulletin 2003; 46(5): 556–561.

[122]  Domínguez-Castanedo N, Rojas-López R, Solís-Weiss V, Hernández-Alcántara P, Granados-Barba A. The Use of Higher Taxa to Assess the Benthic Conditions in the Southern Gulf of Mexico. Marine Ecology 2007; 28 (Supplement 1): 161–168.

[123]  Musco L, Mikac B, Tataranni M, Giangrande A, Terlizzi A. The Use of Coarser Taxonomy in the Detection of Long-Term Changes in Polychaete Assemblages. Marine Environmental Research 2011; 71(2): 131–138.

[124]  Hedges SB, Kumar S., editors. The Timetree of Life. New York: Oxford University Press; 2009.

[125]  Fitzhugh K. Clarifying the Role of Character Loss in Phylogenetic Inference. Zoologica Scripta 2008; 37(5): 561–569.

[126]  Fitzhugh K. Evidence for Evolution versus Evidence for Intelligent Design: Parallel Confusions. Evolutionary Biology 2010; 37(2–3): 68–92.

[127]  Peirce CS. Illustrations of the Logic of Science. Sixth Paper – Deduction, Induction, and Hypothesis. Popular Science Monthly 1878; 13(August): 470–482.

[128]  Peirce CS. Collected Papers of Charles Sanders Peirce, Volume 1, Principles of Philosophy. In: Hartshorne C, Weiss P, Burks A. (eds.). Harvard University Press, Cambridge; 1931.

[129]  Peirce CS. Collected Papers of Charles Sanders Peirce, Volume 2, Elements of Logic. In: Hartshorne C, Weiss P, Burks A. (eds.). Harvard University Press, Cambridge; 1932.

[130] Peirce CS. Collected Papers of Charles Sanders Peirce, Volume 3, Exact Logic. In: Hartshorne C, Weiss P, Burks A. (eds.). Harvard University Press, Cambridge; 1933.

[131] Peirce CS. Collected Papers of Charles Sanders Peirce, Volume 4, The Simplest Mathematics. In: Hartshorne C, Weiss P, Burks A. (eds.). Harvard University Press, Cambridge; 1933.

[132] Peirce CS. Collected Papers of Charles Sanders Peirce, Volume 5, Pragmatism and Pragmaticism. In: Hartshorne C, Weiss P, Burks A. (eds.). Harvard University Press, Cambridge; 1934.

[133] Peirce CS. Collected Papers of Charles Sanders Peirce, Volume 6, Scientific Metaphysics. In: Hartshorne C, Weiss P, Burks A. (eds.). Harvard University Press, Cambridge; 1935.

[134] Peirce CS. Collected Papers of Charles Sanders Peirce, Volume 7, Science and Philosophy. In: Hartshorne C, Weiss P, Burks A. (eds.). Harvard University Press, Cambridge; 1958.

[135] Peirce CS. Collected Papers of Charles Sanders Peirce, Volume 8, Correspondence and Bibliography. In: Burks A. (ed.). Harvard University Press, Cambridge; 1958.

[136] Hanson NR. Patterns of Discovery: an Inquiry into the Conceptual Foundations of Science. New York: Cambridge University Press; 1958.

[137] Harman G. The Inference to the Best Explanation. Philosophical Review 1965; 74(1): 88–95.

[138] Achinstein P. Inference to Scientific Laws. In: Stuewer RH. (ed.) Volume V: Historical and Philosophical Perspectives of Science, Minnesota Studies in the Philosophy of Science. Minneapolis: University of Minnesota Press; 1970. p87–111.

[139] Fann KT. Peirce's Theory of Abduction. The Hague: Martinus Nijhoff; 1970.

[140] Reilly FE. Charles Peirce's Theory of Scientific Method. New York: Fordham University Press; 1970.

[141] Curd MV. The Logic of Discovery: an Analysis of Three Approaches. In: Nickles T. (ed.) Scientific Discovery, Logic and Rationality. Dordrecht: D. Reidel Publishing Company; 1980. p201–219.

[142] Nickles T. Introductory Essay: Scientific Discovery and the Future of Philosophy of Science. In: Nickles T. (ed.) Scientific Discovery, Logic and Rationality. Dordrecht: D. Reidel; 1980. p1–59.

[143] Thagard P. Computational Philosophy of Science. Cambridge: The MIT Press; 1988.

[144] Ben-Menahem Y. The Inference to the Best Explanation. Erkenntnis 1990; 33(3): 319–344.

[145] Lipton P. Inference to the Best Explanation. New York: Routledge; 2004.

[146] Josephson JR, Josephson SG., editors. Abductive Inference: Computation, Philosophy, Technology. New York: Cambridge University Press; 1994.

[147] McMullen E. The Inference That Makes Science. Milwaukee: Marquette University Press; 1995.

[148] Hacking I. An Introduction to Probability and Inductive Logic. New York: Cambridge University Press; 2001.

[149] Magnani L. Abduction, Reason, and Science: Processes of Discovery and Explanation. New York: Kluwer Academic; 2001.

[150] Douven I. Testing Inference to the Best Explanation. Synthese 2002; 130(3): 355–377.

[151] Psillos S. Simply the Best: a Case for Abduction. In: Kakas AC, Sadri F. (eds) Computational Logic: Logic Programming and Beyond. Springer: New York; 2002. p605–625.

[152] Godfrey-Smith P. Theory and Reality: an Introduction to the Philosophy of Science. Chicago: University of Chicago Press; 2003.

[153] Walton D. Abductive Reasoning. Tuscaloosa: The University of Alabama Press; 2004.

[154] Aliseda A. Abductive Reasoning: Logical Investigations into Discovery and Explanation. Dordrecht: Springer; 2006.

[155] Nichols, JD, Cooch, EG, Nichols, JM, Sauer, JR. Studying Biodiversity: Is a New Paradigm Really Needed? BioScience 2012; 62(5): 497–502.

[156] Nixon KC, Wheeler QD. An Amplification of the Phylogenetic Species Concept. Cladistics 1990; 6(3): 211–223.

[157] Wheeler QD, Meier R., editors. Species Concepts and Phylogenetic Theory. New York: Columbia University Press; 2000.

[158] Staley JT. The Bacterial Species Dilemma and the Genomic–Phylogenetic Species Concept. Philosophical Transactions of the Royal Society, Series B 2006; 361(1475): 1899–1909.

[159] Koonin EV, Makarova KS, Aravind L. Horizontal Gene Transfer in Prokaryotes: Quantification and Classification. Annual Review of Microbiology 2001; 55: 709–742.

[160] Lawrence JG. Catalyzing Bacterial Speciation: Correlating Lateral Transfer with Genetic Headroom. Systematic Biology 2001; 50(4): 479–496.

[161] Papke RT, Gogarten JP. How Bacterial Lineages Emerge. Science 2012; 336(6077): 45–46.

[162] Shapiro BJ, Friedman J, Cordero OX, Preheim SP, Timberlake SC, Szabó G, Polz MF, Alm EJ. Population Genomics of Early Events in the Ecological Differentiation of Bacteria. Science 2012; 336(6077): 48–51.

[163] Ochman H, Lerat E, Daubin V. Examining Bacterial Species Under the Specter of Gene Transfer and Exchange. Proceedings of the National Academy of Science of the United States of America 2005; 102(Supplement 1): 6595–6599.

[164] Cohan FM. What are Bacterial Species? Annual Review of Microbiology 2002; 56: 457–487.

[165] Konstantinidis K, Ramette A, Tiedje JM. The Bacterial Species Definition in the Genomic Era. Philosophical Transactions of the Royal Society, Series B 2006; 361(1475): 1929–1940.

[166] Wilkins JS. The Concept and Causes of Microbial Species. History and Philosophy of the Life Sciences 2006; 28(3): 389–408.

[167] Sober E. Reconstructing the Character States of Ancestors: a Likelihood Perspective on Cladistic Parsimony. The Monist 2002; 85(1): 156–176.

[168] Sober E. Reconstructing the Past: Parsimony, Evolution, and Inference. Cambridge: MIT Press; 1988.

[169] Sober E. Evidence and Evolution: the Logic Behind the Science. New York: Cambridge University Press; 2008.

[170] Felsenstein J. Evolutionary Trees from DNA Sequences: a Maximum Likelihood Approach. Journal of Molecular Evolution 1981; 17(6): 368–376.

[171] Felsenstein J. Inferring Phylogenies. Sunderland: Sinauer Associates, Inc; 2004.

[172] Swofford DL, Olsen GJ, Waddell PJ, Hillis DM. Phylogenetic Inference. In: Hillis DM, Moritz C, Mable BK. (eds) Molecular Systematics. Sunderland: Sinauer Associates; 1996. p407–514.

[173] Cleland CE. Prediction and Explanation in Historical Natural Science. The British Journal for the Philosophy of Science 2011; 62(3): 551–582.

[174] Cleland CE. Historical Science, Experimental Science, and the Scientific Method. Geology 2001; 29(11): 987–90.

[175] Cleland CE. Methodological and Epistemic Differences between Historical Science and Experimental Science. Philosophy of Science 2002; 69(3): 474–96.

[176] Cleland CE. Philosophical Issues in Natural History and its Historiography. In: Tucker A. (ed.) A Companion to the Philosophy of History and Historiography. Oxford: Wiley-Blackwell; 2009. p44–62.

[177] Tucker A. Historical Science, Over- and Underdetermined: a Study of Darwin's Inference of Origins. The British Journal for the Philosophy of Science 2011; 62(4): 805–829.

[178] Huelsenbeck JP, Ronquist F. MrBayes: Bayesian Inference of Phylogeny. Bioinformatics 2001; 17(8): 754–755.

[179]  Huelsenbeck JP, Ronquist F, Nielsen R, Bollback JP. Bayesian Inference of Phylogeny and its Impact on Evolutionary Biology. Science 2001; 294(5550): 2310–2314.

[180]  Ronquist F, Mark PVD, Huelsenbeck JP. Bayesian Phylogenetic Analysis using MrBayes. In: Lemey P, Salemi M, Vandamme A-M. (eds.) The Phylogenetic Handbook: a Practical Approach to Phylogenetic Analysis and Hypothesis Testing. New York: Cambridge University Press; 2009. p210–266.

[181]  Faith DP. Conservation Evaluation and Phylogenetic Diversity. Biological Conservation 1992; 61(1): 1–10.

[182]  Faith DP. Phylogenetic Pattern and the Quantification of Organismal Biodiversity. Philosophical Transactions of the Royal Society of London, Series B Biological Sciences 1994; 345(1311): 45–58.

[183]  Faith DP. Quantifying Biodiversity: a Phylogenetic Perspective. Conservation Biology 2004; 16(1): 248–252.

[184]  Vane-Wright RI, Humphries CJ, Williams PH. What to Protect? – Systematics and the Agony of Choice. Biological Conservation 1991; 55(3): 235–254.

[185]  Barker GM. Phylogenetic Diversity: A Quantitative Framework for Measurement of Priority and Achievement in Biodiversity Conservation. Biological Journal of the Linnean Society 2002; 76(2): 165–194.

[186]  Vrba E, Eldredge N. Individuals, Hierarchies and Processes: Towards a More Complete Evolutionary Theory. Paleobiology 1983; 10(2): 146–171.

[187]  Eldredge N. Unfinished Synthesis: Biological Hierarchies and Modern Evolutionary Thought. New York: Oxford University Press; 1985.

[188]  Eldredge N. Macroevolutionary Dynamics: Species, Niches, and Adaptive Peaks. New York: McGraw-Hill Publishing Company; 1989.

[189]  Eldredge N, Salthe SN. Hierarchy and Evolution. Oxford Surveys in Evolutionary Biology 1984; 1: 184–208.

[190]  Salthe SN. Evolving Hierarchical Systems: Their Structure and Representation. New York: Columbia University Press; 1985.

[191]  Miller III W. What's in a Name? Ecologic Entities and the Marine Paleoecologic Record. In: Allmon WD, Bottjer DJ. (eds.) Evolutionary Paleoecology: The Ecological Context of Macroevolutionary Change. New York: Columbia University Press; 2001. p15–33.

# Chapter 4

# An Essentialistic View of the Species Problem

Larissa N. Vasilyeva and Steven L. Stephenson

Additional information is available at the end of the chapter

## 1. Introduction

In this paper, we support the critical attitude expressed by some biologists [1-4] of the fail‐ ure of any *species concept* proposed in the biological literature. Nonetheless, we still believe that species themselves are "entities that [are] actually discovered" in nature and not "merely invented" [5, p. 4]. We don't want a situation of 'questioning species reality' [2] to exist and also don't want a "species-free taxonomy" [1], but to fulfill these two desires we should dis‐ card most of the modern and popular ideas about species.

During the history of species debates, there have been different reasons for believing the concept of species to be faulty, but most serious of these were two general points that have surfaced from time to time. Quite recently, Vrana & Wheeler [6, p. 67, italics added] suggest‐ ed that, rather than species, "*individual organisms... should be used as the terminal entities in phylogenetic analysis*". This view returns us to times of Buffon, who wrote that "in reality individuals alone exist in nature" [7, p. 75], and some authors after Buffon also stated that nothing except a living individual possesses "an indisputable ontological significance" [8, p. 344]. Nevertheless, since "a coherent knowledge of the living world is possible only with the aid of a hierarchic classification" (l.c.), and "*there can be no science about individuals*" [7, p. 37, italics added], the ontology of individual organisms should not exclude the ontology of spe‐ cies, at least as the participants in the *natural hierarchy*, if we can prove that the *taxonomic hierarchy* corresponds to the latter.

Even more frequently, the existence of species, similar to that of higher taxa, has been ques‐ tioned from the whole perspective of the theory of evolution representing 'life' as a dynamic world that is supposedly devoid of any static elements. "An analysis of the situation shows where the difficulty lies. The concepts of the taxonomic categories, as all the taxonomic con‐ cepts, are essentially static" and leave "out of consideration the dynamism which is one of the most essential, if not the most essential attribute of life" [8, p. 353]. It has been concluded from the general perspective of evolution that "the taxonomic categories in general, and species in particular, are not static but dynamic units" (l.c.). In such a statement, one can see the poor discrimination between 'categories' and 'units', and this is the most obvious example of a confused view of species that was developed in Mayr's "replacement" of 'typological thinking' by 'populational thinking' [9]. The false contradiction between 'typological thinking' and 'populational thinking' which should be referred to static and dynamic aspects of taxa, respectively, and which characterize each taxon *simultaneously*, has dominated biological literature for a long time, although these aspects correspond to such notions as *intension* and *extension* of taxa [10]. The failure at developing an understanding of intensional and extensional nature of taxa was the main omission from the history of the 'idea of species' [7], and precisely this failure led to a large number of useless arguments [5, p. 21]. Although the attempt to make the "crucial distinction between the species category and species taxa" [5, p. 18] is related to the distinction between intension and extension, this insight has had little impact, since the primary attention was directed toward the vain search for the vague 'intension' of a general 'species category', whereas *each species* (genus, family, etc.) *has its own intension*. (Of course, this diversity of intensions is akin to the diversity of individuals, and, as such, also demands to be placed in the context of a proper science to be naturally ordered.)

In this paper, we expose one more *general* reason for the failure of almost all species concepts, and, paradoxically, this is rooted in the continuous endeavor to introduce a biological sense into 'species' which has constituted an older problem of ancient metaphysics [5, 7]. The incorporation of biology into the species problem was the emphasis on the relationships *within* species, and those relationships involved, for example, a gene flow between populations (in the so-called 'biological species concepts') or lineages from ancestors to descendants (in many phylogenetic species concepts). However, this incorporation of biology was apparently misplaced, since *none of inner interactions within taxa* are of any consequence for the *discovery* of natural entities that make up "a natural system, a system that carves nature at its joints" [5, p. 16]. It might seem that the exclusion of internal interactions as the basis of a species reality throws the species problem back to pure "metaphysics" with its indiscrimination between living and non-living objects, but this is not the case. First, *organisms* remain in the focus of taxonomy and provide unique patterns of relationships that could not be found outside the organic world. Second, the evolutionary perspective keeps its primary importance, although all of the major aspects of this should be re-emphasized. Thus, we don't believe that 'descent' plays an unnecessary role [6, p. 67] in the understanding of nature of taxa and their monophyly, but, in our view, the 'descent' should not be conceived in terms of ancestors and descendants as suggested by some phylogeneticists who wants to reorganize "the very core of biological taxonomy" [11, p. 309], using an inappropriate approach.

We are sympathetic with the statement that "the species problem is not something that needs to be solved, but rather something that... needs to be *gotten over*" [12, p. 232, italics added]. We want to get over species but in a way that is different from the one taken by O'Hara [12] and the like-minded persons whom he mentioned [13-14]. In contrast to opinions that the 'species concept' is "most fundamental in biology" [8, p. 344], and "a central problem to that science" [5, p. 2], we think that species play a rather modest role in the natural world. They are considered unduly important only by being more numerous than genera or families. Artificially inflated, the 'species problem' is extremely narrow from the perspective of the hierarchy of life. This problem often has overshadowed the 'Natural System', and Mayr [15] even indicated that there is no such thing as the Natural System. Here we are more in line with Darwin, who believed in that it does exist and thought that "descent [was] always used in classification" [16, p. 394]. However, what we should do is to be more specific about what is meant by 'descent', since the *only* kind of 'descent', or the only evolutionary sequence, that can be expressed in the system of organic beings is the successive appearance of new differences during the history of the biota, and precisely this sequence gives rise to the *hierarchical structure* of the system.

## 2. Species problem and evolutionary models

The quest for a correct hierarchy of *equally natural taxa of an individual nature* at different levels was partly blocked by some of the ideas derived from one or more aspects of the narrow 'species problem', as well as from accepting certain general evolutionary models.

One of the most unfortunate—for the construction of a natural system—evolutionary models was the Great Chain of Being (ancient *scala naturae*). As Wilkins [7, p. 53] wrote, "The Great Chain does imply that there is no real distinction between species, as all intermediate gaps are filled." This implication threatened the reality of species, so many old 'species concepts' have required 'gaps' between groups for the latter to be clearly outlined. "To Darwin, species form vague, human-defined, and difficult to discern way stations in evolution. *They are detectable only by means of gaps in variation...*, as contrasted to continuous variation normally found among varieties within species" [17, p. 499, italics added], and "Darwin's view of species as clusters of similar individuals separated by gaps remains relevant today" [17, p. 502].

The idea that gaps should exist and that there are larger gaps between higher taxa than between species [15] is highly popular, and "the size and nature of gaps" [18, p. 118] are thought to be identifiable, although few notions in taxonomy are as senseless as 'gaps'. (Having described numerous new species and genera, we don't recall the necessity to measure any 'distance' from a new taxon to any other already described one. Moreover, if two fungal phyla such as the Ascomycota and the Basidiomycota differ in endogenic or exogenic spore production, two orders within the Ascomycota differ in the presence or absence of paraphyses between asci, two genera differ in the number of septa present in their spores, and two species differ in the presence or absence of ornamentation on spore walls, then who is to say which 'gap' is larger, given that a single character is involved in the delimitation at each level?).

The idea of 'gap' has stemmed from the old transformist thesis, shared by some later philosophers of biology, that species 'transform *into each other*' [19-21]. Biologists have imagined a flow of 'species' with no place for genera and families, and only the breaks in that continuum provided some scraps called 'good species' that were arbitrarily united into genera and families by taxonomists. (For example, supraspecific groupings are "arbitrarily delimited sections of phylogenetically *continuous chains of species*" [22, p. 219, italics added]). Although idea of species transformation 'into each other' was seemingly dropped [23] and "the Great Chain of Being had been broken" [24, p. 476], this concept of world development did not die altogether and has only transformed into seemingly different schemes.

A much more promising model is the "Tree of Life", which is currently predominate among biologists, although this model has its own difficulties with the representation of reality. Two main difficulties relate to (1) the incorporation of the above mentioned 'chain-thinking' into 'tree-thinking' (the latter term was used by O'Hara [25]) and (2) the dichotomous branching of 'Tree of Life'.

The fact of preserving a 'chain-thinking' approach within a tree-thinking approach is well illustrated by the discussions about the *trees at the "species level"* [12, p. 244] and the *natural system at "the species level"* [12, p. 243]. However, a tree as a metaphor for the whole of life and for the entire natural system—in Haeckel's sense [26]—cannot be crammed into the flat plane of a single level. If the poetic mind can think of species as green or withered 'leaves' on a 'tree', it also can conceive that leaves by themselves are disconnected and only united by twigs and branches that are 'larger taxa'. Separate 'leaves' can comprise a disordered heap but not a tree or a system. The most important issue to be considered here is the unpleasant fact, that—similar to the linear sequence of 'species' in the Chain of Being—*"trees at the species level"* rule out the consideration of natural higher taxa. The description of the modern 'processual system' of speciation as a "hierarchy" of species-lineages splitting [27] implies that only single-level relationships are involved, so the *natural system* is out of question.

Additional problem with single-level "trees", or cladograms, is their construction as *dichotomous* branching patterns. In this respect, they are similar to the keys for identification used in taxonomy, and it already has been recognized that keys, stemming from "Plato's method of classification" and "Porphyry's comb of tree", are topologically the same as cladograms [7]. There is a belief that "a major distinction between a phylogenetic tree and Porphyry's comb is that the former is derived from history, while the latter is derived from diagnosis" [7, p. 29], but this distinction is really an illusory one. Both cladograms and keys are mostly constructed on the same information about character distribution, and *no character weighting* (see below) is usually involved in this process, so the results are highly subjective. Moreover, the cladograms do not convey a history, not only because of the chaotic choice of characters, but because they are dichotomous.

Interestingly enough, cladists themselves have begun to argue against the tree-like representation of relationships *because of the dichotomy of trees*. The pioneer of cladistics is acknowledged to be Sinai Tschulok [28], who saw systematics as "a tool that allows the investigating mind to master the biological multidimensional multiplicity" and recognized that "every attempt to group plants on the basis of as many characters as possible... had to result in reticulated relationships" [24, p. 490]. On this basis, there was a conclusion that "the dichotomous Tree of Life cannot be a universal metaphor"; instead, "it is an illegitimate metaphor", as also is the case for "Hennig's description of speciation as a universally dichotomous process of cleavages" [24, p. 488].

We agree that the Tree of Life is hardly dichotomous, but a number of unnecessary opposing points are involved in the above critique. First, although the employment of as many characters as possible actually leads to reticulate representation of relationships, these same relationships also could be expressed in an artificial dichotomous key (and a single-level cladogram), so *taxonomic reticulations* (which are different from hybridization processes) are not enough to reject Hennig's description of speciation as a universally dichotomous process of cleavage. Actually, and secondly, we think that Hennig was right in his understanding of speciation, but he wrongly inferred a dichotomous tree from that understanding. Thus, even in clear cases of fan-like or radial speciation on different islands of some archipelago, when different characters of an ancestor change in several descendants, *each particular event of speciation involves a dichotomy* (divergence) of some character into two states—one is possessed by an ancestor and another one by a descendant. In other words, although there is an opinion that "there is nothing in speciation theory that mandated dichotomous speciation" [24, p. 483], speciation *is* a dichotomous splitting, whereas the Tree of Life is not dichotomous: "The appropriate image for the post-Darwinian system is a much-branched tree" [29, p. 5].

Recently, the Tree of Life as an evolutionary model for the development of organic world was strongly attacked from the perspective of a "rampant lateral gene transfer, sometimes across vast phylogenetic distances" [24, p. 488]. The same line of thought could be traced to various theories of endosymbiosis: "In representations of standard evolutionary theory, branches of "family trees" (phylogenies) are allowed only to bifurcate. Yet symbiosis analysis reveals that branches on evolutionary trees are bushy and most anastomose...; indeed, every eucaryote, like every lichen, has more than a single type of ancestor..." [30, p. 10]. However, the entire matter relating to the opposition of the Tree of Life and lateral gene transfer or hybridization is a confusion of representations. It is well known that there are "many slight differences which appear in the offspring from the same parents" [16, p. 60], i.e. in the progeny *from two 'ancestors'*. The differences can even be not so slight (as is often the case with hybridization), but what is more important is that 'children' *diverge* from each other and from parents in some features. It is exactly *these differences and the estimation of their hierarchical level* that are the main concern of taxonomists, so a divergence is the focus of all efforts to convey relationships in the taxonomic system, hence the representation as a tree. Even with extensive gene transfer leading to novelties, we should, first of all, find out how different the new organisms are from those that existed previously, and on which taxonomic level the resultant differences deserve to be placed.

With due consideration to all of the merits of the Tree of Life as an illustration of a natural system, it is unsatisfactory not only because of the endeavors by biologists to make it strictly dichotomous but also because of the confusion that exists between single-level cladograms (lacking *hierarchical*—in a taxonomic sense—relationships) and the tree representing structure with many levels. It is correctly noted that more work is required "to structure a classification in terms of a *nested hierarchy* that corresponds to a branching tree of life" [24, p. 490, italics added], but the image of a tree does not convey a *nested hierarchy*, since its branches are not included within each other.

Stamos [5, p. 29, italics added], when indicating that taxonomists "either take the metaphor of the Tree of Life seriously or they eschew it altogether, providing nothing in its place", stated that "real history reconstruction, combined with a truly viable species concept, requires a *completely new metaphor for best capturing the nature of the history of life*". He was right but did not provide a new metaphor or, in O'Hara's words [12], a new "generalized representation of the single natural system", and, actually, could not do that, because his focus was only on the 'species problem' which cannot be solved outside the context of a *natural* hierarchy of taxa. He purposefully avoided the discussion of the relationship of higher taxa to the species problem [12, p. 28], and that had a limiting influence on his conclusions.

## 3. Hierarchical model of evolution and its implications for the 'species problem'

We began to discuss the hierarchical model of evolution about 15 years ago [31] and paid greater attention to it later [32-36]. There is nothing new about that model, since it is "the traditional *inclusion hierarchy* set out by the ancient Greeks" or "the Linnaean hierarchy, which is an *inclusive system* of groups within groups" [27, p. 311, italics added].

The main problem with the traditional hierarchical model is that it was proposed without any reference to evolution. Although the idea of such a hierarchy was rather productive for the construction of *taxonomic system*, it was described as an "atemporal hierarchy of sets within sets" [27, p. 312] which is "just rather boring ways of arranging things" [37, p. 9]. To overcome this difficulty and to devise a *natural system* from a rather subjective Linnaean hierarchy, we should identify the points of contiguity of the taxonomic model with evolution. The first step in this direction was made by Darwin [16, p. 73, italics added], who described the evolutionary process by which "larger genera... tend to *break up* into smaller genera. And thus, the forms of life throughout the universe become divided into *groups subordinate to groups*". He also associated the *due subordination* of groups with the word 'genealogy' [16, p. 400], but that led to a great confusion afterwards because of the existence for two kinds of 'genealogies'.

These two 'genealogies' ('phylogenies') correspond to two 'hierarchies' as the latter are discussed by David Hull [27, p. 311; 34]. One of them implies the sequence of ancestors and descendants, whereas the other—the sequence of new characters accompanying the division into "groups subordinate to groups". Unfortunately, the first kind of genealogy (ancestral-descendant lineages) was acknowledged as the true basis for taxonomy [38-39], but, contrarily, the *natural system* can only reflect the second kind of genealogy, or the character sequence in time. The 'entities' in this evolutionary sequence cannot be 'ancestors' and 'descendants' to each other, and the presence of a nucleus as marking a highest level of hierarchy is not more 'primitive' with respect to the presence of a dorsal chord, which is later in appearance and lower in level.

After Darwin, Spencer [40, p. 471] considered the arguments from classification in favor of evolution and wrote that "organisms fall into groups within groups; and this is the arrangement which we see results from evolution". Later, "Tschulok also found the enkaptic [inclusive - L.V. & S.S.] nature of the biological system to provide the strongest evidence for the theory of evolution" [27, p. 313], and Hennig's phylonenetic system was recognized as an encaptic one. It was correctly noted that "Hennig's phylogenetic system can easily be read as a Linnaean hierarchy of sets within sets" and "many of the same sorts of inferences can be made with respect to both" [27, p. 315), so this favorable comparison can allow us to recognize a phylogenetic system in the *properly* constructed Linnaean hierarchy It is not correct that "the Linnaean system... makes no reference to evolution" [3, 39], and it is only necessary to find ways to construct a taxonomic evolutionary model. Extensionally, the taxonomic hierarchy is an inclusive system of groups within groups, and this structure reflects the evolution of groups by differentiation. However, intensionally, the taxonomic hierarchy is the system of levels marked by different sets of characters appearing successively during evolution.

With respect to general schemes, the hierarchical model overcomes the difficulties of the Great Chain of Beings and allows for the *equal naturalness of all taxa* that are characterized by the same differences, from their very origin up to the present day. It does not require the notion of 'gaps', but, instead, it is based upon the *proper subordination of characters*. It does not require "a strictly dichotomous system of encapsulation" [27, p. 314] or "a logical dichotomization of the world" [24, p. 479] either, since the older groups can split into many subgroups simultaneously (cf. 'Cambrian explosion' [41] or 'Ordovician radiation' [42]). The hierarchical model conflates micro- and macroevolution, which often have been divorced in the biological literature [43], since lower and higher taxa could originate *simultaneously* in the same process of splitting.

The hierarchical model turns the Darwin's scheme of evolution upside-down, because the image of "lesser difference between varieties becoming augmented into the greater difference between species" [16, p. 112] is hardly correct, since varieties originate *within* already existing species and differ in their own characters, which are lower in level. The differences between taxa at higher levels—for example, between the animal and plant kingdoms (such as the main types of nutrition [i.e. heterotrophic or phototrophic], food reserve substances in the form of glycogen or starch, the presence or absence of cellulose in the cell walls, etc.)—seemed to separate them at the dawn of their existence and have not been 'augmented' during the millions of years since. These groups were *only differentiating into subgroups* during their further evolution and retain the differences that exist between them as the *same phylogenetic relationship*.

Notably, there are two kinds of innovations, namely *new states* of previously existing characters (as in the sequence of primitive and advanced states in ancestors and descendants), and completely *new characters* (in their own diversity of states) within existing groups. The appearance of new states is accompanied by an increase in diversity at some hierarchical level, but the appearance of new characters makes the organic world hierarchically structured. The increase of species diversity due to evolution of characters does not mean the evolution of species themselves, whereas the differentiation is almost the synonym of the evolution of all taxa. Therefore, the *evolution of species* could be described in similar terms and should be based on their further *disintegration* into numerous populations. Of course, such a view of *species as an evolving entity* would likely conflict with some of the concepts that describe species as the "most integrated" units when compared with higher taxa. Moreover, such terms as 'reality' and 'individuality' of species are mostly discussed in association with their 'integrity', provided by some inner processes, so it might seem that exactly the *'evolutionary species concept'*, representing species as a disintegrating unit, threatens other species concepts that seem to be 'non-evolutionary' ones. To overcome this possible conflict, the whole issue relating to 'reality', 'individuality' and 'integrity' of taxa might be modified by examining the hierarchical model of evolution more closely.

It already has been noted that the *increase of inner diversity* is accompanied by strong tendencies towards disintegration in time and space [44] and that "species cohesion is often guaranteed by non-reproductive factors" [45, p. 186]. Through time, species should gradually lose "one kind of cohesiveness" [46, p. 262] but keep another kind of cohesiveness, and the *"character continuity* is sufficient and testable evidence of such cohesion" [47, p. 220, italics added]. From the hierarchical model of evolution it follows that there is a *persistence of differences*, or *phylogenetic relationships*, between earlier originated groups. That was exactly Darwin's [16] understanding of *the same 'genealogical arrangement'* that exists between groups at each successive period of their modification. The groups themselves could change drastically, and one can compare the world of dinosaurs or the forest of the Carboniferous Period with modern organisms belonging to animal and plant kingdoms, respectively. From their very origin, these groups have become highly differentiated, but they remain the same. In other words, they continue as the same evolutionary units, and the *integrity* of such units *is provided by* those *sets of characters* that distinguish them from closely related taxa. The set of features distinguishing each taxon from its relatives at the same hierarchical level is called *intension*. Another aspect of a taxon is related to its changeable content and is called *extension* [10]. Because of the incessant changes of extensions that cause the same taxa to look so different during evolution, the *cohesiveness, reality, and individuality of historical taxa can be associated only with intensions.*

Moreover, the considerations mentioned above may destroy the *extensional* definition of monophyly in phylogenetic systematics. There is a belief that a monophyletic genus should include an ancestral species and *all* of its descendants [13, 48-49], but even if the genus at the time of its origin was represented by a single species (an ancestor of other species in that genus), it already should have been characterized by intensional monophyly (hidden in characters marking a genus appearance) and had to retain this monophyletic nature during the increase of inner diversity. In addition, the descendants of an initial species may be taxa of different levels in accordance with changes in different characters distinguishing groups of the previously originated hierarchical diversity. Just this case was described by Darwin [16, p. 401] as follows: "Nor can the existing species, descended from A, be ranked in the same genus with the parent A." The unification of all descendants into a single genus would result in an unnatural classification. Therefore, the only acceptable definition of monophyly may be associated with intensions, or defining characters that also show the common descent of members. Moreover, the intensional monophyly causes the notion of paraphyletic taxa completely to be both useless and superfluous. In taxonomy, the phrase "Much of life's history consists of *non monophyletic ancestral taxa*" [50, p. 413, italics added] would mean that the *proper* intensions of 'ancestral taxa' providing their monophyly are not to be found.

Now, what about the debates that surround the species problem and exhaust the energy of biologists? There is no place here to discuss all of them, and we only refer to "a sterile debate about whether species are individuals or classes" [51, p. 191]. We can admit that species are individuals with respect to evolutionary theory [52] and this is in agreement with the hierarchical model of evolution. However the narrow form of "the individuality thesis" as advocated by Ghiselin [53-54] is so inadequate for biological systematics that Ruse [55] believed Ghiselin to be 'dead wrong' in this matter, while Kitcher [56, p. 649] wrote that the 'species-as-individuals' thesis is "one of the least-promising suggestions in recent philosophy of science". It was even suggested [57: p. 456] that modern evolutionary theory does not require species to be individuals and "does not require species at all, only lineages". However, the hierarchical model of evolution requires 'species' as one of lower levels in the genesis of a nested hierarchy of groups, while all evolving taxa are necessarily individuals (with the promising aspect of individuality associated with intensions).

The opposition of individuals and classes usually has been described from the perspective of spatiotemporal location. It is said that biological species have been treated traditionally as spatiotemporally unrestricted classes, but they must be spatiotemporally localized individuals [58]. However this opposition can appear in a different light from the perspective of intension and extension. Intension (as a set of features) makes a class of a species, but intension has its origin in time (at the point of speciation) and keeps together organisms scattered in space, so the class-forming aspect of species is deeply related to spatiotemporal localization of a species. On the contrary, individuality and spatiotemporal restriction would be expected from extensional relationships, but precisely indefinite extensional changes in the number, structure and relations of populations (with the increase of inner polymorphism) make 'rivers' of species flowing through time and space within the 'banks' of the same differences. Thus, *species evolution* is completely extensional (internal), and species cannot be "spatiotemporally restricted' in this aspect (cf. also *'spatiotemporally unrestricted [nature] of evolutionary theory'* [58, p. 354, italics added]).

The popular belief that "classes cannot evolve" [5, 59] is not pertinent to biological taxa, since they are classes of a special kind—they are the constituents of the hierarchically evolving organic world. Their intensions comprise an evolutionary sequence themselves and, simultaneously, *direct* the internal (extensional) evolution of taxa, *defining their internal diversity*. (Surely the main kinds of nutrition—photosynthetic, osmotic, digestive—which early separated such groups as plants, fungi, and animals did not allow the members of those higher taxa to have the same features at lower levels. By analogy with this influence, the internal or external skeleton of animals defines a completely different diversity of forms, say, within the phyla Chordata and Arthropoda). From this perspective, the opposition of single-level relationships (the so-called "hierarchy" of species lineages) and the Linnaean hierarchy (which should be properly constructed to embody the *natural hierarchy of intensions*) as having "directional and temporal dimension" and being "atemporal and non-directional" [27, p.

311], respectively, should be read inversely. The dichotomies connecting species in clado-grams look much more atemporal and non-directional (often simply appearing meaning-less) than the "phylogenetic hierarchy of groups within groups" (l.c.) characterizing the true taxonomic hierarchy. At least the latter allows us to understand the *directional and temporal* influence of older characters (which are higher in level) on the appearance of later ones dur-ing evolution. Moreover, the dichotomous trees constructed for a particular group of species could be numerous, whereas the phylogenetic hierarchy should be a single one.

## 4. Concept of a character and character-based concepts of species

The practice of taxonomy is surely at odds with the suggestions to reject, abandon, or re-place the Linnaean system [38-39, 60-63]. One of the reasons of rejecting the latter is an ob-servation that there is no clear distinction between species and higher taxa, and that species are *non-comparable* entities [64]. Indeed, there is no distinction between species and higher taxa, since the former can evolve (differentiate) into the latter while the organic world devel-ops [16, 23], but, for taxonomists, the deep distinction between categories of the Linnaean system is different characters marking different levels. Moreover, taxonomists can make species (and other taxa) *comparable* entities, and this, again, depends upon dealing with char-acters. In fact, characters are *all* that matters most to taxonomists, despite what we have been told about "the poverty of taxonomic characters" [65]. Of course, characters by themselves can be miserable, and only their correct distribution among hierarchical levels can help us to develop a natural classification. The *characters* are the main "building bricks" in taxonomy, not 'species', although the latter were assigned such an important role [66].

Long ago Darwin supposed that simply being a competent taxonomist was enough to have 'good species', but competent taxonomists often disagree about their species or genera to a greater extent than incompetent ones, since the latter simply follow, often unquestionably so, the concepts proposed by the former. What would be most attractive to a taxonomist in Darwin's views is "his character-based view of species" [17, p. 503], although Darwin himself was hopeless in dealing with characters. Moreover, the *Origin of Species* contains a number of phrases about species that are not based on characters. Thus, he wrote that "if a variety were to flourish so as to exceed in numbers the parent species, it would then rank as the spe-cies, and the species as the variety" [16, p. 68]. However, the number of individuals (exten-sional characteristics of a group) is not a basis for recognizing a group as 'species' or 'variety'.

"What was required to sort out the... multiplicity of characters was a *theory of taxonomic char-acters*" [24, p. 480, italics added], but exactly such a theory is wanting in the science of taxon-omy. We do not have a claim here as to the full development of the character theory, but we can think of two concepts that require the consideration as a first step. These two concepts are (1) the character concept and (2) character weighting (= ranking). Although "characters are among the most fundamental units we use to systematize the things in our world, to-gether with ideas about species" [67, p. 2], there are many notions about characters which are unsuitable for taxonomic theory. "A preliminary definition of a character that could temporarily serve as a guide through the jungle of ideas and observations around the character concept" suggests that "a *biological character* can be thought of as a part of an organism" [67, p. 3. italics added], but this is not the definition of a taxonomic character. The definition given above is considered to be "broader" than "the narrow definition of a character in systematics" (l.c.), but, actually, that "broader" sense, when associated with systematics, was always the source of confusion, since *taxonomic characters are not parts* of organisms. As in the case of the species problem, the biological "burden" in the character problem prevents the construction of a natural system.

The *taxonomic* character concept also has narrow and broad meanings, and the first of these tends to dominate. In its narrow meaning, a 'character' is merely a feature or any attribute of a member of a taxon [15, 68-69], and many biologists have expressed its inadequacy for taxonomy. They have tried to make a shift from 'organismal traits' to 'phylogenetic relationships' [11], but the wider character concept also was proposed long ago, and it even covers 'phylogenetic relationships' as a particular case. In a wider meaning, a character is an independent variable consisting of mutually exclusive character states [59, 70-71]. These states may be displayed simultaneously, as in Mendelian inheritance, or, being apomorphic and plesiomorphic [72], they may follow each other in time. The main consequence of the broader character concept is that it is a *partial basis* for making taxa *comparable* in their hierarchical levels: *Taxa of equal rank should be characterized by different states of the same characters.* [When we say "a species differs", this means "from other species", since a species cannot "differ" from a genus (being a subordinate unit) or a variety (being a super-taxon). This rule excludes the distribution of plesiomorphic and apomorphic states of the *same* characters among different hierarchical levels [24, p. 482].

The character in a broader sense means a *discriminative relation between* taxa and represents a "unity in diversity", whereas the narrow sense of a "character" is related only to similarity *within* taxonomic groups. The broader concept is always of greater importance for any theory, and the broader character concept removes, for example, the opposition between the Linnaean hierarchy supposedly based on 'intrinsic' properties and Hennig's hierarchy based on 'relational' properties [27]. The Linnaean hierarchy is based on the ranking of *characters in all the diversity of their states*, so it is also based on relational properties. If only the narrow character concept is the focus, rather limited species concepts appear, and one of them is a "biosimilarity species concept". We believe the discussion of this concept to be highly productive [5], especially because it admits the *ontology of similarity relations* (contrary to the dismissive attitude to 'organismal traits'). However, this concept is not very helpful for dealing with species diversity in nature. Apart from the fact that it fits a taxon of any level, this concept does not make an important discrimination between intensional and extensional similarity.

The main problem with the extensional similarity of characters comprising an *internal* polymorphism in closely related taxa, for example in genera within the same family, is that it greatly overweighs few differences between those taxa. The most famous examples are Vavilov's studies of variation within genera of several families of vascular plants [73]. Thus, he has counted more than *hundred* of the same features repeating themselves *within* such genera as *Secale, Triticum, Hordeum, Avena* (and others), which differ in very few characters. Vavilov's law tells us that species and genera that are genetically closely related are characterized by similar series of heritable variations with such regularity that by knowing the series of forms within the limits of one species, we can predict the occurrence of parallel forms in other species and genera. The more closely related the species in the general system, the more resemblance there will be in the series of variations. However, despite the fact that extensional similarity can serve as an indicator of close relationships, the employment of similar internal polymorphisms for the unification of organisms would entail the use of artificial groups instead of natural taxa. Exactly such an unfavorable situation exists in many current molecular studies, since the extensional similarity in molecules should be even higher than for morphological features [74]. The discrimination of intensional and extensional similarity of taxa requires the distribution of characters among hierarchical levels, but this is impossible when the focus is on the 'species problem'.

Another character-based concept, rather popular today, represents species (and higher taxa) as natural kinds. In a wider sense, this concept seems to be the familiar class-concept, judging from discussions that have appeared in the biological literature. Thus, classes and individuals have been opposed in the 'species problem' before, but now natural kinds take the place of classes in this opposition. Also, in an analogous way, natural kinds, similar to classes, are thought to be supposedly incapable of undergoing evolutionary change, and it has been noted rather clearly that "species had originally been considered as *classes or natural kinds*" [59, p. 78, italics added]. Here, we want to emphasize that the traditional opposition of 'classes' and 'individuals" was based upon the false opposition of intension and extension; it was false because two different aspects of taxa do not constitute a polarity. As pointed out in the discussion given above, the *intension of each evolving taxon* is the *class-forming and individual-forming* aspect of such a biological unit *simultaneously*, so the intensional consideration does not contain any opposition between classes *or* individuals. In this respect, our attitude deviates from some older one-sided views of biological taxa and, moreover, represents the traditional concept of species as natural kinds in a different light.

What is wrong with an interpretation of biological taxa as natural kinds? The main features of the latter were described as follows [59, p. 79, italic in the original): "(1) All members of a natural kind have the same characteristic properties... (2) The identity and boundary of a natural kind is *metaphysically* determined by an essence... The first condition does not apply to species as there is substantial variation across the members of a species, and even a feature shared by all conspecifics at a time may be modified in evolution." Let us look more closely at the first objection to natural kinds as it is applied to species. This objection does not seem to be valid, because it does not consider intensional and extensional changes during the evolution of species. The fact that each taxon can possess a substantial internal variation in its extension does not exclude the presence of some characteristic properties in its intension, and when intensional features are modified in evolution, new taxa of the same level appear. Now, what about the second of the two "wrong" characteristics of natural kinds? "In the case of the second condition,... an essence has typically been taken to be an *intrinsic* property of a kind member, as in the case of chemical structure. But no intrinsic property (= internal feature) of an organism — be it genotypic or phenotypic — can serve as the definition of its species..., as other species members have or may evolve different features" (l.c., italic in the original).

As indicated, different features in the members of a particular species are observed in an extensional aspect (species content), so this is not the reason for the failure of 'intrinsic' properties in intension. What is more important here is the fact that *all* of the available *properties* of organisms may define taxa at different levels (and be 'essential' at some level), so there is no need to look for special 'intrinsic' properties. The intensions do really define the *identity and boundary* of taxa since the differences between earlier appearing taxa are not disturbed by either their differentiation into subgroups or the appearance of new taxa of the same level. The position in which some features determine a taxon's identity was considered to be a version of essentialism [75-76], and this might have a negative taste for many biologists, but the *science of taxonomy* was born in the framework of essentialism. More productive than the struggle with essentialism is the search for points for its compatibility with the evolutionary development of a biota.

From the considerations outlined above, it follows that the old concept of natural kinds could be—with some reservations (about intensional and extensional aspects)—applied to taxa, but its alleged failure has stimulated the appearance of a new concept of natural kinds, describing them as "homeostatic property clusters" [59, 77-78] and directed towards the biologically meaningful units. At first inspection, the new concept also might seem to remove the old opposition of classes and individuals based on an opposition between intension and extension and to make the shift of individuality to intension, which was discussed earlier. Thus, the *cluster of properties* should determine "the identity and boundary" of a taxon across time and make it a "cohesive entity" [59, p. 80]. However, the discussion of proponents of the new concept have repeatedly slipped back into extensional considerations of cohesiveness. For example, "gene flow" is mentioned as "one of the several features determining a taxon's identity" [59, p. 82], although such processes as gene flow or interbreeding occur in the extensional space of taxa and should have nothing to do with the concept of intension as a cluster of properties. In addition, the "new philosophical notion of a natural kind" has been said to attempt "to reconcile the fact that such kinds are *cannot be defined by necessary or sufficient conditions...*" [59, p. 79, italics added]. However, this is exactly the old objection against natural kinds (or classes) altogether, so the "new" concept does not have any advantages.

There is even an impression that the proponents of the new natural kinds look only for extensional 'cluster of properties (= processes)' and this does not distinguish them from the advocates of the extensional 'individuality thesis'. In other words, they admit the notions of "individuals" and "classes" to encounter *metaphysically* (in Brigandt's [59] words) within extension, whereas we associate both these notions with intension. That's why we agree with Brigandt's view [59] that both approaches to taxa (as individuals and classes) are compatible but on the different grounds. There is also a trend in phylogenetic taxonomy that involves *extensional considerations* when taxa are thought to be defined "*intensionally*". Thus, De Quieroz [11, p. 304] wrote that "intensional phylogenetic definitions" have a focus "on the parts [of taxa - L.V. & S.L.] and the relationships that unite them to form the whole", and the relevant *process* responsible for the unification of populations to form a whole of a particular species is indicated as *interbreeding* (l.c.), whereas higher taxa are supposedly united by the relationships between ancestral and descendant species. However, such relationships cannot "form the whole", since the "whole" is earlier in origin and might exist in an undifferentiated state similar to an egg prior to embriogenesis. That was really a trick to consider the extensional relationships *within* taxa as their "defining properties" [11, p. 305], and that, once again, reconciled the opponents in the debate whether species are classes or individuals. Such a 'phylogenetic' approach [11] to taxonomy coincides with the "new concept of natural kinds", and we have already emphasized the dominance of extensional thinking in biology [35], especially in the 'species problem'.

## 5. Character weighting and ranking

The problem of character weighting in taxonomy is much more important than the species problem, but it is not so popular among biologists. At present, this problem is complicated by an unfounded belief that "molecular techniques have provided a powerful new tool to independently *evaluate the validity* of taxonomic designations" [2, p. 67, italics added], although molecular characters still need to be evaluated themselves. It is really interesting that same authors [2, p. 74] suggest that "abandoning the concept of species" and replacing "the current artificial views of life with a system that describes groups of organisms based on the amount that they differ from other groups" would be an appropriate course of action. Unfortunately, using the 'amount' of difference as the delimitative basis for classifying taxa strongly reminds one of a 'gap' and thus has no meaning in taxonomy. It is not the 'amount' of difference but rather the weight assigned to the differences that matters.

It has been correctly understood that "weighting was, in fact, at the heart of taxonomy" [79], although practical taxonomists were doing weighting rather intuitively, and the theoretical basis for this process was not properly developed. Moreover, the earlier suggestion by Adanson to assign an *equal* weight to characters [29]—which is incompatible with the taxonomic hierarchy that is based on admitting the higher or lower position of characters instead of their equality—was appreciated as the anticipation of the advent of evolutionary theory [19]. This situation, again, shows that biologists failed to perceive the character hierarchy as the *only* evolutionary sequence reflected by a taxonomic system.

Apart from the complete denial of character weighting, which is surely most unproductive, there has been an array of opinions about character values. For example, there was an observation that "any character that varies in a particular case cannot have general value" [29, p. 33]. In Darwin's notebooks, one can find the following sentence: "Definition of species: one that remains at large with constant characters..." [7, p. 131]. The constant characters were always highly evaluated, but we should keep in mind that a character in a broader sense might be always changeable in its states and *always the same as the discriminative relation* between groups, so all characters are variable and constant simultaneously. As for particular character states marking particular taxa, they might be variable or non-variable at different levels, but this does not influence their usefulness. Variable states can be very valuable, because they characterize taxa on the basis of *certain ranges of variability*, which sometimes repeat themselves in different species of the same or different genera within a particular family with a remarkable regularity. So, the constancy or variability of characters (and their states) does not influence their comparative importance for taxonomy.

Further, there is a belief that "unique characters have a real taxonomic weight" and that "we can attach some importance to shared rarity" [80, p. 350, 354]. This conviction was a key to the development of the cladistic notion of 'autapomorphy' (as a unique trait characterizing a given taxon), and even the 'autapomorphic species concept' was suggested [81-82]. However, a unique state of some character can be of weight at the species or genus levels only in those instances in which other states of that character have the same relative importance (see below). Otherwise, unique states lead to the haphazard segregation of taxa that are non-comparable in level. Exactly this chaotic practice is recommended in current molecular taxonomy, and people think that such phrase as "This species is *distinguished by this gene from all others* in the species complex..." [83, p. 325, italics added] is a good discriminative basis. Some biologists are anxious about situations in which there is "the move away from providing character evidence with phylogenies" [84, p. 26], and they state that it is "time to show *some* character". However, "some character" might be of no importance at the level in question. As such, we need numerous characters that should be properly distributed in the taxonomic hierarchy.

"A connection between the taxonomic importance of a character and its function" also has been considered by some biologists [29, p. 82], and very often the fundamental importance of characters to the life of organisms was confused with their taxonomic importance. In taxonomy, the *function of all characters* is to discriminate groups at different hierarchical levels, and the importance of some trait for the functioning of some organism is not relevant. Besides, in this matter, there is some misunderstanding about the nature of taxonomic characters. For example, the fruits of flowering plants are functionally important as the protectors of seeds, but when taxonomists use fruits for comparisons, they use such characters as *the presence or absence* of fruits (in the rough division of seed-bearing plants), *simple vs. compound* (i.e. formed from a single or several ovaries or carpels) fruits, *fleshy vs. dry* fruits, *dehiscent vs. indehiscent* fruits—at lower (and different) taxonomic levels. On one hand, taxonomists use characters that are mostly irrelevant for the functioning of plants, and organisms with dehiscent or indehiscent fruits are likely to be doing equally as well. But even if, on the other hand, the character is as important as the presence or absence of the protection provided for seeds, one can understand that a *taxonomic character* is not a trait of an organism, it is *something* that exists only in opposite qualities and quantities.

The most valuable characters are considered to be those which are indicating relationships [29, p. 122], but exactly *all characters indicate relationships, although at different hierarchical levels*. So, the weight of a character has nothing to do with the description of a character itself (its constancy, uniqueness, importance for an organism, etc.). Instead, *the weight of a character reflects its position in the taxonomic hierarchy*, and the weighting should come to *ranking*. That is the gist of the whole subject of taxonomy, but the problem of character ranking has never been solved completely: "The establishment of *principles* that would allow a proper subordination of characters remained a desideratum" [29, p. 123, italic added]. Strangely enough, a ranking has been considered by some biologists as something that is "simple" and "must be gone beyond" [84]. Hierarchies that fulfill the desideratum are believed to be "hierarchies of embedment in which the components of each level interact in such a way that novel properties are expressed at higher levels" [85, p. 121]. However, this statement cannot be applied to the taxonomic hierarchy as a hierarchical *model of evolution*. Not a single interaction between components of lower level (e.g., species) can define the expression of novel properties at a higher level; on the contrary, properties of higher levels appear earlier in evolution and define the appearance of properties at lower levels, including some of their interactions.

There has been at least one general principle consistently used by biologists in their classifications, namely the 'principle of generality': "the more general characters marking out more inclusive groups, less general characters marking out subordinated, less inclusive groups" [24, p. 479]. Exactly this principle reflects the evolutionary sequence of characters, since groups of lower level with their less general characters appear within groups that appeared earlier. As a result, biologists can have a kind of 'rough' hierarchy reflecting the stepwise development of the organic world. However, really "heavy" characters marking certain steps or levels of evolution did not originate frequently, and after the initial great evolutionary events there was an increase of diversity at lower hierarchical levels along with the development of characters into states and a combination of character states. Even with few characters possessing several states, the combinatorial variation might lead to large numbers of species and genera. This combinatorial variation is the cause of reticulate relationships between taxa at the same level and has nothing to do with "reticulations" caused by hybridization, simply because the combinations appear as the result of an usual divergence of characters [36].

Now, if the evolutionary sequence of characters can help us to construct the axis of a taxonomic hierarchy, the combinatorial variation of states at each level can provide the "principle" of finding numerous *taxa of equal rank*. The problem inherent in all taxonomy is the chaotic employment of the differences that exist between groups, especially between species and genera, and this does not allow them to be comparable in rank. To overcome this difficulty, we have suggested a special method for the weighting of differences that can be considered as *a posteriori weighting* or *test for rank equality* [32]. The analysis begins with the available diversity of *tentative* species or genera segregated by taxonomists chaotically, so that some of them might represent 'good' taxa at their appropriate level, whereas others need to be united or separated. All of the tentatively described groups represent *taxonomic hypotheses* for us. These hypotheses might be deduced from very different *a priori* considerations, including morphological, ecological, biogeographical, even those that are associated with reproduction in case of 'species'. The groups could be *very unequal extensionally* (i.e. they could be small and large), but they should be *equal intensionally* if we call taxa of the same level 'species' or 'genera'.

What is most important is the *estimation of differences* between already segregated groups, and since this estimation is carried out *after* group formation, the testing or weighting is of *a posteriori* nature. In this process, the most "heavy" differences are those that most frequently participate in the delimitation of species within a genus or genera within a family. In our experience, it appears that very few characters are sufficient to outline the majority of already established species and genera, and not many of the previous taxonomic hypotheses need to be re-considered. As a result, *taxa at each level came to be defined by state combinations of the same character set,* and the characters of each level produce the multidimensional combinatorial space of states. This procedure of testing for rank equality can be used for an evaluation of different taxonomic schemes. Each particular classification before *a posteriori* weighting could contain 'good' and 'bad' elements, but in comparing some of them, the approach used can be either eclectic and based on subjective choosing of 'good' elements from all schemes or be synthetic and based on objective *self-coordination* of taxa defined by a *single character set*. Linneaus was completely right when he wrote that "every genus is natural...; it is not to be capriciously split or stuck [to another], for pleasure,... especially *a posteriori*" [86, p. 114, italics added].

One could regard these considerations as 'operational' and not 'ontological', but they have a number of implications for taxonomic and evolutionary theories. First of all, the resulting combinatorial arrangement at each level (after weighting) might possess a very high prognostic power and thus allow us to predict taxa with certain combinations of features. Such predicted combinations are often found to occur in nature, and we have never described a new species or a new genus without checking the existence of an 'empty' place for it in the network of combinatorial space. It should be noted that a prognostic value was required from a *natural* system [87], and this might impart the necessary ontological sense to single-level combinatorial arrangements. At the same time, the employment of a single character set for delimitation of taxa at each level allows us to have a 'fundamentum divisionis' that is *necessary for the logic used in classification* [88]. In the case of small genera where species are not numerous and sometimes cannot be arranged within a combinatorial space, a taxonomist can make a comparison with a large genus within the same family (or order) and evaluate equally the same differences between species within small and large genera. Such an approach will conform to Vavilov's law of homologous variation occurring within closely related genera.

The combinatorial arrangements might help resolve several useless debates encountered in the taxonomic literature. For example, the opposition known as "a single difference vs. many differences" for outlining particular groups is hardly adequate, but even today reviewers may reject some manuscripts because a species or a genus is described on the basis of a single difference. However, the whole set of characters is used to distinguish groups at each level, and the combinatorial space of their states is constructed on the basis of many differences, whereas *each taxon* within that space *differs from a neighboring one* (along the line of character states) *in a single difference*. So, it's quite permissible to indicate only one difference between a new species or genus and its nearest taxon, if the states of a delimitative character belong to a weighted discriminative complex. Thus, characters are of highest value at some level if they are most frequent, and their *unique* states characterizing particular groups also are of great importance. On the contrary, if we encounter some character that occurs very rarely in group delimitation, its unique states are mostly of no value at all. Therefore, one cannot apply thoughtlessly the "autapomorphic species concept" to the analysis of species diversity.

There is also an unfounded appraisal of apomorphic states to the prejudice of plesiomorphic ones. We have explained the origin of combinatorial (reticulate) relationships as the result of the initial divergence of characters into states in different descendants of the same ancestor and the subsequent combination of primitive and advanced states during further evolution [36]. We also have emphasized that the resulting combinations can include only one combination of exclusively apomorphic states and only one combination of exclusively plesiomorphic states (an ancestor), but all other ones are mixtures of apomorphic and plesiomorphic features. Therefore, any unification of these entities into a higher taxon on the basis of 'shared similarity' (apomorphy) will lead to artificial groups.

It has been supposed that the obligate reticulation in relationships "effectively invalidates hierarchical classification" [89, p. 583], but this is not the case. Reticulate relationships are observed at each level of the hierarchical classification and cannot be incompatible with the latter. On the other hand, taxa of the same level also can be connected by a key-like structure, or cladogram. In this respect, reticulations could be considered to be in opposition with *single-level* 'trees', although these arrangements do not exclude each other. The main problem here is that the combinatorial space contains repetitive character states in all directions, and the different ways of folding that space into tree-like structures along any delimitative character might divide very closely related taxa and separate them by some distance in the cladogram, or those taxa that possess the repetitive state appear to be neighbors in the 'tree', and taxonomists think they represent a 'natural clade'. In any case, the construction of cladograms strongly distorts the multiple relationships that exist between taxa on the same level.

## 6. Some words about essentialism

Essentialism as a "philosophical idealism" [90, p. 217] was associated with the typological theory that "stems from Plato and his sources" [91, p. 46] and implies the existence of static 'patterns' shared by all members of a particular group. These 'patterns' or 'types' correspond to Plato's 'ideas' (or 'forms') and those examples of Aristotle's 'essences' that were identified with 'differences' (cf. "the *definition* is the formula of the *essence*" and "the *definition* is the formula which comprises the *differentiae*":*Metaphysics* 7, 1031a, 10-15 & 1038a, 5-10, italics added [92]). Hence, the 'essences' of taxa were designated as sets of *distinguishing* or *defining* characters.

Two problems have arisen with respect to such 'essences' and the consequence of both is the conviction that any classification is completely arbitrary. The first problem related to the uncertainty about 'defining' or 'essential' characters; and such an uncertainty led to the creation of numerous artificial groups of organisms chaotically segregated on the basis of subjectively chosen characters that seemed to be 'essential'. This problem could not be resolved without evolutionary ideas, but the advent of the latter did not help and only created the second problem. Instead of an elaboration of a program with the purpose of searching for essential characters, the very existence of 'essences' was rejected. It was noted that evolutionary theory "necessarily challenged the ontological assertion that species as Forms existed" [19, p. 318]. Consequently, taxonomists were forced "to abandon Aristotelian definitions of taxa names" [19, p. 317] in terms of essential characters that are shared by all members of a given taxon. In fact, the solution to essentialistic "mysteries" could be found in the "practice of weighting some properties more heavily than others because of their varying phylogenetic significance" [19, p. 316], but all the practice and theory went in wrong direction.

It has been supposed that "the concept of distinct and static patterns cannot meaningfully be applied to real groups of organisms, which are parts of an evolutionary continuum and which are always highly variable", and that "typological theory... should have no part in modern taxonomy" [91, p. 50]. Contemporary generations of biologists have been trained to believe that "*typological thinking* is the other major misconception that had to be eliminated before a sound theory of evolution could be proposed" [93, p. 4, italics in original]. In a real sense, this statement has proposed to eliminate systematic foundations for the sake of evolution theory. Later, Mayr [94, p. 145, 156] wrote in an assertive tone that "biological species... have no essence", because "the outstanding characteristic of an essence is its unchanging permanence". Therefore, "it was precisely the variability of species populations that led to population thinking, a dramatic departure from essentialism." [94, p. 156).

Terms such as 'essentialist' or 'typologist' rapidly became disgraceful and abusive, so that systematists are completely intimidated to remain silent. Some scientists have ventured to vote for essentialism [20, 95-100], but their opponents simply ascertained their effort as an "unfortunate attempt to reintroduce essences into systematics" [101, p. 110-111]. The latter were partially right because, very often, non-*classificatory 'essences'* were reintroduce (for example, the 'genetic code' [97]). Such 'essences' were correctly thought to be quite different from "Platonic or Aristotelian essences"; and the main advantage of the former was seen in the fact that they contain "nothing metaphysical about them" (l.c.). This fear of metaphysics seems to be the most striking characteristic of contemporary theorists, but systematics actually needs precisely '*metaphysical* essences' for the ordering of natural diversity.

Sometimes, the defenders of 'essences' are forced to make various reservations to convince readers that their attitude has nothing to do with what Mayr calls 'essentialism', and they do not ignore variation and evolution [102, p. 52]. Others [103, p. 446] have distinguished 'good' essentialism with the fruitful idea that species taxa are classes and 'bad' essentialism, a "virulent version of essentialism" or "typological essentialism". Curiously, 'good' essentialism is associated with the class concept (i.e., with 'types' as sets of properties), while 'bad' essentialism is what has been referred to as 'organismic' essentialism [104] dealing with 'types' as particular things or organisms. The latter 'typology' which considers only type specimens (valuable exclusively for the purposes of nomenclature) sometimes becomes a 'virulent version' of essentialism indeed, because it obscures the advantages of Plato's typology. Lidén and Oxelman [105, p. 183, italics added] wrote that «nomenclatural types allow us to free systematics from *typology*»; thereby, the good practice of the 'type method' is placed in opposition to the very core of classification and becomes an 'enemy'.

Caplan and Bock [103, p. 446] agreed with Mayr that "typology has hindered the development of evolutionary and other theories in biology and that it has no conceptual role in modern biology". Such a criticism is only relevant to the set of ideas surrounding 'typical organisms' but is directed against Plato's typology. Moreover, Caplan and Bock believed that "quite possibly typology as advocated by Plato has no role in any area of science" and that Plato's philosophy is merely "a ghost, long since dead, buried and forgotten". They [103, p. 453] announced "the funeral of Plato's ghost", but one can only be surprised at such a low estimation of Plato's ideas that have been, are, and always will be the one of the most fertile and vital components of the theory of knowledge, which includes the classification of objects in the world. Precisely typology as advocated by Plato has role in any area of science; it cannot be dead, buried and forgotten at all. Biologists have been appreciated for their struggling with older philosophical traditions [106, p. 185], but such an arrogant orientation is most fruitless and disastrous for the development of science.

Plato was a philosopher of primarily universal thinking; his 'ideas' or 'forms' make sense only in the world of estimations, and classification is precisely such a world. *Plato's 'ideas' are 'characters' in the diversity of their particular states* (cf. *Parmenides*, 132c-d [107]) of many levels that determine the place of a particular thing among others. This place is not the place of a thing (an organism) in the 'economic of nature' (for example, a tree in the nearest forest); instead, it is the place of a thing (an organism) in the *whole world of existing differences*. Beyond comparison and reasoning, one cannot see the 'idea' (or 'essence') of a 'thing' (cf. *Theaetetus*, 186d-e: "knowledge does not consist in impressions, but in reasoning about them" [107]). Aristotle's essentialism was more complicated; he preferred particular things but *could not* avoid universals in discussions of *species and genera*. In the latter context, Plato's 'forms' repeatedly appeared in Aristotle's metaphysics, and Aristotle even wrote that "nothing... which is not a species of a genus will have an *essence*—only species will have it" *(Metaphysics* 7, 1030a, 10-15, italics in original [92]).

Although Aristotle "dissected fishes with Plato's thoughts in his head" [108, p. 136], biological systematics assimilated mainly the part of his metaphysics relating to the equation of 'thing' and 'essence' ("Each thing... and its essence are one and the same"; *Metaphysics* 7, 1031b15-20 [92]). Since taxa became to be perceived as 'individuals' (i.e. 'things' of a kind), the equation cited above led to the shift from relationships between taxa to the search for 'essences' *within* isolated taxa, and naturalists tried to figure out what is represented by a 'typical organism'. However, the enterprise was completely hopeless: there were no typical organisms to be found, and when the high diversity of members within a group was observed, a most fantastic 'type' was invented. It was either an 'average' caricature (schematic figure of a plant or an animal) or a chimera combining all the variability of organs in different individual organisms [109]. In both cases, 'types' were unreal, and this kind of 'typology' was justifiably criticized because, being merely an exercise in imagination, it could not help in the construction of the natural system. More than that, those unreal 'types' led to the conviction that the only real 'type' is a type specimen [15]. This created a second variety of 'organismal' typology (although with the same equation of 'essence' and 'thing') that has nothing to do with classificatory typology.

As a result of inventing such an image of 'typology', biologists and philosophers began to criticize their own fantastic notions and, what is worse, the imaginative 'ideal organism' became to be associated with Plato's 'ideas' [110, p. 501]. In addition, some authors accused Plato's essentialists of requiring that "type specimens represent the 'typical' species member" [111, p. 465]. Others, on the contrary, separated the platonic line of thought (or 'class interpretation') and the practice of 'type' specimens and wrote that "on the class interpretation, the role of particular organisms as type specimens is anomalous" [58, p. 353] It is true that specimens cannot be 'types' of taxa, and — to continue Hull's statement — one can read that, on the class interpretation, "the role of lower taxa as types for higher taxa is even more anomalous" (l.c.). Here, a typologist of the platonic line would even more emphatically agree, because lower taxa *cannot* be 'types' for higher taxa indeed. Taxa of all levels have their own types (different sets of characters). Furthermore, Hull (l.c.) put in opposition "class interpretation" and "historical entity interpretation", but, as was emphasized above, the class interpretation allows historical entity interpretation because 'essences' (as sharing traits) are often the only indicators of such entities (how to find 'essences' is another matter).

In fact, all 'pitfalls' of essentialism originate from misunderstood 'essentialism'. Grantham [111, p. 464] wrote that, in his critique of 'essences', he followed Sober's "important insight", but the reading of Sober's papers [112-113] does not leave a negative impression. Sober [112, p. 355] even noted that "the mere fact of evolution does not show that species lack essences" [112, p. 356], and that "if essentialism is simply the view that species have essential properties..., then the doctrine remains untouched (by Hull and Ghiselin)" [112, p. 359]. Nevertheless, Sober [112, p. 370, italics added] believed that "far from ignoring individuals, the *typologist... focuses on individual organisms* as the entities which possess *invariant properties*". In our opinion, this is a quite untenable view of the typologist. The 'invariance' of the *organisms* belonging to a species was not maintained by platonic typologists, since things are "almost always changing and hardly ever the same, either with themselves or with one another...; *they are always in a state of change*" (*Phaedo*, 78e-79, italics added [107]). The 'essentialist theories' about 'essences' have treated the latter as something more than merely properties that are common to some group of objects because 'essences' were supposed to cover the opposite properties, too. Plato's theory of 'forms' repeatedly represented an 'essence' as the 'unity of opposites'.

Thus, 'essences' are not 'invariant properties' in particular individuals; they are 'invariant' (the same) units (unity of opposite properties) and, in addition, they are 'themes' with variations. 'Essences' consist of evolving characters and also could be called 'individuals', taking into account Ghiselin's [114, p. 303] statement that "if something can evolve, it must be an individual". Therefore, nothing supports Sober's [112] view that 'the essentialist's method of explaining variability' failed. Interestingly, Sober [112, p. 381, italics added] believed that "essentialism pursued an *individualistic (organismic) methodology*, which population thinking supplants by specifying laws governing objects at a higher level of organization". This only demonstrates how deeply Aristotle's 'organismic' essentialism is rooted in modern biology and how carelessly it is deduced from Plato's essentialism which could be called 'classificatory essentialism'.

## 7. Conclusions

It has been said that the treatment of the species problem is seriously unbalanced because the biological aspects of the problem are quite strong, but the philosophical aspects are very weak [5, p. 4]. When trying to show that the interactions *within* species (i.e. the very "biological aspect" that was always the primary focus) are of no consequence for the construction of a natural system and for the segregation of natural species themselves, we seem to make a shift to the philosophical aspects. Moreover, trying to show that only the weighting of *differences* between groups matters, and taking into account that 'differences' (or delimitative characters) are *ontological relations* between groups, we return the biological 'species problem' to the 'theory of universals' [52] and the wider field of metaphysics [5]. Ghiselin [114, p. 304] stated that "we are still being metaphysical, albeit opting for a better metaphysics". He is right in that bad 'metaphysical' beliefs that "have no heuristic value" [91, p. 49] dominated biological systematics for a long time, but we may hope for a better metaphysics.

In the light of the discussions outlined herein we can answer some questions in a peculiar way."(1) Should taxonomic classification proceed in terms of descent alone or on the basis of similarity (cladism versus gradism)? (2) If classification rests on clades, are homologies (apomorphies) indicators of history, or are they patterns that are evidence in favor of a historical reconstruction but not themselves a model of evolution? Briefly, this is the distinction between ontological and epistemological notions of classification" [7, p. 208].

First, our classification proceeds in terms that are neither 'cladism' nor 'gradism', and both 'descent' and 'similarity' are represented in intensions distributed among different hierarchical levels along the evolutionary sequence of novelties. Second, classification can be based on 'clades', if 'clades' are defined by combinations of character states at different hierarchical level. Apomorphies, of course, should not be considered "themselves as *model of evolution*", since the latter term is applied only to the whole taxonomic system, whereas apomorphic and plesiomorphic states of the same character are of equal weight (equal rank). Lastly, we believe that our epistemological notions of classification are not divorced from ontological ones.

There has been an opinion expressed that "a species concept is the description of the role a species plays in the household of nature" [115, p. 99], and we have tried to show that the only role of species is to represent the biological diversity at one of the lower levels of the biological hierarchy. The theory of evolution was "invented" to *explain* taxonomic relationships, and the proper formulation of an evolutionary theory requires a good taxonomic theory as a starting point. Therefore, we wish to attract more attention to notions that are important for the development of taxonomic theory, mostly those dealing with characters and their ranking. We also wish to emphasize that the taxonomic hierarchy is *not* "a fictitious grid we place on nature" [50, p. 422]; it is the model of evolutionary differentiation of the organic world, and that evolution proceeds within the framework of differences that originated earlier in time. The notions of extension and intension can help us to trace the development of taxa and also to understand their changeable and unchangeable aspects. Some authors [116, p. 444] wrote that "the world may appear static or dynamic, discontinuous or continuous, hierarchical or linear, as revealed by the taxic versus transformational approach". However, the taxic (taxonomic) approach does not exclude transformation, while the world is simultaneously 'static' (in intensions) and dynamic (both in extensions and intensions because the latter display variation in character states), discontinuous (in characters of different levels) and continuous (in states of characters at the same level), hierarchical (multileveled) and linear (at each level).

When discussing the failure of species concepts, we are not opposed to species in taxonomy and we do advocate their *objective* segregation or their *discovery* in the nature. Before coming up with a definition of the term 'species', biologists should check out whether or not all of their particular 'species' are described properly (i.e. they should test the tentative species hypotheses for a rank equality). After dealing with all natural diversity and testing all taxa in the hierarchy, it may appear that species represent the first level where self-coordination of taxa is possible.

## Author details

Larissa N. Vasilyeva[1*] and Steven L. Stephenson[2]

*Address all correspondence to: vasilyeva@biosoil.ru

1 Institute of Biology and Soil Science, Far East Branch of the Russian Academy of Sciences, Vladivostok, Russia

2 Department of Biological Sciences, University of Arkansas, Fayetteville, USA

## References

[1] Pleijel F. Phylogenetic taxonomy, a farewell to species, and a revision of Heteropodarke (Hesionidae, Polychaeta, Annelida). Systematic Biology 1999;48(4): 755-789.

[2] Hendry AP, Vamosi SM, Latham SL, Heilbuth JC, Day T. Questioning species realities. Conservation Genetics 2000;1(1): 67-76.

[3] Pleijel F, Rouse GW. Least-inclusive taxonomic unit: a new taxonomic concept for biology. Proceedings of the Royal Society London, B, 2000;267(1443): 627-630.

[4] Hey J. On the failure of modern species concepts. TRENDS in Ecology and Evolution 2006;21(8): 447-450.

[5] Stamos DN. The species problem. Lanham: Lexington Books; 2003.

[6] Vrana P, Wheeler W. Individual organisms as terminal entities: laying the species problem to rest. Cladistics 1992;8(1): 62-72.

[7]   Wilkins JP. Species. A history of the idea. Berkeley: University of California Press; 2009.

[8]   Dobzhansky T. A critique of the species concept in biology. Philosophy of Science 1935;2(3): 344-355

[9]   Mayr E. Species concepts and definitions. In: Mayr E. (ed.) The species problem. Washington: AAAS; 1957. p1-22.

[10]  Vasilyeva LN. Does the clash between instinct and science exist in naming nature? Cladistics 2012;28(3): 330-332.

[11]  De Queiroz K. Phylogenetic definitions and taxonomic philosophy. Biology and Philosophy 1992;7(3): 295-313.

[12]  O'Hara R. Systematic generalization, historical fate, and the species problem. Systematic Biology 1993;42(3): 231-246.

[13]  De Queiroz K, Donoghue MJ. Phylogenetic systematics and the species problem. Cladistics 1988;4(4): 317-338.

[14]  De Queiroz K, Donoghue MJ. Phylogenetic systematics and species revisited. Cladistics 1990;6(1): 83-90.

[15]  Mayr E. Principles of systematic zoology. New York: McGraw-Hill; 1969.

[16]  Darwin C. The origin of species. A Mentor Book; 1958.

[17]  Mallet J. Why was Darwin's view of species rejected by twentieth century biologists? Biology and Philosophy 2010;25(4): 497-527.

[18]  Stevens PF. An end to all things? Plants and their names. Australian Systematic Botany 2006;19(2): 115-133.

[19]  Hull DL. The effect of essentialism on taxonomy - two thousand years of stasis. British Journal for Philosophy of Science 1965;15: 314-326.

[20]  Kitts DB, Kitts DJ. Biological species as natural kinds. Philosophy of Science 1979;46(4): 613-622.

[21]  Loevtrup S. On species and other taxa. Cladistics 1987;3(2): 157-177.

[22]  Kinsey AC. Supra-specific variation in nature and classification from the view point of zoology. American Naturalist 1937;71(734): 206-222.

[23]  Gayon J. The individuality of the species: a Darwinian theory? - From Buffon to Ghiselin, and back to Darwin. Biology and Philosophy 1996;11(2): 215-244.

[24]  Rieppel O. The series, the network, and the tree: changing metaphors of order in nature. Biology and Philosophy 2010;25(4): 475-496.

[25]  O'Hara RJ. Population thinking and tree thinking in systematics. Zoologica Scripta 1997;26(4): 323-329.

[26] Haeckel E. Anthropogenie oder Entwickelungsgeschichte des Menschen. Leipzig: W. Engelmann; 1874.

[27] Rieppel O. Hennig's enkaptic system. Cladistics 2009;25(3): 311-317.

[28] Rieppel O. Sinai Tschulok (1875-1945) - a pioneer of cladistics. Cladistics 2010;26(1): 103-111.

[29] Stevens PF. The development of biological systematics. New York: Columbia University Press; 1994.

[30] Margulis L (1991) Symbiogenesis and symbionticism. In: Margulis L, Fester R. (eds.) Symbiosis as a source of evolutionary innovation. Cambridge: MIT Press; 1991. p1-14.

[31] Vasilyeva LN. The hierarchical model of evolution. Zhurnal Obzschey Biologii [Journal of General Biology] 1998;59(1): 5-23. (in Russian with English Summary).

[32] Vasilyeva LN. Systematics in mycology. Bibliotheca Mycologica 1999;178: 1-253.

[33] Vasilyeva LN. Evolutionary and classificatory meanings of the Linnaean Hierarchy. In: Kryukov AP, Yakimenko LV. (eds.) Problems of evolution. Vol. 5. Vladivostok: Dalnauka; 2003. p9-17 (in Russian).

[34] Vasilyeva LN. The hierarchy and combinatorial space of characters in evolutionary systematics. Proceedings of Zoological Institute of Russian Academy of Sciences, 2009;Supplement 1: 235-249 (in Russian).

[35] Vasilyeva LN, Stephenson SL. The Linnaean hierarchy and 'extensional thinking'. The Open Evolution Journal 2008;2: 55-65

[36] Vasilyeva LN, Stephenson SL. The problems of traditional and phylogenetic taxonomy of fungi. Mycosphere 2010;1(1): 45-51

[37] Grene M. Hierarchies and behavior. In: Greenberg G, Tobach E. (eds.) Evolution of Social Behavior and Integrative Levels. Lawrence: Erlbaum Associates; 1988. p3-17.

[38] De Queiroz K, Gauthier J. Phylogenetic taxonomy. Annual Review of Ecology and Systematics 1992;23: 449-480.

[39] De Queiroz K, Gauthier J. Toward a phylogenetic system of biological nomenclature. Trends in Ecology and Evolution1994;9(1): 27-31.

[40] Spencer H. The principles of biology. Vol. I. New York, D. Appleton and Company; 1868.

[41] Marshall CR. Explaining the Cambrian "explosion" of animals. Annual Review of Earth and Planetary Sciences 2006;34: 355-384

[42] Droser ML, Finnegan S. The Ordovician radiation: a follow-up to the Cambrian explosion? Integrative and Comparative Biology 2003;43(1): 178-184.

[43] Gould SJ. The structure of evolutionary theory. Cambridge: Belknap Press; 2002.

[44] Corning PA. Rethinking categories and life. The Behavioral and Brain Sciences 1981;4(2): 286-288.

[45] Stamos DN. Popper, falsifiability, and evolutionary biology. Biology and Philosophy 1996;11(2): 161-191.

[46] Frost DR, Kluge AG. A consideration of epistemology in systematic biology, with special reference to species. Cladistics1994;10(3): 259-294.

[47] Nixon KC, Wheeler QD. An amplification of the phylogenetic species concept. Cladistics 1990;6(3): 211-223.

[48] Mishler, BD, Brandon RN. Individuality, pluralism, and the phylogenetic species concept. Biology and Philosophy 1987;2(4): 397-414.

[49] Ridley M. The cladistic solution to the species problem. Biology and Philosophy 1989;4(1): 1-16.

[50] Ereshefsky M. Darwin's solution to the species problem. Synthese 2010;175(3): 405-425.

[51] Kitcher P. Ghostly whispers: Mayr, Ghiselin, and the 'philosophers' on the ontological status of species. Biology and Philosophy 1987;2(2): 184-192.

[52] Williams MB. Species are individuals: theoretical foundations for the claim. Philosophy of Science 1985;52(4): 578-590.

[53] Ghiselin MT. A radical solution to the species problem. Systematic Zoology 1974;23(4): 536-554.

[54] Ghiselin MT. Species concepts, individuality, and objectivity. Biology and Philosophy 1987;2(2): 127-143.

[55] Ruse M. Species as individuals: logical, biological, and philosophical problems. The Behavioral and Brain Sciences 1981;4(2): 299-300.

[56] Kitcher P. Bewitchment of the biologist. Nature 1986;320(6063): 649-650.

[57] Stamos DN. Buffon, Darwin, and the non-individuality of species – A reply to Jean Gayon. Biology and Philosophy 1998;13(3): 443-470.

[58] Hull DL. A matter of individuality. Philosophy of Science 1978;45(3): 335-360.

[59] Brigandt I. Natural kinds in evolution and systematics: metaphysical and epistemological considerations. Acta Biotheoretica 2009;57(1-2): 77-97.

[60] Hull D. Phylogenetic numericlature. Systematic Zoology 1966;15(1): 14-17.

[61] Griffiths G. On the foundation of biological systematics. Acta Biotheoretica 1974;23(3-4): 85-131.

[62] Griffiths G. The future of Linnaean nomenclature. Systematic Zoology 1976;25(2): 168-173.

[63] Ereshefsky M. Some problems with the Linnaean hierarchy. Philosophy of Science 1994;61(2): 186-205.

[64] Ereshefsky M. Species and the Linnaean Hierarchy. In: Wilson RA. (ed.) Species: new interdisciplinary essays. Cambridge: MIT Press; 1999. p285-305.

[65] Rieppel O, Kearney M. The poverty of taxonomic characters. Biology and Philosophy 2007;22(1): 95-113.

[66] Stuessy T. Plant taxonomy. 2nd ed. New York: Columbia University Press; 2009.

[67] Wagner GP. Characters, units and natural kinds: an introduction. In: Wagner GP. (ed.) The character concept in evolutionary biology. San Diego: Academic Press; 2001. p1-10.

[68] Ghiselin MT. 'Definition', 'characters' and other equivocal terms. Systematic Zoology 1984;33(1): 104-110.

[69] Rodrigues PD. On the term character. Systematic Zoology 1986;35(1): 140-141.

[70] Sneath PHA, Sokal RR. Numerical taxonomy. Nature 1962;193(4818): 855-860.

[71] Colless DH. On 'character' and related terms. Systematic Zoology 1985;34(2): 229-233.

[72] Hennig W. Phylogenetic systematics. Urbana: University of Illinois Press; 1966.

[73] Vavilov NI. The law of homological series in variation. Journal of Genetics 1922;12(1): 47-89.

[74] Sanderson MJ, Donoghue MJ. Patterns of variation in levels of homoplasy. Evolution 1989;43(8):1781-1795.

[75] Keller RA, Boyd RN, Wheeler QD. The illogical basis of phylogenetic nomenclature. Botanical Review 2003;69(1): 93-110.

[76] Rieppel O. The PhyloCode: a critical discussion of its theoretical foundation. Cladistics 2006;22(2): 186-197.

[77] Boyd R. Homeostasis, species, and higher taxa. In: Wilson RA. (ed.) Species: new interdisciplinary essays. Cambridge: MIT Press; 1999. p141-185.

[78] Griffiths PE. Squaring the circle: natural kinds with historical essences. In: Wilson RA. (ed.) Species: new interdisciplinary essays. Cambridge: MIT Press; 1999. p209-228.

[79] Yoon CK. Naming nature: The clash between instinct and science. New York: W.W. Norton; 2009.

[80] Kendrick B. The generic concept in Hyphomycetes - a reappraisal. Mycotaxon 1980;11(1): 339-364.

[81] Rosen DE. Fishes from the uplands and intermontane basics of Guatemala: revisionary studies and comparative biogeography. Bulletin of the American Museum of Natural History 1979;162: 267-376.

[82] Nelson GJ, Platnick NI. Systematics and biogeography: cladistics and vicariance. New York: Columbia University Press; 1981.

[83] Cook LG, Edwards RD, Crisp MD, Hardy NB. Need morphology always be required for new species description? Invertebrate Systematics 2010;24(3): 322-326.

[84] Mooi RD, Gill AC. Phylogenies without synapomorphies – A crisis in fish systematics: Time to show some character. Zootaxa 2010;2450: 26-40

[85] Zylstra U. Living things as hierarchically organized structures. Synthese 1992;91(1-2): 111-133.

[86] Freer S. Linnaeus' Philosophia Botanica. Oxford: Oxford University Press; 2003.

[87] De Hoog GS. Methodology of taxonomy. Taxon 1981;30(4): 779-783.

[88] Stafleu FA. Linnaeus and the Linnaeans. Utrecht: A Oosthoek's Uitgeversm; 1971.

[89] Lawrence JG, Retchless AC. The myth of bacterial species and speciation. Biology and Philosophy 2010;25(4): 569-588.

[90] Johnson LAS. 'Rainbow's end: the quest for an optimal taxonomy. Systematic Zoology 1970;19(3): 203-239.

[91] Simpson GG. Principles of animal taxonomy. New York: Columbia University Press; 1961.

[92] Aristotle (1984) Metaphysics. In: Barnes J. (ed.) The Complete works of Aristotle. Princeton: University Press; 1984. p1552-1728.

[93] Mayr E. Populations, species, and evolution. Cambridge: University Press; 1970.

[94] Mayr E. The ontological status of species: scientific progress and philosophical terminology. Biology and Philosophy 1987;2(2): 145-166.

[95] Caplan AL. Back to class: a note on the ontology of species. Philosophy of Science 1981;48(1): 130-140.

[96] Schwartz SP. Natural kinds. The Behavioral and Brain Science 1981;4(2): 301-302.

[97] Kitts DB. Can baptism alone save a species? Systematic Zoology 1983;32(1): 27-33.

[98] Kitts DB. The Names of species: a reply to Hull. Systematic Zoology 1984;33(1): 112-115.

[99] Bernier R. The species as an individual: facing essentialism. Systematic Zoology 1984;33(4): 460-469.

[100] Grene M. A defense of David Kitts. Biology and Philosophy 1989;4(1): 69-72.

[101] Hull DL. Can Kripke alone save essentialism? A reply to Kitts. Systematic Zoology 1984;33(1): 110-112

[102] Van Valen LM. Species, sets, and the derivative nature of philosophy. Biology and Philosophy 1988;3(1): 49-66.

[103] Caplan A, Bock WJ. Haunt me no longer. Biology and Philosophy 1988;3(4): 443-454.

[104] Vasilyeva LN. [Essentialism and typological thinking in biological systematics]. Zhurnal Obzschey Biologii [Journal of General Biology'] 2003;64(2): 99-111. (in Russian, with English summary)

[105] Liden M, Oxelman B. Do we need phylogenetic taxonomy? Zoologica Scripta 1996;25(2): 183-185.

[106] Kitcher P. Ghostly whispers: Mayr, Ghiselin, and the 'philosophers' on the ontological status of species. Biology and Philosophy 1987;2(2): 184-192.

[107] The collected dialogues of Plato including letters. Princeton: University Press; 2005.

[108] Whitehead AN. Adventures of ideas. New York: Free Press; 1933.

[109] Hammen L. Type-concept, higher classification and evolution. Acta Biotheoretica 1981;30(1): 3-48.

[110] De Queiroz K. Replacement of an essentialistic perspective on taxonomic definitions as exemplified by the definition of "Mammalia". Systematic Biology 1994;43(4): 497-510.

[111] Grantham TA. Beyond 'individuality' and 'pluralism': a review of Ereshefsky's Units of Evolution: Essays on the Nature of Species. Biology and Philosophy 1993;8(4): 457-468.

[112] Sober E. Evolution, population thinking, and essentialism. Philosophy of Science 1980;47(3): 350-383.

[113] Sober E. Sets, species, and evolution: comments on Philip Kitcher's "Species". Philosophy of Science 1984;51(2): 334-341.

[114] Ghiselin MT. Taxa, life, and thinking. The Behavioral and Brain Science 1981;4(2): 303-310.

[115] Mayr E. Comments by Ernst Mayr. Theory in Bioscience 2002;121(1): 99-100.

[116] Christoffersen ML. Cladistic taxonomy, phylogenetic systematics, and evolutionary ranking. Systematic Biology1995;44(3): 440-454.

# Chapter 5

# Biological Species
# as a Form of Existence, the Higher Form

Victor Prokhorovich Shcherbakov

Additional information is available at the end of the chapter

## 1. Introduction

*Our major enemy is not another being but our imperfection*

*"Some say the world will end in fire, Some say in ice."*

Robert Frost

*"Artificial Intelligence description method for deliberative agents functioning on the basis of beliefs, desires and intentions as*

*known in Artificial Intelligence, can be used successfully to describe essential aspects of cellular regulation" [1]*

The Universe consists of discrete entities: elementary particles, atoms, molecules, planets, stars, galaxies. That is there are a limited number of configurations of matter that are fairly stable and lasting, the intermediate ones being volatile. The Universe is structuralized. It means that the world components and the Universe itself resist chaos. They are far from thermodynamic equilibrium. They exist. The existence is resistance to chaos. Many will agree that thermodynamic equilibrium means death for a biological entity, but it is true for any system as well. It may sound funny today that the Darwinian natural selection, acting by accumulation of tiny heritable changes, was initially supposed to produce an even continuum of the living beings. This expectation was never corroborated. The creationists still keep using the absence of this continuum as evidence against biological evolution. But the biological world follows the same global principle: organisms, populations, species are discrete stable entities, the intermediate configurations being volatile. Biological evolution cannot retain and does not retain everything that randomly emerges. The existence of the

Universe depends on the mutual affinity of its constituents, their ability to interact with each other, thus resisting the general aspiration for evenness. This is, however, only one side of the coin. The interaction should prevent the dissipation, but not more than this. Any existence implies a balance between two opposite forces – dissipating and compressing, repulsion and attraction. Too strong interaction leads to collapse, disappearance in singularity.

There are two major forms of existence: inanimate and animate. Any existence is not perpetual. The second law of thermodynamics predicts final dissipation or destruction in collapse of everything in the Universe. What is the dissipating force? Generally speaking, it is energy. Why does the Universe not dissipate immediately? The components of the Universe *interact* with each other, thus retarding the dissipation. The four fundamental physical interactions (weak and strong nuclear interactions, electromagnetic interaction, and gravitation) prevent immediate dissipation of the inanimate entities. These interactions beget numerous new forms of existence, new entities. The Universe as a whole evolves. The prebiotic evolution shares the same major principle with biological evolution [2,3,4]. Various objects continuously arising in the Universe have different longevities, from infinitesimal fractions of a second to billions of years. In the course of evolution, ephemeral forms are replaced by more lasting ones. This principle – "survival of those who survive" – sounds as a tautology, but it is *the great tautology*: Everything genuinely new emerges through this principle. "The only meaningful and objective definition of adaptation would be persistence" [5]. *Longevity is a quantitative measure of existence*. Thus, the evolution of the Universe implies the development of resistance to dissipation, to chaos, to entropy. On the other hand, the longevity implies limitation of the force of interaction, resistance to collapse, to annihilation in singularity. To exist long, the living forms of existence must be able to keep the balance of dissipation and attraction.

## 2. Biology is special

---

*… "…knowledge is a natural phenomenon which originated long before humans."*[6]

---

It gets increasingly evident that life is not a physical process. It is not just extended physics and chemistry. It is absolutely different form of existence, the higher form. The swallow, building its nest in my shed near Moscow, flies to South Africa and returns strictly to the same shed every year. I hope nobody will try to calculate the probability of such event proceeding from the physical causality and stochasticity. It is evident that swallow knows the route, there and back. Just as I know the route from my home to my work and back. And this knowledge is only a trifling part of the total knowledge any organism enjoys. This total knowledge is knowledge how to reproduce itself. I beg pardon of those who cannot stand anthropomorphisms. Human beings are biological entities and they share many defining characteristics with other living entities (and vice versa). Knowledge is one of them. *Knowledge can be defined as ability to accomplish reliably a low-probable action*. The knowledge may be given quantitatively as the inverse of the probability of the action. This ability is acquired during evolution of species and ontogenesis of organism (not only in a high school). Defined this way, knowledge is linked to behavior. The behavior based on knowledge is an expedient behavior. Taking the word "behavior" in the broadest sense, we may speak about the behavior of molecules, organelles, cells, organisms, populations, and ecosystems.

The autocracy of physics ends at the border between the inanimate world and the biosphere, where the world of sense and knowledge begins, and behavior of matter becomes expedient. The words: knowledge, memory, coding, transcription, translation, function, signaling, recognition, decision-making, governing, creation, which are impossible and needless in describing inanimate nature, become not only acceptable but unavoidable in the description of living systems [4].


## 3. Organization as a form of existence

An organization is a complex system that can perform certain functions by virtue of its particular assemblage of parts [7]. Organized systems must be distinguished from the ordered ones. Neither system is random, but the ordered systems are generated according to a simple algorithm and therefore lack complexity, whereas the organized systems must be assembled element-by-element according to an external program or plan. Organization is complexity endowed with function. It is not random due to design or selection, rather than to the necessity of crystallographic order [8]. Living entities (cells, organisms, populations, and species), and only living entities[1], are organizations with the function of survival. Parts of the living entity may hold only particular functions.

"Life is based on semiosis, i.e., on signs and codes" [9], and it cannot be adequately described by means of physics and chemistry. Everything essential in biology is determined not by physical causality but by semantic rules and goal-directed programs. This principle operates on all levels of biological organization. Coding is not limited by the coding of polypeptide sequence by the nucleotide sequence. The entire life cycle is carried out by sequential development of the organic codes and interpretation rules for stepwise self-manufacturing of the entity. In contrast to the objects of the inanimate world that come into being as a result of stochastic interactions, the living entities and their components are *manufactured* on the basis of ontogenetic intention [10]. "All biological objects are artifacts, and ... life is artifact-making" [9]. Coding and instructions involve the use of symbols, but a symbol is connected to the symbolized subject semantically, not physico-chemically. The DNA sequence cannot be deduced from the physico-chemistry of the nucleotides just as the text cannot be deduced from the alphabet. Moreover, this non-deductibility is a necessary stipulation for the capability of DNA to code genetic information. The ambition and hope of molecular biology was to explain phenomenon of life via some new sophisticated physics, in particular, via non-equilibrium thermodynamics. Inadequacy of this approach in explanation of life becomes

---

1 Machines may also be organizations but they are human made creatures.

more and more evident. Living entities survive not because of some special physics or chemistry but owing to their sensible behavior [4,11].

We are used to think that all the entire stuffing of the Universe is presented by two interconvertible essences: energy and substance or matter. Latterly, several bold guys [6,9,12-17] started talking about the third fundamental essence – organic information, which is neither energy nor matter. It is an attribute of life and only life. I think the term "information" may not be the most suitable one. It may be confused with the homonym used in the information theory. Shannon's information is devoid of meaning whereas the meaning is just what we are interested in in context of biology. Most importantly, "the third fundamental essence" is not just information but behavior, which includes internal and external signaling, their interpretation and implementation in the form of organization with a function of survival. I think adequate name for this third fundamental essence would be "mind". Four fundamental physical interactions prevent immediate dissipation of the inanimate world yet they cannot explain existence of living entities. The living world resists chaos by means of *behavior*. Let us define a behavior *sensible* if it is aimed at survival of the behaving entity. Inanimate evolution might be portrayed as self-construction of Nature: matter from energy; biological evolution might be portrayed as self-knowledge of nature: mind from matter [18,19]. The mind is not just epiphenomenon of life. It is *the fifth fundamental interaction* preventing the living world from dissipation and from collapse as well. Mind is a major life-specific instrument for survival. The words "mind", "knowledge", "sense" should not be taken as metaphors. They are good terms. The cells of my body know how to replicate DNA. This knowledge does not differ conceptually from my knowledge of how to read and write, being of course much more important for my survival than the literacy. The total knowledge a biological entity enjoys is the knowledge of how to reproduce itself. Not more and not less. Life is everlasting self-reproduction.

In inanimate nature, all the processes are directed from a less probable state to a more probable state: movement to equilibrium. Life is a movement to a low-probable state. It is a river flowing upward. A stone falling down from a mountain is an example of a physical process; an alpinist climbing up a mountain is an example of a biological process. It needs not only energy, but intention (will) and knowledge; it needs mind. The laws of physics are not violated during this "climbing up". Instead, they are harnessed by the goal-directed programs of life in such a way that low-probable, virtually impossible events become the most probable. Hence, life is a form of existence that differs radically from any form of inanimate existence.

A living entity may disappear for two reasons: it may die or it may change. In both cases the previous entity ceases its existence. To exist means to exist long. Biological species know how to exist long. This is a miraculous knowledge because living entities are improbably complex things. It is not complexity in itself that is miraculous but the fact that the biological complexity is *highly organized*. Complexity implies availability of a large space of the states for the system, while the term "organized" implies that only very few of the states are compatible with vitality. "For the gate is narrow and the way is hard, that leads to life". I.e., living systems are non-equilibrium, low-entropy, low-probable things. We do not say that organisms violate the second law of thermodynamics. Living things are open systems, continuously sharing matter and energy with their environment. Organisms export entropy to environment, thus keeping their internal states far from equilibrium. This is a correct description of the state of affairs, though it is not an adequate explanation of biological organization. Energy is necessary but not sufficient for implementation of life. It needs knowledge. Energy in itself is a chaotic factor. To support organization, energy must be sensibly harnessed. The living entities survive due to their sensible behavior. They know the way that leads to life. By the way, the necessity of knowledge for implementation of life was already evident for the naturphilosophers of the eighteenth century [20].

# 4. Living entities

## 4.1. Prokaryotic cell

The basic form of the biological existence is a cell. Prokaryotic cell is a minimal biological entity. Though any separate thing may be named "entity" in English, it would be heuristically important to name living or biological entity only the organization with the function of survival. The components of a cell (proteins, nucleic acids, membranes, ribosomes, viruses and the like) possess only particular functions, so they are not *biological entities*. A living entity is a monad. It is specifically isolated from the environment and it is a cohesive whole; it exists in time by means of self-reproduction. I am quite aware of possible disagreements with such definition of a living entity; however, I find it not only useful but even inevitable in context of this paper. Cell is a minimal quant of Life. A living entity cannot be less than cell. The components I mentioned above are products of cellular activity. Prokaryotic cells are also the most ancient of the known forms of life. They appeared on the Earth about three billion years ago [20].

## 4.2. Eukaryotic cell

The eukaryotic cell is a much more complex entity. It appeared on the Earth about one billion years ago as a result of cooperation between several prokaryotic cells [22,23]. As such, they created the enormous and abundant world of unicellular eukaryotes.

## 4.3. Multicellular organisms

The high complexity of the eukaryotic cells enabled further cooperation and appearance of multicellular organisms. The multicellular organism is a monad. Providing it is asexual, it is substantive entity (see below) that can reproduce itself acting alone.

## 4.4. Biological species

It occurred that the higher multicellular organisms with large genomes and complex development fail to reproduce themselves reliably across generations. This impediment brought about further cooperation: creation of *multiorganismic entity* as a self-reproducing unit – biological species. Individual organisms comprising the species lack their status as substantive monads. The above scheme is a simplification. On the one hand, the real self-reproducing entity is a generation of interbreeding population – deme. On the other hand, there exist various forms of sexuality, not only the obligatory sexuality on which I am concentrated in this paper. Existential consequences of sexuality are most clearly expressed in the obligatory sexual forms.

### 4.5. Substantive and attributive existence

In accord with the above definition of living or biological entity, I am going to use here the notions of "substantive" and "attributive" existence. Substantive existence implies autonomy and self-sufficiency of the entity in its reproduction and evolution. Substantive entity is a sovereign player on the stage of life. Attributive existence implies existence as a part of a higher rank entity (host). Its survival and evolution is causally linked to survival and evolution of the host. In case of asexuality, an organism is a substantive entity, whereas a sexual individual is an attributive entity that exists as a part of a higher rank entity – deme. The attributive existence is a ubiquitous form of biological existence. For example, hepatocytes are entities that exist as an attribute of the animal organisms. They represent a class of polyphyletic entities. Hepatocytes of the different animals are much more similar to each other than, for example, to neurons or any other cells of the same organism. While demonstrating transspecific "epigenetic consanguinity", they reproduce and evolve as an attribute of the host and for the good of the host. Other good instances of the attributive existence are organelles (mitochondria, plastids). I find the concept of substantive and attributive existence useful for description and understanding of biological organization and evolution. In context of this book, it must be quite important to keep in mind that the existence of a sexual individual organism is attributive.

## 5. Extant and lasting forms of biological existence

As was stated above, evolution of the Universe automatically leads to replacement of ephemeral forms with more lasting ones. This result is inevitable in the ever changing world. Inanimate entities are the most probable configurations of matter at a given situation; their lasting is provided by their physical durability that is provided by the balance of dissipation and attractive power. Organisms are low-probable configurations of matter; they are physically flimsy, extremely complex, low-entropy systems. They cannot withstand entropy growth perpetually. The homeostatic mechanisms cannot be absolutely perfect. They make errors and they lose their robustness. Absolutely perfect homeostasis would require infinite energy expenses. Organisms inevitably die even in the most favorable environment, in the absence of any competition, with an abundance of energy and substance. They perish because of entropy.

It looks like organism as a form of existence reached the thermodynamic limit and is unable to further improve its homeostatic facility. The accuracy of the cell processes is tuned to the point where it is optimal. Both too little and too much accuracy will adversely affect organismal vitality. The energy expenses are concentrated on fidelity of DNA reproduction. And it is really high: one incorrect nucleotide is incorporated only once in $10^8$–$10^{10}$ events. Transcription and translation proceed with a much lower fidelity, with misincorporation rates of 1 in $10^4$ and 1 in $10^3$–$10^4$, respectively [24,25]. With this error rate, significant proportion of newly made polypeptides contains amino acid substitutions [26]. And it is not the whole problem. All biopolymers and supramolecular structures are continuously damaged and the defects are accumulated with time. Accumulation of errors must have self-accelerating dynamics inevitably leading to catastrophe. Living systems bypass the catastrophe by means of reproduction. They reproduce to avoid death. Sometimes, single-cell organisms are referred to as immortal. It is misunderstanding. They also save themselves by reproduction [27]. Even for the apparently symmetrically dividing cells of Escherichia coli, it was shown that the two supposedly identical cells produced during division are functionally asymmetric. The old pole cell should be considered an aging parent repeatedly producing rejuvenated offspring [28].

A characteristic property of life is that its stability is dynamic. Living entities continuously change during lifespan. In essence, this changing is self-regeneration, self-manufacturing, self-renewal. This is the content of life. Organisms and generations of species continuously reproduce themselves through the time. They are transient, renewable forms of existence. The lasting forms of biological existence are lineages and species. An individual organism and a generation of species is a transient link in the existence of lineage or species.

The reproduction may be coupled with multiplication, with increasing the number of organisms. This expansion is an important but contingent factor of species survival. The essence of the reproduction is the replacement of an old, worn-out body by a new one. It may not and, in a standard situation, should not lead to the increasing of the number of organisms to avoid the resource exhaustion. The genuine evolutionary success is stable reproduction [4].

Why does the reproduction, a more complex phenomenon than the simple existence of the individual, prevent entropy from growing up? The point is that reproduction is always coupled with selection. Natural selection is a quality control of reproduction. Imperfect copies are rejected while novelties have a chance to be saved[2] only if they improve or at least do not essentially worsen the homeostasis. So, the outcome of natural selection is largely conservative. Life would not be possible without this conservatism. Evolutionary biologists were mostly concentrated on the generative, inventive side of evolution. But evolution is a dual phenomenon: it generates novelties and it stabilizes them. Moreover, evolution is just a by-product of reproduction, its imperfection. The changes as such are entropy-driven [29] and do not appear enigmatic. Rather they are inevitable. The true marvel is stability, the persistence, the resistance not only to destruction but to further changing as well. Note that only conservative mechanisms are related to sensible behavior (by definition). Reproduction coupled with selection is the very mechanism that is able to provide lasting existence, potential immortality of a living entity. I would like to stress that natural selection is not a special

---

2 This saving is memorizing new knowledge.

goal-directed mechanism invented for the lineage survival. It operates automatically by retaining those that persist and letting go off those that give up. Nevertheless, special mechanisms improving the efficiency of purifying selection are known [30-32].

Ideally, reproduction should be precise; otherwise, the goal of immortalization is not got. Template-directed synthesis was the first and major invention of nature, from which life itself started. It is clear that the precision of DNA replication must be such that most progeny received unaltered genetic information. The real fidelity of DNA replication is remarkably high [33]. For unicellular organisms, the attained fidelity of DNA replication is enough for potential immortality of the lineages. However, in multi-cellular organisms with large genomes and complex development, the number of mutations per genome per generation is unacceptably high, up to three orders of magnitude higher than, for example, in yeast [34-35]). I.e., *genomes of higher organisms are not reproduced with high fidelity*. For example, in man the number of mutations per zygote is 60 or even more. At such rate of mutation, the higher organisms must rapidly degrade because of mutational overload, i.e. because of entropy. They, however, persist. Their longevity needs explanation.

Why did the replication fidelity not evolve to a higher level? The matter is that faithful replication is a costly process. High accuracy needs too much energy. It looks like a further increase in fidelity of genome reproduction was not possible. Hence, higher organisms have to be able to fulfill ontogenesis successfully, and species must be able to persist in time despite never-ending mutational perturbations. This problem has no solution in the frame of asexual ("homeogenomic") lineages. They would rapidly degrade and become extinct or blurred out in the course of the reckless evolution.

Earlier [4], I discussed what other means, besides the high fidelity of genome replication and purifying selection, were invented by evolution to avert or evade the fatal outcome of the mutational deluge. The phenomenon of canalization or robustness [36,37] is directly related to the problem. Robustness is generally defined as a property that allows a system to maintain its functions despite external and internal perturbations. In case of biological systems, it is an ability to perform successful ontogenesis despite environmental and mutational perturbations. Robustness is the retaining of function (meaning) despite changes in structure and environmental impacts. The resources of robustness are derived from all levels of biological organization. Global degeneracy of the link between structure and function is one of the definitions of canalization: there are more genotypes than phenotypes. Function, not structure, is selected during evolution. Different genotypes may correspond to the same phenotype. This principle operates on various levels of biological organization, including operation of multiple pathways leading to the same final result. This is possible owing to the fact that biological processes are determined not by physical causality but by semantic rules and goal-directed programs. Simple organisms, reproducing their genomes with high fidelity, have rather simple semantics with relatively simple hermeneutics. The language of higher organisms is much more complex, with rich synonymy and complex, context-dependent, hermeneutics. This helps to provide resistance of development to mutational and environmental perturbations.

Nevertheless, we have to admit that for the higher organisms with large genomes and complex development all these salutary efforts have appeared insufficient: they fail to reproduce themselves reliably across generations. Their lasting needed another instrument. This instrument was sexual reproduction, the creation of *multiorganismic entity* as a self-reproducing unit.

For a long time, the problem of emergence and maintenance of sexual reproduction attracted little attention from evolutionists. The matter probably seemed too obvious. No one doubted Weismann's idea that sexual reproduction, creating genetic variability, produces material for natural selection and enhances the evolutionary potential of the species. A possibility of the acceleration of evolution at amphimixis was quantitatively substantiated by Fisher [23] and H.J. Muller [24]. The conception of the evolvability is still popular among population geneticists. It is frequently assumed that the capability for rapid and diverse evolution is a positive trait supported by natural selection, while a shortage of the evolutionary potential is fraught with extinction. The notion of evolvability as a selectable trait is in evident contradiction to the known efforts of evolution aimed at creating genetic stability of organisms and lineages [2-4]. It is obvious that the evolvability cannot be easily taken as a species homeostatic mechanism. Direct selection for evolvability is impossible conceptually, so the transition to sexuality needs another explanation, independent of the evolvability. Though sexual reproduction and genetic recombinations are a source of combinative variation in populations, they do not produce new alleles but only new combinations of the extant ones, which are, moreover, doomed to be destroyed in the next generation. If to think that the sexual reproduction was invented for acceleration of evolution (Lamarckian thought, by the way) than the continuous shuffling of the genomes (heedless of their merits) looks more than strange. I think we should not assume special mechanisms for the acceleration of evolution created by evolution. These would be suicidal mechanisms. A species with accelerated evolution would not exist long. All the organisms populating our earth today belong to species resistant enough to further evolution. Evolution is inevitable because the systems created by evolution for protection against evolution, species homeostasis, are not absolutely perfect, and the entropy pressure overcomes them now and then. All the species are capable of evolving just because they originated from the ancestors that were capable of evolving and inherited their imperfection, their "original sin". It is hard to avoid evolution.

## 6. Biological entities as self-reproducing units

There is some complication in delineating the self-reproducing unit in case of sexual reproduction. Two individuals of different sexes are enough to produce progeny. But it is known that the stable reproduction needs a rather large interbreeding population – deme. Small populations have low robustness because of inbreeding that leads to considerable homozygotization. The homozygous individuals usually have a drastically reduced vitality, and the populations they form also have low robustness because of lack of polymorphism and weak genotypic plasticity. Small population size is fraught with the risk of extinction. On the other hand, a species may consist of many demes with rare interdeme genetic exchanges because of geographical impediments or habitual preferences. Different species have various forms of intraspecies organization and sexual relationships. So the borders of a self-reproducing unit must be of necessity fuzzy. In context of this paper, this complication seems not to be principal. All the existential advantages of sexual reproduction are fully realized on the level of deme at any form of sexual relationships. The rare interdeme genetic exchanges may complicate the picture for the evolutionary theorists but not change it principally. Though demes are not completely closed entities, they are closed enough to depend in their survival primarily on the merits of their own.

The early group selection models were flawed because they assumed that genes acted independently, whereas now it is apparent that gene interaction, and more importantly, genetically based interactions among individuals, were an important source of the response to group selection. As a result many are beginning to recognize that group selection is potentially an important force in evolution. So I will try to avoid here the painful discussion related to group selection. I limit myself by the statement that a unit of selection and a unit of substantive reproduction are strictly the same units, the same monads. The notorious replicator/interactor discrimination was invented to save the selfish gene theory which I regard as erroneous. There are two great delusions in the evolutionary theory: gene as an ideal replicator and individual organism as a quintessential unit of selection. Gene is not a substantive entity. I am not even sure that ontologically it is an entity at all because it is definitely not organization. It even lacks any defining characters of an autonomous thing. It is a product of cell activity, even if the very important product. Functionally, it is a piece of text that acquires meaning only in context of the whole organism. It is getting clear that the concept of a selfish gene is not based on real premises. The linkage "one gene - one trait - one selection vector" is not observed: one gene may affect several traits, and most traits depend on many genes [12]. On the other hand, a sexual individual is not a self-replicating entity either, so it cannot be selected as such. Organisms are unique, inimitable parts of species manufactured as piece-goods during species reproduction. The selective meaning of an individual organism is appreciated in context of population via inclusive fitness. Generation of deme is a minimal entity that reliably reproduces itself with a high fidelity (according to the Hardy-Weinberg law). Deme and species are ontologically comparable entities. They differ only in the size and degree of cohesiveness. So we may say that generation of species is a self-reproducing unit. In most contexts, I use "deme" and "species" as equivalent terms.

Thus, substantive sexual entity is a generation of deme or generation of species. The term "generation" may need some clarification. Deme (and species) is a lasting form of existence; generation is a transient (extant) form of a species existence. Now and here we deal only with generation. I think it is heuristically useful to keep in mind that actually living spatiotemporally restricted entity is a generation, not a lineage, which is in fact a historical phenomenon. A lineage could be even comprehended as existing only in our consciousness. However, there must be something more substantial, more existential in the lasing existence than our human perception of the historical reality. "The defining characteristic of a living organism is that it is the transient material support of an organization" [36]. This definition is fully applicable to a species generation as well. The material support is transient. But what is lasting? According to Merser, organization is lasting. I.e., mind is lasting. The knowledge how to reproduce itself is lasting. Exactly and only this knowledge is transmitted from generation to generation. And of course, this knowledge is not just DNA sequence. Only entire substantive entity is a carrier of this total knowledge, not only genes and brains. I would like to stress that the "material support" and the mind are not separate things like the hardware and software of a computer (organism is not a Turing machine) [16]. They are aspects of the same whole [17].

A real extant population may contain organisms of different age, from new-born to mature to old individuals (overlapping generation) or it may be more or less synchronous. The lifespan of a generation is equal to the average lifespan of individual organisms that comprise the generation. The extant entity is an individual organism in case of asexuality and a generation of deme in case of sexual reproduction. When we speak that a species lives million years we may mean that an entity very similar to the extant entity lived million years ago and it is directly connected with the extant entity via sequence of reproductions.


# 7. Meaning of sexual reproduction

Sexual organisms are constituents of a higher rank entity – biological species. The transition to sexuality, like all other major evolutionary transitions, is cooperation. Individual organisms forfeited their ability to autonomous reproduction and autonomous evolution. They exist and evolve as a part of biological species. Their existence is attributive. The transition to sexuality is ascension to a new and a higher quality. Sexual population is a coherent system able to self-reproduction. Species reproduction should not be confused with speciation. Species reproduction is not formation of another (daughter) species. Reproduction is a way of species existence. The reproduction must be precise. In case of stably existing species, the reproduction is really precise. One generation may somewhat differ from another generation in accord with the environmental variations owing to a species' genotypic plasticity and the phenotypic plasticity of the organisms. These changes are reversible manifestations of species robustness. They are not evolutionary changes [12]. By the same token, speciation is not species reproduction, not continuation of the given species. Speciation is a macroevolutionary event that should not be confused with the reproduction. Reproduction is an essentially conservative process. Reproduction is renewal of the same whereas speciation is creation of the new. Similar to a sexual organism, a species is unique: it emerges on one occasion in the history of biosphere and never appears again. Therefore, it is not correct to regard a species as a segment of a species level lineage, as is suggested by De Queiroz [39]. There are no "species level lineages" just as there are no "organism level lineages" in case of sexuality. Lineage is a sequence of generations. In its totality, it represents species ontogenesis [12]. Speciation is not a pre-programmed stage in the species existence. Speaking metaphorically, the species is never "interested" in speciation. For the extant species, begetting a "daughter" species means begetting a competitor. Note that allopatry is a precondition for survival of a new-born species [40].

Reproduction of asexual and sexual organisms differs in many respects (Table). Asexual organism (typically it is a cell) is a self-reproductive unit by itself. Its ontogenesis is relatively simple, typically, the way from the young cell to the mature cell. The reproduction is perfect. It is precise and reliable. Prokaryotic lineages seem to exist billions of years [21]. The ordinary (i.e. reductionistic) viewpoint ascribes this stunning longevity to the high fidelity of DNA replication. According to the ontologically right (holistic) viewpoint, a prokaryotic cell *reproduces itself with a high fidelity*. The ability to replicate DNA belongs to cell, not to DNA. This comprehension is crucially important for adequate portrayal of the life phenomenon. Biological organization is hierarchical, and sensible behavior is a property of the whole system, not of its parts. Though the lifespan of an individual (the time from division to division) is short, the lineage is potentially (proper environment being provided) immortal. So, the lasting form of existence of an asexual entity is an organism level lineage, a linear sequence of individual organisms.

| Feature | Individual organism | Species |
|---|---|---|
| Founding | By fusion if two gametes | Usually, by geographic or ecological isolation of a small group of the individuals of different sex (founder) |
| Probability of abortion | Low | Extremely high |
| Process of ontogenesis | Individual development | Microevolution |
| Genetic basis | Non-evolving genome | Evolving gene pool |
| Contents of the ontogenesis | Embodiment of the ontogenetic intention. | Creation of species robustness |
| Causal mode | Teleonomy. Downward causation. The final result is determined by ontogenetic intention (boundary conditions) within the limit of the norm of reaction of the genotype | Group selection The final result is not determined. The process is limited by the initial conditions (historical constrains) and by a necessity to create perfect species organization |
| Unit of self-reproduction | Self-reproduction is impossible | Generation of a deme |
| General attractor | Adult organism | Stasis (ceasing of evolution) |
| Ending of ontogenesis | Obligatory death (probably programmed) | Extinction. Potential immortality is not excluded |

**Table 1.** Comparative characteristics of ontogenesis of a sexual individual and a biological species.

Many give an import to the fact that asexual organisms can exchange genetic material now and then. However, the biological sense of such exchanges is quite different. There is sex but no sexual reproduction. Asexual entities do not form biological species sensu Mayr-Dobzhansky, an entity of a higher rank. They have no need in this complication just because they reproduce themselves with high fidelity acting alone. If a mutated individual survives, it initiates a new lineage that may compete with the previous lineage and may swap it.

Metaphysics of a biological species as an individual is an intricate philosophical and epistemological problem and its discussion has rather long history [12,39-55]. Here, I am interested in the ontological aspect of the problem: species as a form of existence. As the basic definition of species, I take that of Ernst Mayr [49]: "Species are groups of interbreeding natural populations that are reproductively isolated from other such groups". Both traits "interbreeding" and "reproductively isolated" are obligatory.

Many, including me, regard biological species as an ontological individual. This view was clearly formulated by Michael Ghiselin [43]. Similar to an individual organism, a species has ontogenesis: birth, infancy, adolescence, maturity, aging, and death [12]. This lasting existence is carried out as a sequence of generations. This sequence is commonly known as microevolution that may be confusing because the essence of the species ontogenesis is not an evolution. In case of success, a species' ontogenesis culminates in stasis (i.e. cessation of evolution) that may last dozens of millions of years and more [4]. Theoretically, the potential immortality cannot be excluded. Generation of deme is a self-reproducing entity. It means that generation of deme, not the deme, is the ontological individual. So a deme is equivalent to an asexual individual organism in its role in survival and creation of the lasting entity, which is a lineage, sequential row of self-reproducing entities, sequential raw of generations of deme.

Species is organization. Bonding of the intra-species components (individual organisms) is carried out by means of behavior. I suggested the term "behavioral bond" to designate the interaction between organisms by analogy with ionic, covalent, hydrogen, etcetera bonds [12]. Behavioral bonds provide cohesiveness of species. Species-specific behavior implies operating of special connections between the individuals, which transform the species into organization with the function of survival. Primarily, these are the connections accountable for the interbreeding and reproductive isolation, which make a species a genetically closed monad. Reproductive isolation is determined by the mutual affinity of organisms. The affinity is not limited by choosing a mating partner; it includes all the intraspecies interactions as distinct from the interspecies ones. Reproductive isolation and preventing inbreeding are two opposite "forces", analogous to the attraction and repulsion, the proper balance of which is a necessary precondition for a species existence. Both inbreeding and promiscuous sexual behavior are destructive for a species. What is "the proper balance"? What is the final state this balancing is aimed at? It is an optimal species gene pool.

Genetically, a species as a whole is a closed system. Parts of the species (groups, demes) are potentially capable for substantive existence in nature. This capability is analogous to the capability of plants and lower animals to regenerate the whole body from the parts. As such, this is not speciation. The absence of the physical skin hampers us to grasp a species as a unity. But it is only a matter of habit and imagination. Behavioral bonds are common in biology. They unite families, groups, tribes, armies, companies, states, and humankind.

The transition from genome to genetic pool is far from being the whole story. Genetic pool, similar to genome, is not a substantive entity. It is reproduced by a generation of the deme as its part, its attribute. The necessity to reproduce a genetic pool drastically changes the biological status of the individual organisms. An asexual organism is a self-sufficing sovereign player on the stage of life. It is a monad. It can reproduce itself through generations acting alone. A sexual individual is a law-obedient citizen of the multi-organismic realm. It has to cooperate. This cooperation is not limited by the finding of a sexual partner and rearing a progeny. The final goal is transmitting an *optimal gene pool*. The entire species organization, including behavior of individual organisms, is submitted to this final goal. What gene pool is optimal? The most general answer is like this: the pool which provides reliable survival of the next generation. Of course, this answer is too general. The function of survival belongs not to a gene pool but to the organization as a whole, i.e. to the generation of deme. A property characterizing perfect organization is known under the label of robustness. The robustness is defined as an ability to perform successful ontogenesis despite mutational and environmental perturbations. The resources of robustness are derived from all levels of biological organization. At the species level, the main factor creating robustness is diversity of organisms comprising the species. This factor influences species survival in two ways. Immediately, diversity of organisms is the material for creation of perfect species organization. The same diversity is the prerequisite for creation of an abundant gene pool bestowing species with genotypic plasticity. The diversity of this type is an emergent trait of a species and it must be selected. Not all diversities are of equal merit. However, it may be postulated that the absence of diversity would result in rapid extinction of the species just because such homogeny annuls all the advantages of sexual reproduction.

The above consideration makes it clear that Darwinian selection of individual organisms, the conquerors in the intraspecies competition, does not work in case of sexuality. There are a lot of objections to "the selection of the best", and this one is one more: such selection would lead to the virtual annulling of the diversity.

A crucial feature of sexual reproduction is manufacturing individual genomes by picking them over from the continuously shuffled population gene pool instead of the direct copying of the ancestor's genome. The main advantage of this way of reproduction is quite evident. Though large genomes cannot be precisely replicated, there always exists a possibility to manufacture one errorless genome from the two with errors. Moreover, degeneracy of the link structure-function implies that the functionally robust genomes may have various sequences. *I.e.*, the intrinsic property of sexual reproduction is creation of great diversity of genomes and hence individual organisms. The emergent species property that follows is the *genotypic plasticity* – the ability to change *reversibly* the population gene pool configuration in different environments. This is a powerful factor of species stability.

This is the basic level of cooperation – genetic. Though the genome of a sexual organism cannot be replicated with an adequate accuracy, a genetic pool can be reproduced with an adequate accuracy. All the well-known complications and troubles of the sexual reproduction are justified by this capacity for an accurate reproduction because only accurate reproduction provides longevity.

The "picking over" mechanism of genome reproduction supports the diversity of genomes and individuals. It is evident, however, that this diversity inevitably includes a high proportion of genomes (and corresponding individuals) of low vitality doomed to perdition. Unfortunately, this was interpreted in the classic Darwinism (and in Neo-Darwinism as well) in the spirit of Malthus's idea of exponential growth of populations leading to the competition for resources: struggle of everyone against everybody. The very idea of natural selection was based on the assumption that organisms produce more offspring than can survive. Organisms, therefore, have to compete with each other. This competition was construed as a moving force of biological evolution leading to continuous perfecting of the biological entities. It was stressed that a most fierce struggle must be between the individuals of the same species because they have identical needs.

This misleading interpretation begot most malignant forms of social Darwinism and, as a counterbalance, an antievolutionary attitude of many intellectual and spiritual leaders. In reality, the seemingly "extra" progeny is a compensation for the poor fidelity of genome reproduction and for random death of the organisms[3]. In both cases, the differential survival of individual organisms is not a result of competition for resources. First and foremost, they survive or die accordingly to the merits of their own, irrespective of the presence of other individuals. As a rule, they are not killed or starved to death by their fellows or rivals.

The stability of the biotic entities is determined not merely by their physical durability but by their expedient behavior especially. They are organizations with the function of survival. The Universe evolves via the interaction and cooperation of the entities, whence its complexity and hierarchical structure come from. The major transitions in biological evolution (prokaryotic cell → eukaryotic cell → multicellular organism → biological species) are the steps of cooperation. Though a complex entity consists of the other simpler ones, it is not just an aggregate of the included entities. It is a qualitatively new form of existence; it is an organization of a higher rank. Hierarchy in biology doesn't mean just complexity or heterogeneity. It implies a functional predestination of their parts for the sake of the whole. Survival of the parts crucially depends on survival of the whole. Hence, constituent entities are to be included into the higher entities only in an appropriately transformed configuration. The operating principles of the organization of the higher rank are not necessarily related to or derivable from the properties of the parts or to their internal operating principles. That is the principles organizing an upper rank are novelties. They are not necessarily predictable from the rank below. On the other hand, the organizing restrictions of the living entities, being emerged as a frozen chance, cannot be deduced from any general principle or law. They can be understood only retrospectively, in the context of their history. The above statements imply that the evolution of a higher entity cannot be adequately presented as self-sufficing evolution of its constituents. The prosperity of the whole is the vector of selection for the constituent entities.

---

3 At r-strategy of reproduction the random death, i.e. associated neither with competition nor with genetic defects, may be massive.

# 8. Altruism

Neo-Darwinism defines altruistic behavior of an individual organism as a behavior that diminishes its own fitness and enhances the fitness of other individuals. In its turn, fitness is defined as a relative fecundity of the individual. According to the same paradigm, "Evolution is based on a fierce competition between individuals and should therefore reward only selfish behavior. Every gene, every cell, and every organism should be designed to promote its own evolutionary success *at the expense of its competitors*." [56]. If so, altruism must be impossible. Altruism, however, is ubiquitous. One more inconsistency is the fact that sexual organisms cannot reproduce themselves. To cope with this snag, G.C. Williams [57] and then R. Dawkins [58] announced the gene as a unit of selection. The gene seemed to be an ideal immortal replicator. The hypothesis of kin selection [59,60] and of reciprocal altruism [61] gave a formal explanation of how evolution could favor altruism despite a fierce competition between individuals. It did not prove of course that an individual organism or a gene can really serve as a unit of selection. Most biologists, being not trained in mathematics, had been pressed to take for granted this logical trick under the label of "gene centered view". Regretfully, it got global success among population geneticists and molecular biologists and buried the holistic understanding of life and evolution for dozens of years.

Meanwhile, an altruistic behavior, which is ubiquitous among people and other animals, keeps being a headache for the evolutionary biologists. Hamiltonian pill has helped to alleviate the headache but the phenomenon remains enigmatic and gives food to unending and mostly fruitless discussions. It is really difficult not to see the numerous and various forms of cooperation, mutual aid, friendship, and love at every turn. For this, one must specially train his/her imagination in the reductionistic logics or reject the phenomenon ironically as did Michael Ghiselin [44]: "Scratch an 'altruist,' and watch a 'hypocrite' bleed". Meanwhile, the problem of origin and maintenance of altruism is just a seeming problem begotten by the gene-centered point of view and reductionistic philosophy. The reductionistic methodology is not an adequate tool for operation with the hierarchically organized world of life.

The idea of gene as a replicator is bewildering. Gene is not a living entity. It is not a self-replicator. It is replicated. It is a replica or a template. Genes are manufactured by cell, just like all the other cell constituents: RNAs, polypeptides, organelles. Only self-reproducing substantive entity can serve as a unit of selection. By this I say of course in favor of deme (or group) selection as the only meaningful level of selection for obligatory sexual organisms. Deme is the lowest substantive entity that reproduces itself with a high fidelity. Opponents of the group selection reject it as a too slow process[4]: lower-level selection easily trumped higher-level selection. First, the group selection may be rapid enough: a generation of population (the unit of reproduction) is of the same longevity as an individual organism; second (and uppermost), the lower-level selection in itself is the destructive side of the overall process. If it is not trumped by the higher-level selection, the group simply will not go through. The mechanism of species evolution (microevolution or species ontogenesis) presumes a

---

4 A successful species ceases to evolve [4]

group selection. Successful groups may prosper, while less successful shrink and extinct. And again, competition and struggle between the groups may play little role in their fate: they parish or prosper primarily because of their own merits, e.g. because of the prevailing of selfish or altruistic behavior of the constituent entities. For a comprehensive discussion of the problem of group selection, see [62].

The idea of multilevel selection has now received a substantial support [63]. It is a step in the right direction. However, I think that here remains some inconsistency. Given that the higher level of selection operates, the selection at the lower levels must be forbidden because it can produce nothing but casualties like a parasitic DNA or a malignant cell. The so called ultra-selfish genes are factual parasites with a net harmful effect on the host. They, along with other parasites and harmful mutations, are representatives of the destructive force of nature. Evolution in action is an unending struggle against this force. And the most productive way of this struggle is cooperation. Biological species is organization, which is the cooperation of individual organisms. A hierarchical organization presumes submitting behavior of parts in favor of the whole. Selection presumes the selection of genes but the vector of selection is "for the good of species", not "for the good of gene"! Just because a gene is not a living entity, not organization with the function of survival! It is strange for me to insist on such a self-evident statement. These two goods coincide. If they not coincide, the gene will be rejected "for the good of species". The reverse (rejection of species for the good of gene) is nonsense. For more discussion, see [12].

Once we took a population as a self-reproducing unit, once we have grasped the biological species as an individual of a higher rank, we see no enigma in altruistic behavior. It is simply inevitable. We do not wonder why cells of our body, e.g. those of skin, living only several days, do not fight for unlimited proliferation. We know very well what follows if they do. Let us define altruistic behavior as behavior of parts for survival of the whole. Sexual organism lost its status as a substantial biological entity. It places its genes into the common gene pool hereby demonstrating the hundred-per-cent altruism at the most basic genetic level. Organisms are unable to reproduce themselves. They are reproduced by species as a class of entities. So they just have to be altruistic or they will disappear along with the whole. This statement may look contra-intuitive. The vernacular understanding of altruism as disinterested aid to other organisms hampers us to see the altruism as a multifaceted biological phenomenon. Faces of the altruism are numerous and they may look unexpected. I remind that an altruistic behavior is the behavior for the good of group. It may sometimes look unfriendly, hostile, and cruel in relation to the other individuals, still being altruistic.

This nontrivial comprehension helps to interpret inter-individual relationships as largely altruistic. True selfish behavior of an individual organism would be a mistake, similar to the behavior of a cancer cell. Normal behavior of the parts is always aimed at survival of the whole. Let me present one example of apparently selfish behavior that is actually behavior for the good of group. Fighting for leadership is often presented as an example of a fierce struggle. Though the picture is slightly spoiled by the ritual character of such battles, the scene remains to be impressive. But is this fighting really selfish behavior? It is hard and dangerous. The transmitting of genes to the next generation does not look as a final cause.

Every genome is unique and it is not transmitted as a whole. The semantic content of genetic information depends on the combination of genes. But the combination is not transmitted. Meanwhile, the meaning of the fighting is quite evident. A proper leader is extremely important for survival of the group. And if we give up our human envy to leaders, we will be able to recognize that the life of leaders is fairly altruistic. It is completely devoted to the group survival and often has a sacrificial character. The essence of interindividual conflicts in population is not the fighting for power but the verifying of the relative status. The correct status is an extremely important parameter for a proper organization of the population. An individual that lost in this fighting still did not lose in life. The correct status is important for getting a *right* position. A wrong position, even if it is the higher one, would be failure for the individual and for the population. It is important and comfortable to occupy a proper position in the group.

Hence, altruism, as defined above, is not something special. The balance of two counter-forces is a prerequisite of any existence. In case of species, these two forces are behavior of the parts for survival of the whole and behavior of the parts for their own survival. They are two aspects of the sensible behavior, of which a disinterested aid of one individual to another is only a particular type of altruism. Classic Darwinism proclaimed the struggle for existence. And it was implied that it is mainly the struggle between individual organisms of the same species for resources. But a lack of resources is not the common cause of organism's death. The universal enemy of life, which acts everywhere and always, is entropy. And the only force that helps to resist it is the sensible behavior.

What do we mean when we speak about the behavior of such a complex entity as a deme? The overall content of this behavior is self-reproduction. The deme is a substantive entity. Its behavior is aimed at its own survival. So it is selfish (by definition).The behavior of all its parts, including individual organisms, which is aimed at survival of the whole, is *altruistic* (by definition). The living world is organized hierarchically. Though only deme is a substantive entity, other (attributive) components may also be *relatively* autonomous in their existence. So, their behavior must be also aimed at their own survival. Moreover, the survival of the parts is absolutely necessary for survival of the whole. The behavior of the relatively autonomous entities is dual. It may look selfish and competitive in a certain respect being altruistic as a whole. This trivial consideration just shows that altruism, being complex and important phenomenon, does not look strange and enigmatic for the holistic perspective.

### 8.1. The faces of altruism

Suicide is the most common form of altruistic behavior. It operates on all the levels of biological organization, from molecules to organisms. One example of molecular altruism is the DNA damage repair by enzyme $O^6$-methylguanine-DNA methyltransferase, which transfers the methyl group of the damaged base to one of its own cysteine residues in a suicide reaction [64]. Numerous and diverse forms of cellular suicide are well known.

The evolution of the "picking over" mechanism of genome formation was necessarily coupled with the evolution of the intrinsic or internal selection. The intrinsic selection is a purifying selection. It begins to operate long before the organism is tested by the environment or came across the other members of population. Moreover, it starts operating even before the appearance of the individual, during the formation of generative cells (sperm and eggs). The overwhelming majority of the generative cells and their predecessors undergo programmed death. This mass suicide is aimed at selecting robust generative cells [31,65-68]. For example, in the testis of mice, the mutation rate declines five-fold during spermatogenesis: the heavily mutated cells commit suicide. During ontogenesis of multicell organisms, cells with damaged DNA also commit suicide (apoptosis). It is really suicide, not a killing. This behavior prevents malignisation.

One more phenomenon, inconceivable from the individual-centered view, is phenoptosis, the programmed death of organisms. In the most expressive form, it occurs in salmon: death of the adult individuals after spawning. It looks probable that an aging is a slurred form of the phenoptosis. The existence of a programmed altruistic ageing and death was suggested in [69]: "The similarities between the molecular pathways that regulate ageing in yeast, worms, flies and mice, together with evidence that is consistent with programmed death in salmon and other organisms, raise the possibility that programmed ageing or death can also occur in higher eukaryotes".

The different longevity of the individual life is a manifestation of the same phenomenon. Both, mouse and man are mammals. Why does a mouse live only two years, while a man lives up to hundred years? The answer is: individual longevity must be optimal for the species survival. And this optimum is the integral constituent of the general strategy of species survival. The phenomenon of frustration may also be construed as a type of altruistic phenoptosis. From the individual point of view, frustration looks strange: it is evidently a programmed reaction to a stress, and it is definitely contra-adaptive, especially the destruction of the immune system. May be it is also a case of altruistic suicide, a form of intrinsic selection, self-elimination of the individuals with inadequate reaction to stress.

The intrinsic selection keeps operating over entire ontogenesis: in the process of fertilization, during embryo implantation, embryogenesis, at birth, during infancy, adolescence, maturity, aging. I would like to stress that this intrinsic selection, though it may look cruel and relentless, may have little relation to the competition between individual organisms. The doomed entities die primarily because of their own imperfection.

The intrinsic selection controls robustness of the individual organisms, their healthiness. This is the first, immediate quality control at the level of an individual. The final quality control is carried out at the level of generation of deme where robustness of the generation as a whole, perfection of the deme organization is checked. It is not easy to define specifically what a "good organization" is. The most evident parameters are diversity and genetic plasticity, which are crucially important for the stability of the deme reproduction. Real success is not maximal but optimal fecundity. So that even infertile individuals may occur useful for the population survival.

The general tendency in progressive evolution is diminishing fecundity. Ideally, one female should provide two healthy offspring, not less and not more. Only during a relatively short initial period of the species founding (during creation of the species robustness and territorial expansion), the exponential Malthusian increase of population in the number makes sense. A matured species needs stable reproduction. The principal "the more the better" stems from the capitalistic psychology that is well known to lead to economic crisis. Natural selection is wiser than the human leaders. Biological species know how to control their numerical strength and thus exist long. The numerical strength should not exceed the resources. The reproductive rate in most species had evolved through group selection to ensure populations remained below the threshold of over-exploitation of resources [70].

While the competition between asexual lineages or between different demes and species may be sometimes a real combating, such combating between the organisms of the same deme would be self-destructive. I do not discard competition, but I only think that its biological meaning should be reconsidered: it is an instrument for creating, fine-tuning and maintaining species organization, which is cooperation. A species is organized hierarchically. The hierarchy is continuously checked. This checking may look as a conflict or struggle for survival yet it is not. To make emphasis on the fierce struggle means to create the problem in theory that does not exist in reality. Scratch antagonism and you find *the good of group*. On the sidewalk under my window, I see a boy jumping on the skateboard. Very dangerous exercise! He risks breaking his neck. Why? What is he fighting with? With the desk? With the gravity? Not at all. He fights with his own imperfection, which is the major enemy of everybody.

Other forms of altruism may look much more attractive: parental care, friendship, mutual aid, and other examples of the uninterested aid. They are well-known. I only would like to raise an objection against an opinion that an altruistic individual always fails in conflicts with a selfish one. This was postulated in the "dove-hawk model" by Price and Maynard Smith [71]. I stress that it was not based on empirical observations. It was assumed. "Selfishness beats altruism within groups. Altruistic groups beat selfish groups. Everything else is commentary" [72]. It is a good phrase in favor of group selection. But in my opinion, the authors over-appreciate the selfish individuals. Why to think that an altruist is always a looser and an egoist is always a winner? I think opposite. A weak individual just cannot afford altruistic behavior. He needs a help itself. Let me cite a rhyme by Theodor Sologub (in my word for word translation):

---

*"It is pitch-darkness in the field.*

*Somebody is calling: "Help!"*

*– What can I do?*

*I am scared and petty.*

*I am dead tired.*

*– How can I help?"*

---

Who could dare to respond to this call in the dead of night? Who will leave his warm and safe dwelling and help? Loser? By no means. Hero or Saint. They may not leave progeny of their own, yet they are certainly not losers, not weak and cowardly. Monks do not have children by definition; however, they are stably produced by the human populations during many centuries; quite similar to the stable production of, for example, hepatocytes or neurons, or worker bees though these entities never cross the frontier of generations. Altruists are stably reproduced across generations even if they happen to have no offspring of their own. It looks most probable that the altruistic/selfish phenotype of an individual is determined by numerous genes, and a population is characterized by a broad continuum of individuals, from the "pure altruistic" to the "pure selfish". This distribution is a "species trait". Owing to the gene pool shuffling, it is totally transmitted to the next generation, even if the extreme altruists do not produce their own offspring while extreme egoists have too little concern for their offspring. During evolution, the form of this distribution is optimized for the species survival. Of course, it is species-specific and must be coordinated with the *general strategy of species survival*. Altruists would be stably reproduced across generations even if they had no offspring of their own. But why? Normally, they have offspring. Women love heroes.

What about competition and struggle between the groups? Group is a substantive entity and its behavior is selfish. As such, this does not presume the survival at the expense of other groups of the same species. Groups also prosper or shrink in accord to the merits of their own. However, competition and struggle is possible and sometimes it may be really fierce. Unfortunately, evolution of Homo sapiens included such struggle in the most extreme forms. The history of humankind was the fighting of the tribes that often acquired a character of genocide. But this is quite another story.

Years ago, I asked once my fellow student about meaning of sexual reproduction. She was a romantic person and she quickly replied: "possibility of the love". And we both laughed at the joke. But now, being an old and wise man, I take it quite seriously. The love is a rather good term for designation of the intraspecific interactions not only between the sexual partners or the parents and children, but for intraspecific interactions in general. The sexual reproduction is a real cooperation not only at the level of gene pool, but at the level of entire inter-individual relationships. The apparent hostility and competition should not hide the basically cooperative character of intraspecific bonds that we would not expect for the asexual organisms that are self-sufficing sovereign players on the stage of life[5]. Above, I defined mind as the fifth fundamental interaction, which is life-specific. In the particular case of intraspecific bonds it could be named "love".

## 9. Conclusion

There are two major forms of existence: inanimate and animate. No existence is perpetual. The second law of thermodynamics predicts final dissipation of everything in the Universe.

---

5 Lover's tiffs end in kisses.

The Universe does not dissipate immediately because its components *interact* with each oth-er. The Universe as a whole is an evolving entity. The four fundamental physical interac-tions produce numerous entities, thus retarding immediate dissipation of the Universe. Various objects in the Universe have different longevities, from infinitesimal fractions of a second to billions of years. In the course of evolution, ephemeral forms are replaced by more lasting ones. This principle – survival of those who survive – sounds as a tautology, but it is *the great tautology*: Everything genuinely new emerges through this principle. Longevity is a quantitative measure of existence. Inanimate entities are the most probable configurations of matter at a given situation; their lasting is provided by their physical durability. Organisms are low-probable configurations of matter; they are physically flimsy, extremely complex, low-entropy systems. Their existence needs a special explanation. We are used to think that all the entire stuffing of the Universe is presented by two interconvertible essences: energy and substance. Latterly, some people started talking about the third fundamental essence – organic information, which is neither energy nor matter. It is an attribute of life and only life. The term "information" may not be the most suitable one. It may be confused with the homonym used in the information theory. Shannon's information is devoid of meaning whereas the meaning is just what we are interested in in context of biology. An adequate term for the third fundamental essence would be "mind". The living world resists chaos by means of *sensible behavior*. I define a behavior *sensible* if it is aimed at survival of the behav-ing entity. Inanimate evolution might be portrayed as self-construction of Nature: matter from energy; biological evolution might be portrayed as self-knowledge of nature: mind from matter. The autocracy of physics ends at the border between the inanimate world and the biosphere, where the world of sense and knowledge begins. All and only the living enti-ties are organizations with the function of survival. The longevity of a living entity is pro-vided by self-reproduction. The sensible behavior is based on knowledge. The entire knowledge a living entity enjoys is the knowledge of how to reproduce itself. In asexual or-ganisms, the self-reproducing entity is an individual organism. The minimal living entity is a prokaryotic cell. More complex living entities are presented by eukaryotic cells, multicell organisms and biological species. There are two different ways of reproduction: asexual and sexual. Asexual organisms (typically a prokaryotic cell) reproduce themselves with a high fidelity that is sufficient for the potential immortality of the lineage. Higher organisms are too complex to be able to reliably reproduce acting alone. They cooperate and form a higher rank multiorganismic entity – biological species. In case of asexuality, an organism is a sub-stantive entity, whereas a sexual individual is an attributive entity that exists as a part of the higher rank entity – deme. A generation of the deme is a self-reproducing unit. A crucial fea-ture of the sexual reproduction is the formation of genomes of individual organisms by ran-domly picking them over from the continuously shuffled gene pool instead of the direct replication of the ancestor's genome. This process inevitably produces individual organisms with different abilities to survive. Generally, they survive or die according to the merits of their own, irrespective of the presence of the other entities. This is a moment of purifying or intrinsic selection. Evolutionary success of a species depends on the perfection of species or-ganization, which includes cooperative interaction between the individual organisms. This cooperation is one of the manifestations of the fifth life-specific fundamental interaction.

## Author details

Victor Prokhorovich Shcherbakov

Institute of Problems of Chemical Physics, Moscow Region, Russia

## References

[1] Jonker. CM., Snoep JL., Treur J., Westerhoff HV., Wijngaards WC. Putting Intentions into Cell Biochemistry: an Artificial Intelligence Perspective. Journal of Theoretical Biology 2002; 214(1) 105-134.

[2] Shcherbakov VP. Evolution as Resistance to Entropy. Evolution is Aimed at a Halt of Evolution. 2004. RIO IPChPh RAS (in Russian).

[3] Shcherbakov VP. Evolution as Resistance to Entropy. I. The Mechanisms of Species Homeostasis. Zhurnal Obshchey Biologii 2005; 66 195-211 (in Russian).

[4] Shcherbakov VP. Stasis is Inevitable Consequence of Every Successful Evolution. Biosemiotics 2012; 41-19.

[5] Wake DB., Roth G., Wake MH. On the Problem of Stasis in Organismal Evolution. Journal of Theoretical Biology 1983;101(1) 211-224.

[6] Sharov AA. Role of Utility and Inference in the Evolution of Functional Information. Biosemiotics 2009; 2 101-115.

[7] Denbigh K. A Non-conserved Function for Organized Systems. In: Kubat L. Zeman J. (eds.) Entropy and Information in Science and Philosophy. Elsevier; 1975.p83-92.

[8] Wicken JS. The Generation of Complexity in Evolution: A Thermodynamic and Information-Theoretical Discussion. Journal of Theoretical Biology 1979;77 349-365.

[9] Barbieri M. Biosemiotics: a New Understanding of Life. Naturwissenschaften 2008; DOI 10.1007/s00114-008-0368-x

[10] Shcherbakov VP. Evolution as Resistance to Entropy. II. A Conservative Role of the Sexual Reproduction.Zhurnal Obshchey Biologii 2005; 66 300-309 (in Russian).

[11] Shcherbakov V P. Biological Species is the Only Possible Form of Existence for Higher Organisms. Evolutionary Meaning of Sexual Reproduction. Biology Direct 2010; 5. http://www.biology-direct.com/content/5/1/14

[12] Noble D. The Music of Life: Biology beyond Genes. Oxford, New York: Oxford University Press; 2008.

[13] Noble D. Biophysics and Systems Biology. Philosophical Transactions of the Royal Society London A2010; 368(1914) 1125-1139; Doi: 10.1098/rsta.2009.0245

[14] Abel DL., Trevors JT. Self-Organization vs. Self-Ordering Events in Life-Origin Models. Physics of Life. Reviews 2006; 3211-228.

[15] Weber A. The Book of Desire: Toward a Biological Poetics. Biosemiotics 2011; 4(2) 149-170.

[16] Battail G. An Answer to Schrödinger's What Is Life? Biosemiotics 2011; 4(1) 55-67.

[17] Coen E. The Art of Genes: How Organisms Make Themselves. Oxford: Oxford University Press; 1999.

[18] Teilhard de Chardin P. The Phenomenon of Man. New York: Harpers & Brothers; 1959.

[19] Delbruck M. Mind from Matter? Palo Alto: Blackwell Scientific Publications, Inc.; 1986.

[20] Roe SA. Biology, Atheism, and Politics in Eighteenth Century France. In: Alexander DR., Numbers RL. (eds.) Biology and Ideology. From Descartes to Dawkins. Chicago and London: The University of Chicago Press; 2010.p36-60.

[21] Cavalier-Smith T. Cell Evolution and Earth History: Stasis and Revolution. Philosophical Transactions of the Royal Society London B2006; 361 969-1006.

[22] Margulis L. Origin of Eukaryotic Cells. New Haven: Yale University Press; 1970.

[23] Maynard Smith J, Szathmary E: The Major Transitions in Evolution. Oxford: Oxford University Press; 1995.

[24] Zaher HS., Green R. Fidelity at the Molecular Level: Lessons from Protein Synthesis. Cell 2009;136(4)746-762.

[25] Lynch M. The Cellular, Developmental and Population-Genetic Determinants of Mutation-Rate Evolution. Genetics 2008; 180(2) 933-943.

[26] Ehrenberg M., Bilgin N. Measurement of Ribosomal Accuracy and Proofreading in E. coli Burst Systems. In: Martin R. (ed.) Protein Synthesis: Methods and Protocols. Methods in Molecular Biology, V.77 1998.p 227-241.

[27] Mortimer RK., Johnston JR. Life Span of Individual Yeast cells. Nature 1959; 183 1751-1752.

[28] Stewart EJ., Madden R., Paul G., Taddei F. Aging and Death in an Organism that Reproduces byMorphologically Symmetric Division. PLoS Biology 2005; 3 e45.

[29] Brooks DR, Wiley EO. Evolution as entropy. Chicago and London: The University of Chicago Press; 1986.

[30] Lynch M., Walsh, JB. Genetics and Analysis of Quantitative Traits. Sinauer Associates Inc.; 1998

[31] Baum JS., St. George JP., McCall K. Programmed Cell Death in the Germline. Seminars in Cell & Developmental Biology 2005; 16(2) 245-259

[32] Levine AJ. p53, the Cellular Gatekeeper for Growth and Division. Cell1997; 88 323-331.

[33] Drake JW., Charlesworth B., Charlesworth D., Crow JF. Rates of spontaneous muta- tion. Genetics 1998; 1489(4) 1667-1686.

[34] Kondrashov AS. Modifiers of Mutation-Selection Balance: General Approach and the Evolution of Mutation Rates. Genetic Research 1995; 66(1):53-69.

[35] Sniegowski PD, Gerrish PJ, Johnson T, Shaver A. The Evolution of Mutation Rates: Separating Causes from Consequences. Bioassays 2000; 22(12)1057-1066.

[36] Schmalhausen I I. Factors of Evolution: The Theory of Stabilizing Selection. Philadel- phia: Blakiston; 1949 (Reprinted Chicago: University of Chicago Press; 1987).

[37] Waddington CH. (1942). Canalization of Development and the Inheritance of Ac- quired Characters. Nature 150(3811) 563-565.

[38] Merser EH. The Foundations of Biological Theory. New York: Wiley-Interscience; 1981.

[39] De Queiroz K. Species Concepts and Species Delimitation. Systematic Biology 2007; 56(6)879-886.

[40] Wiley EO. On Species and Speciation with Reference to the Fishes. Fish and Fisheries 2002;3 161-170.

[41] Dobzhansky T. Speciation as a Stage in Evolutionary Divergence. American Natural- ist 1940; 74(753) 312-321.

[42] Simpson GG. The Species Concept. Evolution 1951; 5(4) 285-298.

[43] Ghiselin MT. A Radical Solution of the Species Problem. Systematic Zoology 1974; 23(4)536-544.

[44] Ghiselin MT. The Economy of Nature and the Evolution of Sex. The University of California Press; 1974.

[45] Ghiselin MT. Metaphysics and the Origin of Species. Albany: State University of New York Press; 1997.

[46] Ghiselin MT. Species concepts: The basis for controversy and reconciliation. Fish and Fisheries 2002; 3 151-160.

[47] Ghiselin MT. Metaphysics and Classification: Update and Overview. Biological Theory. Summer 2009; 4 253-259.

[48] Mayr E. The Growth of Biological Thought: Diversity, Evolution, and Inheritance. Belknap Press of Harvard University Press; 1982.

[49] Mayr E. What is a Species, and What is not? Philosophy of Science 1996; 63262-277.

[50] Wiley EO. The Evolutionary Species Concept Reconsidered. Systematic Zoology 1978; 29(1) 17-26.

[51] Wiley EO. Is the Evolutionary Species Fiction? A Consideration of Classes, Individuals and Historical Entities. Systematic Zoology 1980; 29(1) 76-79.

[52] de Queiroz K. The General Lineage Concept of Species and the Defining Properties of the Species. In: Wilson RA (ed.) Species: New Interdisciplinary Essays. Cambridge, Massachusetts: MIT Press; 1999.p49-89.

[53] de Queiroz K. Species Concepts and Species Delimitation. Systematic Biology 2007;56(6)879-886.

[54] Mayden RL. On Biological Species, Species Concepts and Individuation in the Natural World. Fish and Fisheries 2002;3 171-196

[55] Wilkins JS. What is a Species? Essences and Generation. Theory in Biosciences 2010; 129141-148.

[56] Nowak MA. Five Rules for the Evolution of Cooperation. Science 2006; 314(5805) 1560-1563.

[57] Williams GC. Adaptation and Natural Selection: A Critique of Some Current Evolutionary Thought. Princeton: Princeton University Press; 1966.

[58] Dawkins R: The Selfish Gene. Oxford: Oxford University Press; 1976.

[59] Hamilton WD. The Evolution of Social Behavior I. Journal of Theoretical Biology 1964; 7(1)1-16.

[60] Hamilton WD. The Evolution of Social Behavior II. Journal of Theoretical Biology 1964; 7(1) 17-52

[61] Trivers R. The evolution of reciprocal altruism. Quarterly Review of Biology 1971; 46(1) 35-57.

[62] Wilson DS. Truth and Reconciliation for Group Selection. http://www.huffingtonpost.com/david-sloan-wilson

[63] Okasha S. Multi-Level Selection and the Major Transitions in Evolution. Philosophy of Science 2005; 721013-1028.

[64] Friedberg EC., Walker GC. Siede W. DNA Repair and Mutagenesis. Washington, DC: ASM Press; 1995.

[65] Mori C., Nakamura N., Dix DJ., Fujioka M., Nakagawa S., Shiota K., Eddy EM. Morphological Analysis of Germ Cell Apoptosis During Postnatal Testis Development in Normal and Hsp 70-2 Nockout Mice. Developmental Dynamics 1997; 208(1) 25-36.

[66] Tilly JL. Commuting the Death Sentence: How Oocytes Strive to Survive. Nature Reviews Molecular Cell Biology 2001; 2838-848.

[67] De Felici M., Klinger FC., Farini D., Scaldaferri ML., Iona S., Lobascio M. Establish-
      ment of Oocyte Population in the Fetal Ovary: Primordial Germ Cell Proliferation
      and Oocyte Programmed Cell Death. Reproductive Biomedicine Online 2005;
      10182-191.Doi:10.1016/S1472-6483(10)60939-X.

[68] Liu Z., Lin H., Ye S., Liu Q., Meng Z., Chuan-Mao Zhang C., Yongjing Xia Y., Margo-
      liash E., Rao Z., Liu X. Remarkably High Activities of Testicular Cytochrome c in De-
      stroying Reactive Oxygen Species and in Triggering Apoptosis. Proceedings of the
      National Academy of Sciences USA 2006; 103(24) 8965-8970.

[69] Longo VD., Mitteldorf J., Skulachev VP. Programmed and Altruistic Ageing. Nature
      Reviews Genetics 2005; 6866-872.

[70] Wynne-Edwards VC. Animal Dispersion in Relation to Social Behavior. Edinburgh:
      Oliver and Boyd; 1962.

[71] Maynard Smith J., Price GR. The Logic of Animal Conflict. Nature 1973; 246 15-18.

[72] Wilson DS., Wilson EO. Rethinking the Theoretical Foundation of Sociobiology.
      Quarterly Review of Biology 2007; 82(4) 327-348

# Transitioning Toward a Universal Species Concept for the Classification of all Organisms

James T. Staley

Additional information is available at the end of the chapter

## 1. Introduction

One purpose of this paper is to briefly discuss the various stages in the development of microbial taxonomy to illustrate how new technological developments have influenced our understanding of prokaryotic species. This will be followed by a discussion of why a universal species concept for all organisms is a critical need for biology as a scientific discipline. In particular, arguments will be made for the adoption of the phylogenomic species concept as a Universal Species Concept (USC). The final section will provide an outline of how biologists might implement a USC.

It is worth noting here that the idea of a 'species' is a human idea, not a naturally determined or 'god-given' designation. In nature there are many examples of animals such as the dog-wolf group in which multitudinous variations due in part to human influences have produced a very complex array of descendants, all of which comprise a single 'species' according to the Biological Species Concept. Similar examples can be found in plants as well.

## 2. History of bacteriological treatment of species

Before we proceed, it would be helpful to understand how microbiologists, in particular bacteriologists, have dealt with the species issue from a historical perspective. The classification of prokaryotic organisms, i.e., Bacteria and Archaea, has undergone many changes since microbial life was first discovered. All of these changes have been brought about by technological advances.

The history of microbial taxonomy can be broken down into four periods:

a.  Discovery of microorganisms,

b.  Advent of pure cultures and phenotypic features,

c.  Introduction of molecular analyses and

d.  Gene sequencing and genomics.

The brief historical treatment below illustrates how important the introduction of new in-struments, techniques and analyses were in the development of the field of microbiology and, in particular, microbial systematics.

*Period 1. Discovery of microorganisms 1673 to 1850*

One can appreciate that a long time elapsed between the discovery of microorganisms by Antonie van Leeuwenhoek in the late 1600s and the early attempts to describe and name bacteria and other microorganisms. Microorganisms would not have been discovered had it not been for the microscope. Initially they were known only from observations of their mor-phology and these were hampered because of the poor quality of microscopes that were available at the time, the small sizes of the organisms and the paucity of distinguishing mor-phological features. A few species were named that were quite distinctive morphologically, however, there were not many. A more thorough treatment of this period and more recent history can be found in [15].

*Period 2. Advent of pure cultures and phenotypic features 1850 - today*

A critical innovation in the study of microorganisms was the development of procedures to isolate them in pure culture. Because microorganisms are so small, it was initially very diffi-cult to determine their features, aside from their cell size and shape. The cultivation of micro-organisms on solid media enabled individual cells to be separated from one another to allow them to grow into colonies that could be studied as 'pure cultures' of a single type or spe-cies. This critical breakthrough was first achieved in the mid-1800s in Robert Koch's laboratory.

These pure cultures could be readily grown in abundance in the laboratory thereby enabling the characterization of their cellular chemical composition, physiology, metabolism and life cycles. In time a variety of phenotypic tests were developed that could be used to character-ize each individual species. A testament to the success of phenotypic characterization is il-lustrated by the publication of [2], the first edition of which was published in 1923. It is now in its 9th edition (1994). Phenotypic tests were and still are strongly relied upon for the iden-tification of species.

More recently the 2nd edition of Bergey's Manual of Systematic Bacteriology (BMSB), which is the most encyclopedic treatment of all described prokaryotic species (Archaea and Bacte-ria) has recently been completed with the publication of Volume V [3]. Thousands of species of bacteria have been named based primarily on phenotypic features. In addition, new spe-cies are still being discovered through traditional agar plate isolation approaches as well as novel modifications that enable the growth of previously un-isolated microorganisms. Im-portantly, the taxonomy of the Bacteria and Archaea in this most recent edition of BMSB re-lies on the 16S rRNA gene sequences of organisms from the domain level to the genus. So, it is a phylogenetic classification for all taxonomic levels except the species. This point will be elaborated on below.

The availability of phenotypic and genotypic features coupled with the abundance of strains that were available of some microbial groups such as the enteric bacteria provided fertile ground to use the phenetic approach for their classification beginning in the 1950s. Computers were also becoming available in the last half of the 20th century so comparisons of genera that had large numbers of species and strains could be analyzed using similarity coefficient analysis and related procedures [16]. However, the phenetic approach has taken a back seat since gene sequencing became available.

*Period 3. Introduction of molecular analyses 1960 - 1990*

Once DNA was discovered, its features began to be used to distinguish among microorganisms. A classical example of the impact of this approach was the determination of the DNA base composition, i.e., the mol% G + C content of DNA (mol G + C/mol G + C + A + T) x 100. By conducting this molecular analysis it was determined that one group of coccus-shaped bacteria, the *Staphylococcus* genus and related bacteria could be readily distinguished from another group of coccus-shaped bacteria, the *Micrococcus* group. The mol% G C of the former was very low (ca. 30-35 mol%) compared to that of the *Micrococcus* group (ca. 70-75 mol %). Clearly with differences in DNA content that great, the two groups of cocci must be classified in different groups. Currently, based on 16S rRNA gene sequencing, *Staphylococcus* and related cocci are placed in the Firmicutes phylum whereas *Micrococcus* and related organisms are in the Actinobacteria phylum.

The other technique that was developed in the 1960s was that of DNA-DNA hybridization (DDH). In this procedure, if one wishes to compare how similar two organisms are to one another, the DNA is extracted and purified from each of them. This double-stranded DNA is then 'melted' to single strands by heating and the separated solutions are mixed with melted DNA fragments from a test organism. These are allowed to re-anneal by slowly reducing the heat of the solution. The degree to which they re-anneal with one another to form double strands is then analyzed. An organism's own DNA is used as a control which is stated at 100%. Strains that exhibit <70% re-annealing with another strain by this procedure are considered to be members of a separate species, whereas those exhibiting >70% are considered to be members of the same species.

The combination of phenotypic features and DDH gave rise to the 'polyphasic species definition'. In 1987 a committee of prominent microbial taxonomists adopted the polyphasic definition [25]. This dictum which still stands today is that all members of a bacterial species must have a unique phenotype and exhibit greater than 70% DDH by standardized procedures [23]. More recently, it has been found that organisms that are greater than 97-99% similar by 16S rRNA gene sequence must be different species thereby relieving the necessity of carrying out DDH analyses except for the most closely related strains [17, 8]. This finding was helpful because at this time only specialized laboratories are equipped to carry out DDH analysis.

*Period 4. Gene sequencing and genomics 1990 - today*

In the 1990s gene and protein sequencing became readily available to biologists. This technological advance has fundamentally changed taxonomy because these sequences can be used to trace the evolutionary history of a lineage. Two major early impacts of this on taxonomic were the discovery of the three domains of organisms, the Bacteria, Archaea and Eukarya and the development of the Universal Tree of Life based on 16S and 18S rRNA gene sequencing of representatives from all three domains [27].

Another ad hoc meeting of an international committee of expert bacterial taxonomists which was held in 2002 resulted in a significant modification of the polyphasic species concept [22]. The major change was an allowance to permit the use of multiple locus sequence analysis (MLSA) in which the sequences of typically 5-7 genes are concatenated together and then analyzed phylogeneticaly. This could be used in place of DDH in the polyphasic species definition. This is significant because it is an evolutionary approach that uses sequence based phylogenetic analyses that have been successfully applied already for the identification of bacterial species in some genera such as *Streptococcus* as well as others [e.g., 7]. Since the process of speciation is an evolutionary process a sequence based, phylogenetic approach is well suited for the classification of species.

Since the availability of genome sequences, their analyses have shed considerable light on what comprises a current bacterial species [9, 10]. For example, the determination of ANI (average nucleotide identity) of genomes can be used to replace DDH as a means of determining the boundaries of a bacterial species.

There still remain major drawbacks to the current polyphasic bacterial species definition. First, the current bacterial species definition is not a single concept but a dual concept, a combination of two concepts, one phenotypic and the other molecular. Furthermore, some variations of the concept are not evolutionary. For this reason, the current bacterial species definition is extremely unlikely to become a universal species concept.

Because of these issues, a genomic – phylogenetic (or phylogenomic [1]) species concept was proposed for the Bacteria and Archaea in 2006 [19, 20, 21]. Because genomes contain all the genetic information of an organism, it is provides ideal and sufficient evolutionary information for a species classification based on genome sequences. A partial genome sequence may not necessarily always be sufficient.

One of the rationales for the phylogenomic classification is that it is consistent with the Tree of Life which is based on 16S rRNA and 18S rRNA gene sequences. Although the Tree of Life is phylogenetic, there is insufficient resolution in the 16S rRNA gene to distinguish among many prokaryotic species [Fox et al., 1992; 22]. Therefore, less highly conserved genes must be used at the species level. Hence the phylogenomic species concept relies on sequence analyses of less highly conserved genes [19] as used in the MLSA approach [e.g., 7]. Using the phylo-

---

[1]The term, 'phylogenomic' was introduced by [5] and seems appropriate to replace to the term 'genomic – phylogenetic' in the name of the species concept [19] because it is less clumsy. Therefore, phylogenomic was adopted by [20, 21] for the name of the species concept.

genomic species concept ensures the reliance on phylogenetic analyses from domain to species in the Bacteria and Archaea and could also be applied to the Eukarya.

Phenotypic characteristics will always remain important in taxonomy not only for bacteria but for other organisms. However, they should be confined primarily to the identification of organisms based on known distinctive features of the organism. Phenotype may also be used in nomenclature as many unique phenotypic features can be aptly and often colorfully applied in coining names for novel species. However, phenotypic features should never be used as a basis for classification because, unlike gene and protein sequences, they cannot be analyzed phylogenetically.

## 3. Why should biologists develop a Universal Species Concept?

Most biologists are not taxonomists as their work is quite separated from that of the taxonomist. Also there is already is a classification of the organisms in their fields so they see little need for a universal species concept. Therefore, many would not regard the development of a universal species concept to be very important. Nonetheless, there are two strong arguments in support of a universal species concept.

First, biologists, as well as all other scientists, use terms about the basic units of their science that are critical to their thinking and comprehension. Perhaps an analogy with chemistry is apropos here. In chemistry the basic units of their science are defined very definitively. Thus, a compound is a chemical with a definite formula that can be written out on a piece of paper. Can you imagine if organic chemists used a different concept for the definition of a compound than the inorganic chemists? This is clearly absurd. However, in biology there is no uniform or definitive idea for how the basic unit of life, the species should be classified. I strongly believe that in order for biology to become a true and rigorous science, a concept for a species that would apply universally to all organisms is a basic requirement. The term 'concept' is used here to indicate that the same *methodological approach* should be used to classify a species across all disciplines. Therefore, the dual species concept for microorganisms in which both phenotype and DDH are used should be abandoned in favor of a single concept, e.g. the phylogenomic species concept. This does not mean that phenotype would not play an important role in taxonomy. It would still play a major role in the identification and naming of species, but not in their classification.

The adoption of a Universal Species Concept would not mean that each species from bacterium to plant is constrained by some human-imposed, artificial boundary but that each species would be determined by the same conceptual approach, such as the phylogenomic approach. The result would be that all species would be classified with the same methodology. For taxonomic purposes experts in each discipline would have to decide on the intraspecific constraints for each species.

Second, now is a propitious time for the adoption of a universal species concept in biology because we have all the information that is needed. For example, like chemists, we can now actually write out the chemical formula for the genome of many species. Indeed, [26] has recently proposed that a genome sequence should suffice for the naming a bacterial species irrespective of any additional information. Furthermore, he concludes that this is consistent with the International Code of Nomenclature for Bacteria [11] and the phylogenomic species concept. Of course, associated with that formula are genes and proteins as well as all the characteristics that comprise the innate properties of the organism. This includes not only the genes that are expressed only under certain conditions, but those too, that are not always expressed.

Many microbiologists and perhaps other biologists may not immediately flock to adopt a USC although they may agree that this is an important goal for biology. The reason is that it will take some time to make changes. For example, systematists of many individual groups of organisms will need to identify an appropriate set of genes for MLSA and then invest resources and time in order to properly classify the species they are most interested in. For example, the 16S rRNA gene sequence, which has very low resolution, seems to be an inappropriate gene to include in such analyses because of its highly conserved nature. Eventually, I believe that most biologists interested in taxonomy will adopt the phylogenetic approach and the phylogenomic species concept as more appropriate genes and more genomic information become available.

## 4. The quest for a universal species concept

The Biological Species Concept (BSC) proposed by [12]was a breakthrough in taxonomy. This simple concept states that a species consists of a group of organisms in which a male and a female member may breed to produce progeny which are also fertile.

At that time, bacteriologists were working from a completely phenotypic perspective without any thought that sexuality in bacteria was possible. However, in the late 1940s and 1950s Joshua Lederberg's laboratory demonstrated that enteric bacteria such as *E. coli* could carry out conjugation (Lederberg, 1947; 1957), in which genes and in some cases, the entire genome from an $F^+$ cell could be transferred to an $F^-$ cell. It was then that microbiologists began to think that perhaps bacteria could also fit into the Biological Species Concept.

Arnold Ravin was a bacterial geneticist who explored the possibility of including bacteria in the BSC [13, 14]. He argued that there was sufficient evidence to conclude that bacteria speciated through evolutionary processes, which even then were regarded as the hallmark of speciation. Moreover, it was clear from the experiments on bacterial conjugation that genes could be transferred from an $F^+$ cell to an $F^-$ cell and these could be expressed in the recipient cell. Therefore, the major elements needed to fulfill the BSC were available for at least some bacteria. However, the difficulty remained that the process could not be readily demonstrated on a species by species basis among such a highly diverse group of organisms that reproduced primarily by asexual reproduction. The idea seemed impractical even if bacterial sexuality was more widespread. For those reasons bacteriologists have abandoned the idea of using the BSC for bacteria.

## 5. What should the universal species concept be?

Following closely on the heels of gene and protein sequencing advances and genomic analyses, a number of microbiologists have argued for a new concept for the bacterial species [1, 4, 6, 15, 19, 24].

Likewise there are several other species concepts that can be applied to organisms. However, if the BSC cannot become a Universal Species Concept, which one should be used? Several of these are dual concepts and therefore have little chance of becoming a universal species concept. The phylogenomic species concept was recommended as a universal species concept [21] because it can be applied not only to microorganisms, but to all other organisms as well. Further, it analyses the evolutionary relatedness among organisms which is a key factor in speciation. Other species concepts seem deficient in comparison.

It should be noted, however, that it is more important for biologists to adopt a Universal Species Concept than for a particular concept be adopted [21]. If there is one that is really better than the Phylogenomic Species Concept what is it?

## 6. Implementation of the USC by challenge

Now that a universal species concept has been proposed, how should it be implemented? Ideally it would be wonderful to have biologists meet together, discuss the issue and then vote on it. However, this is very unlikely to happen during this current global financial climate and perhaps not anytime soon thereafter.

An alternative approach is to conduct phylogenomic analyses on species that have a questionable classification. If sufficient evidence can be found, based on phylogenomic data that indicate the current taxonomy is flawed, this could be published to challenge the current classification. In contrast, if the analysis confirms the current classification, it would also provide additional validation of the PSC as a USC. In this manner the phylogenomic species concept could be considered the *de facto* species concept for the classification of that and all other species.

## Acknowledgements

I appreciate the comments of Robert Cleland that have helped clarify several important points in the article. However, he may not agree with all of the views expressed here.

## Author details

James T. Staley*

Address all correspondence to: jtstaley@u.washington.edu

Department of Microbiology, University of Washington, U. S. A.

## References

[1] Achtman, M., & Wagner, M. (2008). Microbial diversity and the genetic nature of microbial species. *Nat. Rev. Microbiol.*, 6, 431-440.

[2] Bergey's Manual of Determinative Bacteriology. (1994). *J. G. Holt, N. R. Krieg, P. H. A. Sneath, J. T. Staley and S. T. Williams, eds). Williams and Wilkins, Baltimore, MD.*

[3] Bergey's Manual of Systematic Bacteriology,. (2012). 5, *M. Goodfellow, P. Kämpfer, H. J. Busse, M. E. Trujillo, K.i. Suzuki, W. Ludwig and W. B. Whitman, eds) Springer New York, N.Y.*

[4] Cohan, F. (2002). What are bacterial species? *Annu. Rev. Microbiol.*, 56, 457-487.

[5] Eisen, J. (1998). Phylogenomics: Improving functional predictions for uncharacterized genes by evolutionary analysis. *Genome Research*, 8, 163-167.

[6] Gevers, D., Dawyndt, P. P., Vandamme, P., Willems, A., Vancanneyt, M., Swings, J., & De Vos, P. (2006). Stepping stones towards a new prokaryotic taxonomy. *Phil. Trans. R. Soc. B*, 361, 1911-1916.

[7] Hanage, W. P., Fraser, C., & Spratt, B. G. (2006). Sequences, sequence clusters and bacterial species. *Philos. Trans. R. Soc. B*, 361, 1917-1928.

[8] Keswani, J., & Whitman, W. B. (2001). Relationship of 16S rRNA sequence similarity to DNA hybridization in prokaryotes. *Int. J. System. Evol. Microbiol.*, 51, 667-678.

[9] Konstantinidis, K. T., & Tiedje, J. M. (2005). Genomic insights that advance the species definition for prokaryotes. *Proc. Natl. Acad. Sci. U.S.A.*, 102, 2567-2572.

[10] Konstantinidis, K. T., Ramette, A., & Tiedje, J. M. (2006). The bacterial species definition in the genomic era. *Phil Trans R Soc B*, 361, 1929-1940.

[11] Lapage, S. D., Sneath, P. H. A., Lessel, E. F., Skerman, V. B. D., Seeliger, H. P. R., & Clark, W. A. (1992). International Code of Nomenclature of Bacteria (1990 Revision). *American Society for Microbiology Washington, D. C.*

[12] Mayr, E. (1942). Systematics and the origin of species. *Columbia University Press, New York, N.Y.*

[13] Ravin, A. W. (1960). The origin of bacterial species. *Bacteriol. Rev.*, 24, 201-220.

[14]  Ravin, A. W. (1963). Experimental approaches to the study of bacterial phylogeny. *American Naturalist*, 97, 307-318.

[15]  Rosselló-Mora, R., & Amann, R. (2001). The species concept for prokaryotes. *FEMS Microbiol. Rev.*, 25, 39-67.

[16]  Sokal, R. R., & Sneath, P. H. A. (1963). Principles of Numerical Taxonomy. *San Franciso W. H. Freeman.*

[17]  Stackebrandt, E., & Goebel, BM. (1994). Taxonomic note: a place for DNA-DNA reassociation and 16S rRNA sequence analysis in the present species definition in bacteriology. Int. J. Syst. Bacteriol. , 44, 846-49.

[18]  Stackebrandt, E., Frederiksen, W., Garrity, G. M., Grimont, P. A. D., Kampfer, P., Maiden, M. C. J., Nesme, X., Rosselló-Mora, R., Swings, J., Truper, H. G., Vauterin, L., Ward, A. C., & Whitman, W. B. Report of the ad hoc committee for the re-evaluation of the species definition in bacteriology. *Int. J. System. Evol. Microbiol.*, 52, 1043-1047.

[19]  Staley, J. T. (2006). The bacterial species dilemma and the genomic-phylogenetic species concept. *Phil. Trans. R. Soc. B*, 361, 1899-1909.

[20]  Staley, J. T. (2009a). The Phylogenomic Species Concept for Bacteria and Archaea Microbe. 4, 362-365.

[21]  Staley, J. T. (2009b). Universal species concept: pipe dream or step toward unifying biology? *J. Ind. Microbiol. Biotechnol.*, 36, 1331-1336.

[22]  Staley, J. T., & Gosink, J. J. (1999). Poles apart: biodiversity and biogeography of sea ice bacteria. *Ann. Rev. Microbiol.*, 53, 189-215.

[23]  Vandamme, P., Pot, B., Gillis, M., de Vos, P., Kersters, K., & Swings, J. (1996). *Microbiol. Rev.*, 60, 407-438.

[24]  Ward, D. A. (1998). A natural species concept for prokaryotes. *Current Opinion in Microbiology*, 1, 271-277.

[25]  Wayne, L. G., Brenner, D. J., Colwell, R. R., Grimon, P. A. D., Kandler, O., Krichevsky, M. L., Moore, L. H., Moore, W. E. C., Murray, R. G. E., Stackebrandt, E., et al. (1987). Report of the ad hoc committee on reconcilliation of approaches to bacterial systematics. *Int. J. Syst. Bacteriol.*, 37, 463-464.

[26]  Whitman, W. B. (2012). Intent of the nomenclatural Code and recommendations about naming new species based on genomic sequences Bull. *BISMiS*, 2, 135-139.

[27]  Woese, C. R., Kandler, O., & Wheelis, M. C. (1990). Towards a natural system of organisms: proposal for the domains Archaea, Bacteria and Eucarya. *Proc. Natl Acad. Sci. USA*, 87, 4576-4579.

# Chapter 7

# Species, Trees, Characters, and Concepts: Ongoing Issues, Diverse Ideologies, and a Time for Reflection and Change

Richard L. Mayden

Additional information is available at the end of the chapter

## 1. Introduction

Consider ideology in reference to species and their conceptualization and discovery. Ideology is a set of beliefs by which a group or society orders reality so as to render it intelligible; speculation that is imaginary or visionary.

An extensive library of writings exists on the topics of species, species concepts and the intersection of these with phylogenetics. Why then another paper on the topic? Following the dialogue on these topics in literature and in discussions of systematics, the discovery and descriptions of species, and arguments and actions of authors of master lists of species, it is clear that some specific areas relating to species warrants additional attention. Herein, I offer a condensed, episode relating to a series of topics that, in my opinion, are not generally understood by some working today with biodiversity or practicing systematics. It is hoped that the brief addresses to particular topics are perhaps more palatable to those not focused on these issues regarding species but, in my opinion need to be fully aware of their intentions and interpretations regarding the nature of species and the use of terms and concepts within a critical theoretical and philosophical context.

In the last two decades we have witnessed great advancements in our conceptual, theoretical and operational understanding of biological species, as well as varied views of species within each of these areas. All of these contributions are, of course, developed and clarified within the context of the standing theory of evolution, or descent with modification and speciation. Should this theory not apply to the origins of biodiversity then our advances with this complex issue must be reframed, discarded, and/or derived anew. Varied "conceptual views" of species have long been proposed or used in discussions and arguments of diversity and have been reviewed [1-17] and whole volumes have been dedicated to the concepts and practices regarding species wherein various authors outlined their preferred concepts and critiqued other concepts [18-20].

Because of the nature of species being individuals, aspects of their discovery and interpretations of their evolution are complex and sometimes require considerable thought and reflection. Unlike those things that humans feel more comfortable with, sets, classes, etc. that have clear operational definitions - is or is not, black or white, species and interpretations of their evolution are much more complex. Species, unlike supraspecific taxa stand alone in many contexts and terms used in systematics. With a general lack a real interest in discussions relating to species, especially among systematists and taxonomists (as well as others in comparative biology) there remain critically important issues regarding the basic understanding of species. As such, in practice, when interpreting data and analyses relevant to the existence of species those not informed can make decisions about the question of "What is a species?" or the species diversity simply inappropriate. Likewise, regarding the above responsibilities of these research communities staying abreast of the general issues regarding "What is biodiversity?", the bottom line for studies by such investigators simply lead to many inappropriate outcomes. For example, misinterpretation or indifference to deciphering natural patterns related to species, leads to a sequence of events that I would think that the comparative biology (in the broadest of sense) would not want to see – inaccurate assessments of diversity, inappropriate evolutionary conclusions, misinterpretations as to the existence of character distributions or their evolution, inaccurate conclusions as to processes in ecology, speciation, biogeography, and lack of official protection and conservation of species. Lineages that are ignored cannot be included in comparative biology analyses – that being revealing natural patterns of descent with modification, appropriate comparisons, and seeking explanations/ processes having lead to these patterns.

The issue of species simply boils down to many practitioners both inside and outside of the informed realm of the underlying theoretical and philosophical issues on species easily slipping into a more comfortable world of "black and white", that being a world of "science made easy", and a sense of self ease and tranquility regarding their own scientific reputation - view species as class constructs with simple definitions (includes most of the diversity of concepts, see below and [15-17]). If something in nature being studied does not fit their concept (actual black and white definition surrounding a concept), then it cannot be a species. Or, there is the unfortunate view of many involved in molecular biology that a species must have a "high enough" p distance to be valid, when the genes that we study today represent an infinitesimally small sampling of the genomes of the taxa. Or, despite the intentions of the discovery of naturally existing lineages, some actually believe that they can only be discovered and diagnosed on the basis of morphologically differentiation. For many, if morphological characters are not easily examined by eye, with a hand lens, or with a microscope, they do not exist unless differentiated enough genetically from other such things.

Adding to the confusion surrounding the topic and the artificial assessments and conclusions by some, many well-intentioned and informed practicing taxonomists/systematists still unfairly judge other's research on diversity and their recognition of species on their artificially contrived basis of degree of difference or "value" of diagnostic morphological characters, as if anyone understands the genetic coding of any morphological characters used in a taxon-specific discipline. Sadly, even today the degree of morphological differentiation is considered as a surrogate to the degree of genetic divergence. Falsifying evidence of this long-standing hypothesis abounds, but as is argued in reference [21], paradigms are not usually shifted, changed, or evolve because to many scientists evaluating examples and data typically spend most of their time seeking examples to simply corroborate hypotheses. Hence, much of the "new synthesis" thinking remains, and with this is the traditional thinking of biological species.

Not all contributors to the recent advancements on species view species as real and individuals that are mutable and are lineages diversifying over time. Others, again for convenience and safety of their own scientific reputation, prefer to consider species as class-like constructs with safe definitions (not the same as diagnoses). Interestingly, this long held argument of propo-nents of concepts that in reality do not consider time as a component of the conceptualization and interpretations of species, a logic incompatible with the current paradigms, theories, and hypotheses underlying evolutionary biology still exist. With these definitions of varied class constructs, logically derived from all concepts except the Evolutionary Species Concept, species thought to participate in processes associated with evolution and used in reconstruc-tions of evolutionary relationships are in fact immutable, thus precluding their existence in reference to any aspect of evolutionary biology.

Regardless of the valuable, important, and numerous advancements in the discipline that have ultimately lead to a critically important view of consilience among concepts with evolutionary biology, several issues and terminology remain unclear to many practitioners of the discipline. These are discussed below. This clarification and a call for a more informed systematic and taxonomic community investigating all of biodiversity will hopefully better inform and unify scientists within this broad field on species so they can be logically investigated within the current paradigm of evolutionary biology. The topics covered herein have been selected as ones that I interpret as a series of issues in this broad discipline wherein practitioners and users remain misinformed or have not dealt with or had the time to deal with the complexities of the issues surrounding species.


## 2. Taxonomy and systematics

Historically, these general areas of the study of biodiversity were considered to be or were referenced as separate but linked disciplines. Taxonomy has traditionally been viewed as the area of science related to the discovery, naming, and classifying species. Systematics was traditionally viewed as the study of relationships of species and supra-specific taxa. Under this organization these two disciplines remained inappropriately dis-connected. Prior to the revolution of systematics by Hennig a sound method for reconstructing genealogical relationships did not exist and the field was dominated by the disciplines of Evolutionary Systematics and Numerical Taxonomy, both disciplines lacking a logical means of providing natural biological classification reflective of ances-tor-descendant relationships [22-25]. Hennig's methods provided a means with which one could reconstruct defendable hypotheses of relationships, as well as a time-dimen-sional discussion of species. Simpson and Wiley's works on species concepts provided the necessary connection between phylogeny reconstruction, character evaluation and in-terpretation of homologies, and species as lineages [1-4, 26].

Thus, today, these two once separate disciplines must be combined under the title Systematics. Neither of the formerly recognized disciplines can today work independently. The natural order is that phylogenetic interpretations of characters and their evolution, as well as phylog-eny reconstruction leads to the discovery of new, previously unrecognized diversity. Phylog-enies provide a framework for properly classifying diversity. This merger provides a more holistic view of a necessarily structured discipline focused on biodiversity. Below, I will use the term systematics to refer to this collective scientific study of biodiversity.

## 3. Life-long education, critical thinking, and biological species

As scientists our work should involve a process of continued education of specialized and many associated fields. Without continued life-long education, our practices and hy-potheses, hence, scientific practices, become outdated and eventually irrelevant. Such "scientists" are anchored in the past but are frustrated and irritated as they watch the world change around them. Those who practice in the disciplines associated with the dis-covery and study of biological species have the responsibility to remain up to date on the theoretical underpinnings relevant to biological species, as well as the various opera-tional mechanism used in the discovery of species, regardless of the types of characters used and their interpretations (see below under "Characters: their real function and why all types are important!" and "Concepts for Species as Lineages versus Operational Defi-nitions"). Admittedly, there are volumes of publications relating to dialogues on species concepts but these are critical discussions for one to better understand the theoretical and philosophical underpinnings of the study of biodiversity as species in the natural scien-ces. They, along with an understanding of the difference between plesiomorphic and apomorphic characters (whether they be discrete characters, frequencies of characters as alleles/nucleotides), also need to have a phylogenetic perspective to evaluating diversity. In the science of biodiversity today, these are essential to understand and effectively par-ticipate in a more accurate depiction of the planet's biodiversity and its evolution. Only through the understanding of these topics (all reviewed in very good papers in the last 20 years) can one make an informed decision on species diversity and decide whether species are lineages and mutable (=participate in such basic process theories as anagene-sis and cladogenesis and are individuals) or are simply artificial and immutable con-structs that humanity has developed to simply organize nature (classes). These are the two basic propositions that one has to evaluate in the central argument to the species is-sue and one's ability to conduct science.

## 4. Participation in educational and experience-based dialogues on species?

In Mayden's discussion of species and species concepts several theoretical and philosophical issues were discussed in this area, as well as a number of practical, experienced-based observations regarding species and their investigation in a time-inclusive perspective [15]. My experience with biodiversity, species, and the lineage/phylogenetic perspective lead me to "paint myself into a corner" on the theoretical, philosophical and operational views on species in nature. This evaluation and discussion lead to the acceptance of species as individuals and the determination that other concepts were really not underlying concepts to be used to provide a theoretical underpinning for use in the organization of thought allowing for an accurate discovery of all types of diversity evolving as lineages. Thus, both theory and real-life experience surrounded my conclusions as to the discovery of the elements of the tree of life for informed study. Hull evaluated concepts and compared them with others discussing species and discussed the theory of monism and pluralism as associated with theories or perspectives on an issue [10]. He also addressed the issue of life experiences with biodiversity regarding multiple dimensions involving both and laboratory studies, both, in my opinion, being fundamental to understanding, discussing, and choice of underlying conceptualizations of species. Experience with naturally occurring entities and their variation in characters is no substitute for the theory and philosophy of species but both are necessary and sufficient to successful and meaningful study and dialogue pertaining to the discipline.

Scientists discovering and studying biodiversity and its evolution learn about the entities, known as species, through studies of theory and philosophy, as well as real-life experiences, on species and guidance from mentors and colleagues. A background on the readings of the theory and philosophy of species alone, and with no practical experience honed over time and with continued study is not enough, in my opinion, to prepare one to either discuss species as individuals or classes, appropriately viewing the reality of species, or judging the validity of any described species. As we all know, one does not become an expert in science overnight – it comes through study and mentoring. The theory and philosophy purported by some without these one of classes of experience is not enough to serve as an expert on this topic. The same is also appropriate for those restricting themselves to only phylogenetics or population genetics of entities hypothesized to be species (see section below entitled "The Nature of Biological Species: Assertions as to the Monophyly of Species?). Likewise, the same applies to efforts in evaluation and hypotheses of homologies of character states or characters, regardless of the type of character system involved. Continued experience, studying literature and mentoring from one or more persons experienced in these areas of diversity cannot be substituted by the limits of theory and philosophy of a topic. This combination of experiences is essential in these and other areas of the natural world to fully understand process and diversity, as many more "things" in the natural world are actually individuals (like species; etc. individual organisms, populations, communities, ecosystems, planets, etc.) and cannot be defined; it is inappropriate to view them as simple class constructs artificially restricted by rigid definitions of membership. The repetitive study, examination, and tests of levels of tokogenetic versus phylogenetic relationships of biological diversity, of characters, and other individual-type entities is a critical area of study in organismal biology that is uniquely occupied by specialists in systematics.

## 5. Concepts for species as lineages versus operational definitions

Wiley and Lieberman provide a very lucid summary for an important introductory discussion laying the foundation for understanding the issues and importance of topics surrounding species concepts [27]. As discussed by these authors two critical factors have been discussed for many years include:

1. Species-as-taxa are individuals and,

2. A species concept for the category species of the Linnaean classification is a kind concept – by intention.

This important distinction is likely the most common reason for many being confused in dialogues and writings about species. This has also lead to the long-time debate on concepts, the delays in biodiversity discovery, and likely the extinction of biodiversity by neglect and ignorance in our interpretation of characters through time. The above distinction as to species-as-taxa and the species category is essential to understanding the central issues discussed by Mayden [15-17], de Queiroz [13-14, 28], Wiley and Mayden [29-31], Naomi [32], and Wiley and Lieberman (27). Without a clear understanding of this basic premise, future discussion and understanding will be difficult for the systematist and/or evolutionary biologist to integrate and be consistent with underlying principles of evolution given current paradigm as to descent.

In specific writings by Mayden [15-17] an explanation for the growing number or large number of historically developed concepts was related to 1) the lack of understanding or view of species as lineages versus classes and 2) concepts being developed based on the mentor-student lineages of researchers working with specific types of characters for specific taxa that ideologically think are the most important characters for the recognition of species – a legacy that should continually be evaluated by the researcher via life-long learning. All of the concepts, save the Evolutionary Species Concept, have arisen by this means. However, concepts underlying entities that are individuals must be developed under the premises through which they supposedly originated – or process-based concepts. As Wiley and Lieberman [27, pg. 29] correctly assert "Process-based concepts attempt to characterize species in a manner predicted by "covering laws" thought to explain processes occurring in nature." One must examine each of the concepts individually to determine if they are concepts derived out of the currently held theory of descent with modification and related hypotheses and assertions. Thus, without an understanding of process-based concepts and kind concepts relevant to biological species, what constitutes a species across biological diversity derives from a historically "blinded," non-lineage, definition-based view of species. Such concepts of naturally occurring entities make it impossible in comparisons, contrary to a lineage-based concept viewing species as individuals. What results from this limited insight into species and an underlying process-based concept of what we are looking for will be discoveries or decisions based on limited thought as to lineages and time, and misunderstanding of the evolution of homologous characters. Hypothesized homologous character transitions are simply historical and heritable markers or tags to serve as evidence to formulate hypotheses as to independent lineages, entities comprised of organisms that maintain their identities from other such entities through time and over space and which have their own independent evolutionary fates and historical tendencies visa-via the Evolutionary Species Concept. Species are christened with proper names and are always testable hypotheses, not fact, as is the same for phylogenetic trees.

The important issues at risk here for evaluating and selecting definitions purporting to be equally viable for species are partially outlined below from Wiley and Lieberman (27, p. 28).

---

As kind concepts, they are defined by intention, each concept having properties that should provide necessary and

sufficient conditions for "speciesness."...

---
---

We might also expect some concepts to be nominal kinds, kinds that do not have direct connections with evolutionary

theory but which are thought to be useful in some manner by those who invented them.

---
---

As kinds, we might expect that some species concepts are candidates for being natural kinds. Such concepts would " fall

out " of evolutionary process theories (Quine, 1969 ). We might also expect some concepts to be nominal kinds, kinds

that do not have direct connections with evolutionary theory but which are thought to be useful in some manner by those

who invented them....

---
---

Wiley and Mayden (2000a – c) suggest that systematists form species concepts in a manner that reflects their ideas of how

these concepts function in systematic and evolutionary theory. Some systematists form concepts that allow them to

discover what they think are species. Others form concepts based on how they think species function in the evolutionary

process. We suggest that this is nothing but the familiar debate on operationalism that surfaces from time to time in

science (Wiley, 2002 ). Some systematists think that it is important to form " operational " concepts. (Wiley and Lieberman,

2011, 28)

---

As Wiley and Lieberman [27] further discuss, this issue is not one that should be easily dismissed. In my opinion (Mayden 15-17], without a clear understanding of these distinctions the "fog" that surrounds one may have about species and concepts, and their inappropriate applications, will plague the outcomes of a researcher and any clade of diversity studied by this researcher. I do not speak for Wiley or Lieberman with respect to my proposition of such a bewildered view of evolutionary diversity. However, this general issue was emphasized by these authors in a similar context [27, pgs. 28-29]:

---

It is exactly this distinction that led Frost and Kluge (1994), Mayden and Wood (1995), Mayden (1997), and de Querioz

(1998) to suggest that a distinction can be drawn between what might be termed " general " concepts and " operational "

concepts. General concepts provide an ontology from which " operational " concepts (= testable concepts) may be

applied....

---

---

Process-based concepts attempt to characterize species in a manner predicted by "covering laws" thought to explain

processes occurring in nature. Evolutionary theory (specifically that part of evolutionary theory concerned with the origin

of species) provides such a " covering law, " which posits that there are two kinds of entities one might expect to find in

nature that are of primary interest to systematists, species and monophyletic groups. Given that evolution appears to

result in a hierarchy of entities, the monophyletic group seems to be a natural kind that falls out of the general theory of

descent with modification, coupled with the general theory of speciation....

---

---

Because all particular examples of the kind "monophyletic group" have a beginner, a common ancestral species, this

suggests that species are also a necessary kind. In other words, if "monophyletic group" is a natural kind, then that from

which particular monophyletic groups are derived ("species") might also have the status of being a natural kind (Wiley

and Mayden, 2000a ). The results form the basis for expecting, as Hennig (1966) suggested, that organisms have two sorts

of relationships. Tokogenetic relationships are formed on the basis of reproduction and obtain among individual

organisms. Phylogenetic relationships are formed by severing reproductive ties to the extent that two tokogenetic systems

are formed out of what was once a single tokogenetic system (or, in the case of speciation via hybridization, two

tokogenetic systems form a third through tokogenesis between the two systems). For sexual organisms, tokogenesis is

nonhierarchical while phylogenesis is hierarchical (speciation via hybridization is the exception).

---

Given this insight into concepts, both process-based and operational (sensu Bridgeman's operationalism [33]), the recent syntheses on species by Mayden and de Queiroz lead to a singular conclusion as to a needed revision of the whole issue of "Species Concepts". Not all existing of species-as-taxa serve to facilitate our understanding of processes and patterns underlying the evolution of biological diversity. Only one, the Evolutionary Species Concept, is the appropriate process-based concept facilitating the discovery of species as individuals and natural kinds. Herein, I argue that the terminology associated with all other previously referenced "species concepts" (except the General Lineage Concept that is a synonym of the Evolutionary Species Concept) should be changed to their use as only definite criteria (sensu Wiley and Lieberman [27]). They serve as none other than simple criteria for the discovery of lineages or surrogates for the discovery of lineages, using any and all types of heritable characters, that are consistent with the processes of evolution under the process-based Evolutionary Species Concept. Thus, these historical concepts of species have new names as criteria, not concepts, and are referenced differently (Table 1).

| | |
|---|---|
| Agamospecies Criterion | Internodal Criterion |
| Biological Criterion | Morphological Criterion |
| Cladistic Criterion | Non-dimensional Criterion |
| Cohesion Criterion | Phenetic Criterion |
| Composite Criterion | Phylogenetic Criterion |
| Ecological Criterion | Polythetic Criterion |
| Evolutionary Significant Unit Criterion | Recognition Criterion |
| Genealogial Concordance Criterion | Reproductive Competition Criterion |
| Genetic Criterion | Successional Criterion |
| Genotypic Criterion | Taxonomic Criterion |
| Hennigian Criterion | |

**Table 1.** Revision of terminology associated with a new process-based conceptual view of biological species. Above listed historical concepts of species (previously considered concepts equal to the Evolutionary Species Concept) represent definitions of class constructs and are herein consider definite criteria or criteria for short. See Mayden [15-17] and Wiley and Mayden [29-31] for further clarification of why these "concepts" are not equivalent or any are to the Evolutionary Species Concept as the underlying process-based concept of species.

A non-lineage perspective held by a researcher also leads to significant problems in accurately estimating biodiversity by not understanding the origins of characters in different populations or species. Very often two or more species are reduced to one or more simply by the non-dimensional interpretation of the existence of characters distributed across the species and intervening populations thought to be one of these species yet in an active and ongoing process of gene exchange through hybridization and intergradation (simple violation of the historical Biological Species "Concept"). Character interpretation even within populations of species or across the geographic landscape of one or more species must be evaluated from a phylogenetic perspective, as illustrated in the simple example in Figure 1. In this instance one geographic grouping of organisms (Species II: population, set of populations?) is interpreted as a group of intergrades between Species I and III. In the world of species concepts without a dimensional component this may be a valid interpretation of the situation. However, as one can see in this example, the observation that Species II has both traits of Species I and II has a more parsimonious interpretation of possessing the unique combination of both plesiomorphic and apomorphic conditions relative to Species I and III. Species II would likely then be considered simply intergrades between the Subspecies I and III – thus a reduction of biodiversity from three likely valid species to only one!

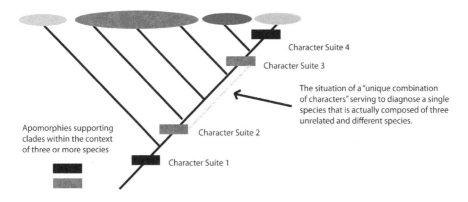

**Figure 1.** Hypothetical example of multiple geographic groupings of hypothesized taxa wherein the centrally located grouping, Species II, is polymorphic for the two character states found separately in Species I and II. A non-dimensional interpretation of this polymorphism would likely conclude the existence of continued gene flow and therefore the modification of diversity to only one species and two or more subspecies. The lineage perspective, however, provides a clear interpretation of Species II simply sharing one plesiomorphic allele with Species I and one apomorphic allele with Species III. No active process is necessary to invoke and the difference in biological diversity is 1 versus 3 species between these two approaches to interpreting the processes involved in possession of characters (alleles, morphology, etc.)."

## 6. Synonymy of the overall theoretical concept of biological species

With the great attention given to species and their conceptual framework over the last 20-30 years there has been tremendous progress in clarification and resolutions on the issue and the education of various communities of researchers and the general public. One of the most significant accomplishments, in my opinion, has been the realization that not all "concepts" of species have an equal status for use as guiding principles enabling one to discover biodiversity on planet Earth. Two basic ideas have been consistently repeated historically in theoretical or practical discussions of species in nature. Those that argue that species are of different kinds and as such require different concepts (pluralism; Hull [10]). Or, there are those researchers that argue that this is an incorrect and poorly formulated logic as to species, regarding the result of natural selection, anagenesis, and cladogenesis or speciation (= individuals). In this second framework, the discipline should have an overarching, non-operation concept of species as lineages and a series of conceptual criteria for discovery of said species (monism). I argue the latter as these principles are consistent with what we currently understand or think we know about evolutionary biology. These two opinions have resulted in an enormous number of papers where researchers, philosophers, or both have simply not understood one anothers arguments. This is exactly the same underlying issue that has plagued the debates over which characters and analyses are "best" or "should be used" in phylogeny construction. Most of the special issue in *Zootaxa* [38] on molecular and morphological characters involved authors that had a poor understanding or were ignorant of the underlying and generally sound methods of what they were criticizing. These two examples have, however, been to the benefit of science and lead to many important discussions that forced people to clearly articulate ideas and ideologies and evaluate principles and process-driven concepts. Unfortunately, however, those systematists, as well as reviewers and editors of papers, not up to date on these issues have simply delayed the discovery, authoritative decisions that have serious consequences. Among many other critical outcomes, there is the delay or denial of christening lineages of biodiversity with proper some names, some before they have gone extinct as their habitats are lost, there is species not being "authorized" for descriptions because of a prejudicial ideology of species as individuals and a process-based concept, there is squandered time for conservation interventions of endangered diagnosable entities due to ignorance and character bias, and there is a result in inappropriate or inaccurate process- and pattern-based studies of species, natural selection, and speciation.

Through these various discussions and debates, some researchers have focused on a higher-level understanding of the issues at hand and how to resolve ongoing incongruences of logic of some, conflation of many things about the vast writings on species and species concepts, the real underpinnings of the various species concepts, understanding process-based concepts and non-process-based concepts, etc. These recent papers were all done by experience professionals (field, laboratory, and theory) with real-life experiences with species and species issues, and were formulated in an effort to better present the history of this critical topic to all of the natural sciences in a more lucid, logical, and cohesive manner [1-4, 7, 11-17, 28-32, 27]. Common amongst these series of writings is that the issue is not a question of there being different types of species in nature that necessitate different concepts to realize their existence as natural biodiversity. Rather, they argue that species are the same theoretical entities and realized by a common overarching, process-based concept of these lineages, and that their discovery relies upon the other existing "concepts" or criteria for different defining qualities for recognition of these independent lineages consistent with the process-based concept.

As outlined by Mayden [15-17] and discussed by Wiley and Lieberman [27] the Evolutionary Species Concept is really a logical analog to the natural kind monophyletic group as neither of these are operational concepts. Critically important to this, however, is that *this is strength of a concept and not a weakness* [5, 29-31]. And, as Wiley and Lieberman [27] discuss, "a concept can embody testable consequences without being operational sensu Gilmour [34]."

The writings of Mayden and those of de Queiroz parallel one another. However, as discussed in the detailed comparisons and review by Naomi [32] both have integrated frameworks of species concepts and consider operational methods of discovery. Each view is ultimately thought to lead to the same consequences as to the discovery and descriptions of species in nature. de Queiroz chose to develop a new title for his line of thinking of a species concept, the General Lineage Concept, to set it apart from the Evolutionary Species Concept. However, as aptly reviewed by Naomi (32, p. 1) the "existing and revised versions of the integrated framework of species concepts all are not new species concepts, but versions of the evolutionary species concept, because they treat the evolutionary (or lineage) species concept as the concept for species category. This was also concluded by Wiley and Lieberman [27]. As such, the General Lineage Concept is herein argued to be a synonym of the Evolutionary Species Concept.

## 7. Definition of the category species in the linnaean classification

---

An evolutionary species is an entity comprised of organisms that maintains its identity from other such entities through time and over space and which has its own independent evolutionary fate and historical tendencies (Wiley and Mayden [29]).

---

**Synonymy:** General Lineage Concept of de Queiroz [14]

## 8. What is a species? Reality, fiction, hypothesis, artifact, special creation?

It is my opinion that species are naturally occurring lineages that exist in nature irrespective of our abilities to find, study, or diagnose them. As such I am a realist. In concert with this, following the premises of scientific investigation, I consider discovered and named species that we place in the Linnaean Category Species to be testable hypotheses. Evolution operates at the level of the species lineage as this is the upper-most and most comprehensive level of tokogenetic relationships that exist prior to speciation. Species are individuals and are mutable and as such take on all types of apomorphic qualities. Their evolutionary relationships can be inferred, as well as any other historical or contemporary patterns and processes as tenants of evolutionary biology. Species are not artifacts or things of special creation that are immutable; if they are then they cannot be investigated or used in investigations of descent with modification that inherently involves anagenesis and speciation.

## 9. Characters: Their real function and why all types are important

*Homo sapiens*, as a species, are inherently compromised in doing an effective job at understanding and finding homologous characters and species as individuals. We, as a species, do not have access to or even know what the appropriate "markers' are to corroborate either a lineage or homologous characters independent of other such lineages/homologues.

Abundant discussion exists as to the question of "What is a character?" and several have provided discussions and overall definitions [3, 27]. The same is true for the question of homology of characters. In fact, without a doubt the two issues are often thought to be independent. This type of thinking about species readily leads a serious logic problem for species discovery and phylogeny reconstruction, and possibly conclusions resulting from related biodiversity investigations.

Simply put, characters are deciphered and hypothesized attributes of species unraveled to their most basic element and then sorted out as to hypotheses of various hierarchical levels of or independent instances of homology. No set type of characters is inherently better than any other, except where a hypothesized character is not heritable or is simply too variable to be evaluated given current algorithms or existing thinking on the topic. However, homoplasy can be very useful in phylogeny reconstruction to resolve nodes wherein homoplasy occurs. Actually, homoplasy is a contrived term for something that actually does not exist in nature. Homoplasy, in the systematic literature, derives simply from our inability to properly identify variation in homologues in our deciphering of character transformations.

In the world of molecular systematics, aligned genes, restriction sites, and frequencies of alleles of a gene are all analyzed. No data are excluded, even though variability exists in any of them. This variability is handled through specially designed algorithms that are formulated to deal with polymorphism of heritable traits. Morphological systematics, however, for the vast majority of studies is immune to these issues as users discard characters that they deem as too variable, in reality it is much more subjective and difficult to decipher the homology of characters, and nothing is known as to the genetic control (simple or complex) of any of the characters. All character types suffer from some sort of issue and they can be argued to be equally problematic. However, as a person that has used all of these types of characters, it is my opinion that hypotheses of homology are more difficult with morphological characters without very detailed ontogenetic evaluations or any additional investigation of any other type of morphological character.

In reality, characters are simply markers or tags that are left behind through anagenesis accompanying speciation. These tags or character transformations, when accurately interpreted in a phylogenetic context, must be evaluated in a context to ascertain its relative state of being either plesiomorphic or apomorphic. If anagenesis does not keep pace with speciation then no resolution of relationships of three or more taxa will be possible; although this depends entirely upon the character being examined as some not examined may have undergone anagenesis and kept pace with speciation. Phylogeny reconstruction is dependent upon the rate of anagenesis of any type of hypothesized heritable character either keeping pace with or occurring faster than the development of independent lineages, or speciation. Thus, in summary, everyone should be careful in evaluations and critiques of characters used in phylogeny reconstruction and diagnoses of species. After all, today we use a variety of types of data, though mostly molecular and morphological data, for reconstructions or inferences of shared common ancestors. These synapomorphies, of varied types, provide evidence for the existence of a species that was an ancestor to two or more descendants. Yet, in the conservative world of those describing species, morphology rules and species are not, for example, diagnosed strictly using molecular apomorphies. This inconsistency in logic is unfortunate as species as lineages and individuals are not being recognized, conserved, and studied – simply because of a conservative character bias that has no underlying metaphysical justification.

## 10. Gene trees versus species trees?

Wiley [3] and Wiley and Lieberman [27] advocate that there is only one tree to explain the diversity of life, regardless of our abilities to discover it. In systematics, among at least phylogeneticists (not transformed cladists or many considering themselves cladists wherein phylogenetic hypotheses do not represent evolutionary trees), there has always been an underlying premise a reconstructed phylogeny represents a hypothesized tree of genealogical relationships for the taxa concerned. Such an inference of relationships is a hypothesis as to the best estimate of species (or species representing supraspecific taxa) relationships - based on synapomorphic, homologous characters employed in a phylogenetic analysis (model based or not). These inferred trees are hypotheses of speciation events (representing a diverse lineage by selected species or single species, subspecies, or other taxon used by the author(s)). It is a hypothesis of a genealogical tree, not a series of branching events that can be reconstructed on anything (nuts, bolts, cars, motorcycles, etc.). Why? Cladistic relationships, as advocated by some, for the above inanimate objects has no underlying reality in assessments of homology. While the objective of those that consider themselves cladists or transformed cladists (literature of the 1980s) is to avoid invoking any evolutionary bias in a reconstructed tree of relationships, this cannot be achieved because a simple use of any type of characters in a matrix invokes the evolutionary process of character homology. Otherwise, why not randomly distribute the character information across columns? Therefore, these arguments are pointless as not all nuts or bolts or cars or motorcycles originated in a unique common ancestor (of each type) wherein through time the character transformations may or may not have changed through time via anagenesis and that may or may not have been accompanied with speciation. When one purports to reconstruct a series of phylogenetic or cladistic relationships the underlying hypothesis of homology is an evolutionary process and cannot be avoided in any science-based study; thus, those wishing to divorce their study from any evolutionary-based assumptions are incorrect.

Prior to the generation and eventual landslide of genetic data in phylogeny reconstruction, species trees were discussed in the above context as descent and ancestor-descendant relationships supported only by homologous and derived characters. With advancements in capturing molecular data (allozymes, restriction sites, sequences) for hypothesized synapomorphic homologues things changed with respect to the idea of trees. References to such genealogical hypotheses involved either gene and/or species trees. Thus, one could discuss gene trees, species trees, or both. This plurality of discussion of trees arose when researchers discovered that different genealogical information (synapomorphies) used separately revealed different patterns of genealogical relationships [35]. This topic has recently been thoroughly covered in the compiled volume edited by Knowles and Kubatko [36]. Hence, these efforts, along with sophisticated algorithms to analyze the types of data resulted in the introduction of a new terminology and the idea of a group of species can actually have different gene trees. The differences amongst these gene trees result from lineage sorting issues, long-branch attraction, and other factors resulting in the evolution of individual genes. While this theory of gene trees is very convincing and is a real issue in systematics today, some solutions to these varied patterns of descent have been and are being developed as noted in Knowles and Kubatko [36].

The question then becomes "Exactly what is a species tree? The term species tree is widespread in the literature, usually in reference to a discussion of gene trees. However, there is limited coverage as to what exactly people hypothesize is a species tree. It could be envisioned to be a variety of things but rarely does one define their meaning of a species tree, a lapses exactly similar to those describing species but do not reference as to what concept they follow in discovering new diversity. The general thought of a species tree is that it is an estimate of all of the data combined, with some ad hoc explanations for some character transformations. However, the idea of gene trees versus species trees is usually in reference to molecular data sets and resulting phylogenies. Thus, while the issue of genes providing different hypotheses of descent of taxa examined is a valid and important issue, it is important to keep in mind the underlying theory of a species tree and not have the latter biased into only discussions relevant to genes.

Thus, what is a species tree? Is it a tree based on all gene sequences available or all molecular types of data? Is it a tree based on mitochondrial, plastid, or nuclear gene sequences? Is it a tree based on some combination or restriction of molecular data in the options above? Where do other types of characters, like morphological, behavioral, ecological, physiological, etc. fit into the discussions on species trees? Are they to be included or are species trees really those reconstructions based on morphological characters as some would argue (Mooi and Gill [37], and see series of papers relating to this issue in *Zootaxa*, Carvalho and Craig [38])?

I argue that when considering the idea of and discussion of a "species tree" one must keep in mind that such considerations should not be in reference to any particular type of data and this should be clearly addressed. As Wiley [3] and Wiley and Lieberman [27] advocate, and I concur with, there is only one tree of the genealogical history of life on earth, regardless of our ability to discover it. It is also an issue of a hypothesis-driven question and outcome. All reconstructions and inferences are biased in one way or another – from the obvious of weighting characters to a model based analysis (all analyses are based on some sort of underlying model, including simple parsimony) to one's selection of characters or those characters that are available to the researcher (aspects of morphology, primers for and ability to generate gene sequences, behaviors, and a theoretically infinite number of character types).

Thus, a species tree is simply the one genealogical history of relationships of diversity of life. All of our analyses and attempts are simply estimates of this tree. However, the ways of combining or dividing characters and the types of characters employed are seemingly infinite and likelihood of converging on the species tree of life must always be considered. Trees are hypotheses subject to testing either through additional types of analyses, altering assumptions to even differing combinations of character transformations.

## 11. Which species tree is the best?

A common misconception amongst those practicing systematics or using systematic information (phylogenies) is that the latest published phylogeny of a group of taxa is the best and therefore represents the most probable set of relationships. This is a gross misconception. A phylogeny of taxa (species) is a hypothesis and multiple published phylogenies on a group of taxa may be equally probable/parsimonious hypotheses and stand alone as multiple hypotheses of relationships for the same taxa. It is not uncommon for the phylogenetic relationships of any group of taxa to change as more characters or taxa are added or taxa or characters are removed. This phenomenon is known as bias via taxon and character sampling [39-42]. It is predicted that with continued addition of samples of taxa or taxa alone as well as more character information on these taxa, the tree will stabilize as to relationships based on synapomorphies. Thus, for a particular group the phylogenetic relationships may change over time with new additions of hypothesized synapomorphies and/or addition or deletion of taxa. However, one must realize that, contrary to assertions of Mooi and Gill [37, 43], they are all competing hypotheses, just like alternative hypotheses of homology. They stand until appropriate statistical tests are used to evaluate their robustness. The future holds many uncertainties and possibly alternative interpretations of genealogical descent. After all, this is science and engages the process of refuting or corroborating hypotheses, not an exercise set out to prove a set of relationships engraved in stone.

## 12. The nature of biological species: Assertions as to the monophyly of species?

The term monophyly is not like many terms in the literature on evolutionary biology or derived from the new synthesis, wherein multiple interpretations or definitions exist. Monophyly was clearly defined by Hennig. So, why ask this question if it is a clearly defined term and many researchers use it in the literature, in descriptions, or in dialogue? Good question, and one that should not have to be discussed if practitioners in systematics and population genetics where, in fact, up to speed on their understanding of the theoretical and philosophical concepts surrounding species and supraspecific taxa, and the difference between these two frequently used or tossed-around terms.

So, what is the deal with the argument pertaining to monophyletic species? Is this the appropriate interpretation when one finds that not all geographic samples of a species form a lineage independent of other such lineages? No! Unfortunately, this represents another instance of practitioners of the above listed fields misunderstanding and misusing the terminology and misunderstanding of concepts of monophyly, paraphyly, and polyphyly. To ask whether or not a species is monophyletic or argue that a species is monophyletic, paraphyly, and polyphyly is simply a *non sequitur*. The argument of one of the above three terms for a species is fallacious because of the disconnection between the initial premise and the resulting conclusion. By their very nature, species as entities, individuals, and lineages in nature or as taxa, one's preference in use, are natural. They are independent lineages that maintain their identity through time, even in face of factors that could compromise their identities, and are restricted in time and space. As Hennig [22-25] and Wiley and Mayden [29-31] have pointed out, the terms monophyly, paraphyly, and polyphyly do not relate at all to the entity species, but are correctly used in discussions relating to only supraspecific taxa of three or more species. The reinforcement of this misunderstanding and misuse of the term most likely stems from the definition and discussions relative to the "Phylogenetic Criterion (see Table 1)." This use of the term confuses issues – species discovery and naturalness of supraspecific taxa. It is a term for one of many frequently used criteria for discovering and diagnosing species in nature on the basis of apomorphic character(s) or the "monophyly" of all representatives of a hypothesized species. Monophyly and related terms refer to supraspecific taxa.

It is true that one may misinterpret variation and that multiple unrelated lineages as species (all but one undiscovered at the time) are recognized under one binominal. However, this will remain such until evidence supporting the contrary and corroborating such a hypothesis of greater diversity exists in the complex. "Problematic" instances that very likely point to such cases of diversity are those species that are diagnosed "on the basis of a combination of characters). How is this so? This type of diagnosis has the potential of using a mixture of plesiomorphic and apomorphic character states that in a unique combination diagnose two or more species but can also result in "cryptic diversity" that will remain to be discovered though a lineage-based interpretation of characters (Fig. 2; also see "Characters: their real function and why all types are important!").

Many researchers and philosophers may possess the theoretical and philosophical background relating to this topic but lack experience in the broad-based diversity and complexities of nature that biodiversity scientists, or systematists have to deal with in their studies. Likewise, those lacking the theoretical and philosophical understanding of species will likely error in their assessment of lineages representing species. The most appropriate solution to this dysfunctional situation is that the student and practitioner of systematics, biodiversity science, or whatever one may call the specialty of discovering and describing species and reconstructing relationships, must be educated in both areas and must learn through apprenticeships. This is how one learns the practice in most areas of the natural sciences. Only through such an experience that is part and parcel to a general education can one be more confident that they have the experience to evaluate a broad range of situations of diversity and generate informed hypotheses. It is analogous to what Hennig called reciprocal illumination. In varied systematic groups the specialists also know what characters tend to be highly variable and are less likely to be useful in corroborating a hypothesis of lineage independence. This only comes through

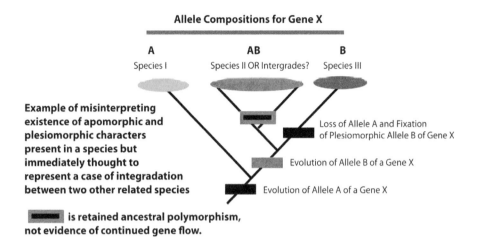

**Allele Compositions for Gene X**

| A | AB | B |
|---|----|---|
| Species I | Species II OR Intergrades? | Species III |

Example of misinterpreting existence of apomorphic and plesiomorphic characters present in a species but immediately thought to represent a case of integradation between two other related species

Loss of Allele A and Fixation of Plesiomorphic Allele B of Gene X

Evolution of Allele B of a Gene X

Evolution of Allele A of a Gene X

is retained ancestral polymorphism, not evidence of continued gene flow.

**Figure 2.** Hypothetical example of a known evolutionary history of a clade of organisms and the existence of multiple independently evolving and unrelated lineages diagnosed and interpreted as a single lineage on the basis of "a combination of characters."

experience with diversity, not through theoretical and philosophical readings, discussions, or debates (although these are fundamental to some types of interpretations – see above). Everyone benefits from apprenticeships – both the student and mentor. Apprenticeships are critical to a developing professional and such experiences should be continuous throughout the process of life-long learning and practicing in the discipline of science.

## Acknowledgements

This contribution has been possible only through the readings of great papers by and discussions on these topics, regardless of their opinions, with many extremely bright scientists, theorists, philosophers, and systematists kind enough to share their time with me. There are too many to be acknowledged here. I particularly thank David Hull, E. O. Wiley, D. Hillis, and D. Frost for their valuable time and ideas as mentors that have helped in molding my perspectives on this broad topic. I also thank A. Ilverson for her assistance with this paper. Finally, I thank the US citizens and taxpayers for their funding of the National Science Foundation, and NSF for research funding to conduct biodiversity and systematic studies, studies that without the accompanying experiences and colleagues would have failed to close a much needed void (others remain I am sure) in my understanding of species and systematics (NSF grants to Mayden EF 0431326, DEB-0817027, DEB-1021840, DBI-0956370, and others predating these).

## Author details

Richard L. Mayden*

Address all correspondence to: e-mail:cypriniformes@gmail.com

Department of Biology, Laboratory of Integrated Biodiversity, Conservation, and Genomics, Saint Louis University, St. Louis, Missouri, USA

## References

[1] Wiley, E. O. (1978). The evolutionary species concept reconsidered. Systematic Zoology 1978; , 27-17.

[2] Wiley, E. O. (1980). Is the evolutionary species fiction? A consideration of classes, individuals and historical entities. Systematic Zoology 1980; , 29-76.

[3] Wiley, E. O. Phylogenetics. The Theory and Practice of Phylogenetic Systematics. New York: Wiley-Interscience; (1981).

[4] Wiley, E. O. Kinds, individuals, and theories. In: Ruse M (ed.) What the Philosophy of Biology Is. Dordrecht: Kluwer Academic; (1989). , 289-300.

[5] Wiley, E. O. On species and speciation with reference to the fishes. Fish and Fisheries (2002). , 3-1.

[6] Wiley, E. O. Species concepts and their importance in fisheries management and research. Transactions of the American Fisheries Society. (2007). , 136(4), 1126-1135.

[7] Templeton, A. R. (1989). The meaning of species and speciation: A genetic perspective. In: Otte D., Endler JA. (eds.) Speciation and Its Consequences. Sunderland: Sinauer Associates; 1989. , 3-27.

[8] Frost, D. R, & Kluge, A. G. A consideration of epistemology in systematic biology, with special reference to species. Cladistics (1994). , 10-259.

[9] Frost, D. R, & Hillis, D. M. Species in concept and practice: Herpetological applications. Herpetologica (1990). , 46-87.

[10] Hull, D. X. The ideal species concept- and why we don't get it. In: Claridge MF, Dawah HA, Wilson MR. (eds.) Species, the Units of Biodiversity. London: Chapman and Hall; (1997). , 357-380.

[11] Mayden, R. L, & Wood, R. M. Systematics, species concepts, and the evolutionary significant unit in biodiversity and conservation biology. American Fisheries Society Symposium. (1995). , 17-58.

[12] Sites, J. W, & Crandall, K. A. Testing species boundaries in biodiversity studies. Conservation Biology (1997). , 11(6), 1289-1297.

[13] De Queiroz, K. The general lineage concept of species and the defining properties of the species categories. In: Wilson RA (ed), Species: New Interdisciplinary Essays. Cambridge: The MIT Press; (1999). , 49-89.

[14] De Queiroz, K. Species concepts and species delimitation. Systematic Biology (2007). , 56-879.

[15] Mayden, R. L. A hierarchy of species concepts: The denouement in the saga of the species problem. In: Claridge MF, Dawah HA, Wilson MR. (eds.) Species, the Units of Biodiversity. London: Chapman and Hall; (1997). , 381-424.

[16] Mayden, R. L. Consilience and a hierarchy of species concepts: Advances towards closure on the species puzzle. Journal of Nematology (1999). , 31-95.

[17] Mayden, R. L. On biological species, species concepts and individuation in the natural world. Fish Fisheries (2002). , 3-171.

[18] Claridge, M. F, Dawah, H. A, & Wilson, M. R. eds.) Species: The Units of Biodiversity. London: Chapman and Hall Ltd; (1997).

[19] Howard, D. J, & Berlocher, S. H. eds.) Endless Forms: Species and Speciation. Oxford: Oxford University Press; (1998).

[20] Wheeler, Q. D, & Meier, R. eds.). (2000). Species Concepts and Phylogenetic Theory. A Debate. New York: Columbia University Press; 2000.

[21] Kuhn, T. S. The Structure of Scientific Revolutions. Chicago: University of Chicago Press; (1962).

[22] Hennig, W. Grundzuge einer Theorie der phylogenetischen Systematik. Berlin: Deutscher Zentralverlag; (1950).

[23] Hennig, W. Kritische Bermerkungen zum phylogenetischen System der Insekten. Beiträge zur Entomologie (1953). , 3-1.

[24] Hennig, W. Phylogenetic systematics. Annual Review of Entomology (1965). , 10-97.

[25] Hennig, W. Phylogenetic Systematics. Urbana: University of Illinois Press; (1966).

[26] Simpson, G. G. Principles of Animal Taxonomy. New York: Columbia University Press; (1961).

[27] Wiley, E. O, & Lieberman, B. S. Phylogenetics. The Theory and Practice of Phylogenetic Systematics. Second edition. New York: Wiley-Blackwell; (2011).

[28] De Queiroz, K. The general lineage concept of species, species criteria, and the process of speciation. In: Howard DJ., S. H. Berlocher SH. (eds.). Endless Forms: Species and Speciation. Oxford: Oxford University Press; (1998). , 57-75.

[29] Wiley, E. O, & Mayden, R. L. The evolutionary species concept. In: Wheeler QD., Meier R. (eds.) Species Concepts and Phylogenetic Systematics. A Debate. New York: Columbia University Press; (2000a). , 70-89.

[30] Wiley, E. O, & Mayden, R. L. A critique from the Evolutionary Species Concept perspective. In: Wheeler QD., Meier R. (eds.) Species Concepts and Phylogenetic Systematics. A Debate. New York: Columbia University Press; (2000b). , 146-158.

[31] Wiley, E. O, & Mayden, R. L. A defense of the Evolutionary Species Concept. In: Wheeler QD., Meier R. (eds.) Species Concepts and Phylogenetic Systematics. A Debate. New York: Columbia University Press; (2000c). , 198-208.

[32] Naomi, S. I. On the integrated frameworks of species concepts: Mayden's hierarchy of species concepts and de Queiroz's unified concept of species. Journal of Zoological Systematics and Evolutionary Research. (2010). , 49(3), 1-8.

[33] Bridgman, P. The Logic of Modern Physics. New York: McMillian; (1927).

[34] Gilmour JSLTaxonomy and philosophy. In: Huxley JS (ed.) The New Systematics. Oxford: Clarendon Press; (1940). , 461-474.

[35] Maddison, W. P. Gene trees in species trees. Systematic Biology (1997). , 46-523.

[36] Knowles, L. L, & Kubatko, L. S. (2010). Estimating species trees: Practical and Theoretical Aspects. New York: Wiley-Blackwell; 2010.

[37] Mooi, R. D, & Gill, A. C. Phylogenetics without synapomorphies- A crisis in fish systematics: time to show some character. Zootaxa (2010). , 2450-26.

[38] Carvalho MRDCraig MT. Overview. Zootaxa (2011). , 2946-5.

[39] Hillis, D. M. Taxonomic sampling, phylogenetic accuracy, and investigator bias. Systematic Biology (1998). , 47-3.

[40] Zwickl, D. J, & Hillis, D. M. Increased Taxon Sampling Greatly Reduces Phylogenetic Error. Systematic Biology (2002). , 51(4), 588-598.

[41] Hillis, D. M, Pollock, D. D, Mcguire, J. A, & Zwickl, D. J. Is sparse taxon sampling a problem for phylogenetic inference? Systematic Biology (2003). , 52-124.

[42] Mayden, R. L, Tang, K. L, Wood, R. M, Chen, W. J, Agnew, M. K, Conway, K. W, Yang, L, Li, J, Wang, X, Saitoh, K, Miya, M, He, S, Liu, H, Chen, Y, & Nishida, M. (2008). Inferring the tree of life of the order Cypriniformes, the Earth's most diverse clade of freshwater fishes: implications of varied taxon and character sampling. Journal of Systematics and Evolution 2008; , 46(3), 424-438.

[43] Mooi, R. D, & Gill, A. C. Why we shouldn't let sleeping dogmas lie: a partial reply to Craig. Zootaxa (2011). , 2946-41.

# Chapter 8

# Species Delimitation: A Decade After the Renaissance

Arley Camargo and Jack Jr. Sites

Additional information is available at the end of the chapter

## 1. Introduction

A decade ago Sites and Marshall [1] described the empirical practice of species delimitation as "a Renaissance issue in systematic biology". At the time there was an odd disconnect between the two frequently stated empirical goals systematic biology: the discovery of: (1) monophyletic groups (clades) and relationships within these at all hierarchical levels above species; and (2) lineages (species); compared to the actual practice of the discipline. While much of systematic biology had been devoted to the first goal, the second goal had until recently been largely ignored [2], despite the fact that species are routinely used as the basic units of analysis in biogeography, ecology, evolutionary biology, and conservation biology [3,4]. However, Sites and Marshall [1] noted "signs of a Renaissance" at the time of their review, which was precipitated in part by others emphasizing the need to distinguish between a non-operational, ontological definition of species, versus the empirical (operational) data needed to test their reality [5-7]. De Queiroz [7] (p. 60) noted that "All modern species definitions either explicitly or implicitly equate species with segments of population level evolutionary lineages." De Queiroz also noted that this was a revised version of Simpson's "evolutionary species concept", which defines a species as "a lineage (an ancestral- descendent sequence of populations) evolving separately from others and with its own evolutionary role and tendencies" ([8], p. 153), and called this a General Lineage Concept (GLC) of species ([7], p. 65). De Queiroz [9] further emphasized that the multiple empirical criteria simply reflect the many contingent properties (differences in genetic or morphological features, adaptive zones or ecological niches, mate-recognition systems, reproductive compatibility, monophyly, etc.) of diverging populations associated with different evolutionary processes operating in various geographic contexts [10,11]. Sites and Marshall [1] noted that the emerging consensus among systematists and evolutionary biologists was based on the utility of this distinction (ontological definition vs. empirical species delimitation [SDL] methods), and as also noted by de Queiroz [12], due to the contingencies of speciation processes, any single criterion or data set will artificially reduce the complexity of evolving lineages.

The subject matter of these and other reviews [12,13] focused strictly on methods of detecting various lines of evidence for lineage independence (reproductive isolation, ecological distinctiveness, diagnosability, monophyly, etc.), and since then new methods continue to be described [14], as do studies comparing the performance of some of these [14,15]. In 2006, the Society of Systematic Biologists (SSB'06) organized the first symposium dedicated to the topic of species delimitation [2]; 11 papers were presented and six of those published, including an update by referenced de Queiroz [16], which emphasized the distinction between the GLC as "separately evolving metapopulation lineages, or more specifically, with segments of such lineages", versus secondary biological attributes or properties of organisms that can be quantified to empirically test for species status. This is a crucial distinction because it clearly separates the conceptual issue of defining the species category from the methodological issues of delimiting species; previously these had been conflated with the result that properties used to infer species boundaries (the empirical test) were also sometimes regarded as necessary for defining a species (a conceptualization issue). The advantage of the unified GLC is that no specific biological attributes of a species are considered necessary properties – species may exist as segments of metapopulations lineages regardless of our ability to empirically delimit them. Prior to this clarification and the realization that many different properties are relevant to the issue of species delimitation [17], the alternative species "concepts" in which various biological attributes had accumulated in diverging lineages required these same attributes to be necessary properties of species. This led to a confusing situation in which a different property was considered necessary under each alternative concept (22 such "concepts" were identified by Mayden [6]), and a long and ultimately non-productive debate about species definitions. Now most of these earlier "concepts" can be viewed as secondary species criteria that provide evidence of lineage separation.

Recently, Hausdorf [17] argued for an up-dated ontological species concept, based in new insights into speciation processes, particularly evidence that reproductive barriers are semipermeable to some gene flow, and that speciation may occur despite ongoing gene flow between diverging populations [18-23]. Two other lines of evidence are relevant to the point of re-visiting the GLC: (1) findings of polyphyletic species of animals, due to parallel speciation in which similar traits conferring reproductive isolation arise separately in closely related populations [24,25], or in plants, due to recurrent polyploidization in different populations of the ancestral species [26,27]; and (2) discoveries of uniparental organisms that can be characterized as distinct units resembling species of biparental organisms [28]. We cannot resolve all of these larger issues here, but we return to some of the general points raised by Hausdorf [17] in the discussion.

Empirically, species delimitation continues to be a topic of increasing interest in evolutionary biology. A reference search in the ISI Web of Science with the keyword 'species delimitation' retrieved 227 articles published since 2000, of which 60% were published after 2008. Less than 10 articles per year were published between 2000 and 2005; subsequently 10-20 articles per year between 2006 and 2008, and after 2008 the publication rate reached ~ 40 articles (Figure 1A). These increases include papers describing new SDL methods, or using existing methods with novel data sets and/or applications to new taxa. Because new SDL methods apply the same coalescent models developed for species tree estimation and usually lead to the discovery of morphologically 'cryptic' species, we also searched for references with the keywords 'species tree' and 'cryptic species'. During the same period of time, papers about 'species trees' were few until 2007, increased between 2008 and 2010 to 5-10 articles per year, and nearly doubled to >20 papers last year (Figure 1B). Publications referring to 'cryptic species' show a constant increase from 20 papers/year in 2000 to 90 papers/year in 2011 with the larger annual increase between 2010 and 2011 (Figure 1C). These publication trends suggest that the recent paradigm shift in phylogenetic systematics to incorporate species trees (29) is having a positive impact on the development of new SDL methods, which are gradually being incorporated into integrative taxonomic practices for the discovery of cryptic species diversity [30].

# 2. Body

## 2.1. Short history of some early methods

Sites and Marshall [1,13] separated SDL methods into non-tree and tree-based approaches, and included among the former (1) pairwise genetic distances that could be tested for either correlations with reproductive isolation [31,32], morphological distances [33], or geographic distances [34]; (2) gene flow statistics to estimate the extent of gene flow across hybrid zones [35]; (3) fixed alternative character states as an indicator of no gene flow in a "population aggregation analysis" (PAA; [36]); (4) the presence of heterozygous genotypes as an indicator of a "field for recombination" [37]; and (5) genotypic clusters [38].

Early tree-based methods included: (1) three versions of the phylogenetic species "concept" based on apomorphy, or lineage splitting, or node-based criteria, following the terminology of Brooks and McLennan [39]; (2) cladistic haplotype aggregation [40]; (3) molecular-morphological assessments using dichotomous flow charts [41]; (4) genealogical exclusivity [42]; and (5) an extension of the nested clade analysis [43] that includes tests of species boundaries [44]. The data sets in these early studies most often included genotypes resolved from multilocus isozymes [15], morphological (usually meristic) characters, and with few exceptions [45,46], mitochondrial DNA (mtDNA) sequences. An innovative phylogenetic method described by Pons *et al.* [14] was based on a likelihood analysis of the mtDNA gene tree that estimates the inflection point between species-level (speciation-extinction) and population-level (coalescent) evolutionary processes, and demonstrated that groups delimited by this approach were generally concordant with geographic distributions and morphologically recognized species. This was one of a small number of early studies comparing the performance of multiple SDL methods (see also [15, 40, 45,46]).

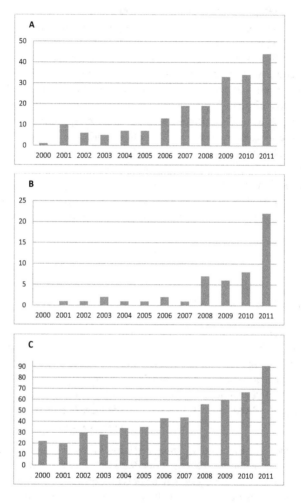

**Figure 1.** The number of papers with (A) "species delimitation"; (B) "species tree", or (C) "cryptic species" in the title, published from 2000 – 2011.

The published contributions of SSB'06 symposium [2] included several novel SDL methods, the first method [47] described a coalescent approach to estimating species boundaries based on multiple unlinked gene trees, and that does not require species to be characterized by reciprocal monophyly. This is an explicitly model-based approach that accommodates stochastic variance of the gene sorting process by linking estimates of two key parameters, a range of estimates of effective population sizes relative to possible divergence times. This type of gene tree-coalescence approach also directly links population genetic SDL methods to phylogenetic inference at deeper levels of divergence, which has been identified as a "new paradigm" in systematics [29]. In this same issue, Shaffer and Thomson [48] introduced a population genetic SDL based on large sets of single nucleotide polymorphisms (SNPs), which would be most suited to delimiting very young species. Finally, this volume included two more novel SDL methods, both in this case using ecological and distributional data in novel ways to model "niche envelopes" that can augment molecular or morphological data in species delimitation [49-51].

## 2.2. Recent progress

### 2.2.1. New methods & new theory

New empirical SDL methods continue to be developed, based on multiple lines of evidence and multiple statistical methods. Among some of these is the approach of Bond and Stockman [52] that is especially relevant to highly geographically-structured populations in which traditional sequence-only data sets are likely to recover large numbers of well-defined, well-supported, and geographically concordant/genetically divergent-but-morphologically cryptic populations (species). These authors describe a framework for testing potential genetic and ecological exchangeability as a means of delimiting cohesion species [53], and present an example in trapdoor spiders of the *Aptostichus atomarius* complex. A completely different approach [54] is based on statistical tests of both population structure [48] and genealogical exclusivity [54] of nuclear loci, to test species provisionally identified from well-supported mtDNA haploclades; the focal taxa in this study were Malagasy mouse lemurs (55; genus *Microcebus*). As a third example, Puillandre et al. [56] described a four-step approach to "generating robust speciation hypotheses" in mollusk family Turridae (genus *Gemmula*) based on: (a) collection of the COI DNA barcode gene for GMYC [14] and ABGD (Automatic Barcode Gap Discovery; [57]) analyses; coupled with (b) nuclear gene (rRNA 28S), morphological, geographical and bathymetrical data, to redefine species boundaries in this clade. This protocol more than doubled the previously known species diversity in *Gemmula*, and may be useful for large-scale SDL in hyperdiverse groups. A few additional examples include genotype-based methods for dominant and co-dominant multi-locus markers [58], combined estimates of divergence times and gene flow to discriminate intraspecific from interspecific patterns [59], and an extension of the R package GENELAND to include genetic, phenotypic (morphometric), and geographic data for delimitation of populations and species [60].

The recent merge of coalescent theory with phylogenetics has driven a new generation of SDL methods and a new paradigm in systematics [29]. This new theoretical framework, and its derived analytical applications, was in part required as a solution for accommodating the observed conflict among genealogies from multiple loci (gene trees) with the underlying population-level genealogies (species trees) [61]. A multi-species or 'censored' model was formulated to account for this discordance by considering each branch of the species tree as a separate coalescent model and by connecting them into a population-level genealogy following the topology of the species tree [62,63]. Under this new approach, two major key innovations over the classic phylogenetic methods were achieved. First, multiple individual samples can be assigned to a single species and the estimated phylogeny represents the speciation history of ancestral and descendant species-level lineages, in contrast to the gene genealogies estimated with individual samples. Second, because the coalescent process of each gene tree is dependent upon parameters of its containing species tree, this approach can co-estimate gene and species tree simultaneously, by-passing the task of calculating a consensus tree or estimating a phylogeny from a concatenated dataset. This new theoretical framework allows prediction of the probability distribution of gene trees given the species tree, and consequently, several methods were developed for estimating species trees from a collection of multiple gene trees under different algorithms [64,65]. Based on these new methods, a generation of fully-coalescent SDL methods was introduced that consisted of selecting the best species-tree model from a set of alternative models that represent different hypotheses of species limits. For instance, one approach finds the maximum-likelihood for the full species tree (all species are hypothesized as separate lineages) and for alternative species trees (two or more species at a given node are collapsed into one), and then selects the best model using Akaike information criteria, assuming fixed gene trees and constant population sizes along the species tree (SpeDeSTEM; [66]).

Another SDL method consists of sampling from the Bayesian posterior distribution of species delimitation models using reversible-jump Markov chain Monte Carlo (rjMCMC) with the program BP&P 2.1 [67]. This approach accomodates gene tree uncertainty and variable population sizes, but a "known" species tree must be provided *a priori*. In addition, heuristic and/or semi-parametric approaches have been developed for: resolving the boundary between coalescent and speciation processes using single gene trees (generalized mixed Yule-coalescent, [14]), finding both the optimum species tree and species limits via minimization of gene tree conflict and intraspecific structure (Brownie; [68]), and selection of SDL models using approximate Bayesian computation (ABC) [69]. Other tree-based [54] and non-tree-based [58] SDL methods that can handle multiple loci with limited variation have been applied with success. In addition, there has been also a resurgence of morphology-based SDL using multivariate techniques in a hypothesis-driven statistical framework [60,70].

## 2.2.2. New kinds of data

The development of new multi-species/multi-locus SDL methods was also in part due to the demand of efficient analytical tools to handle the rapidly increasing amounts of molecular data collected with modern techniques. New SDL methods should be able to handle tens of loci for multiple individuals derived from the development and screening of anonymous nuclear loci (ANL), introns, and protein-coding loci using genomic resources [71-73]. However, these new SDL methods are inadequate to analyze the influx of whole-genome data that have started to be collected for non-model organisms via next-generation sequencing (NGS) technologies ([74-76]; e.g, genome of the lizard *Anolis carolinensis*; [77]). NGS technologies have been recently applied to development of thousands of gene regions spanning multiple divergence times [78], or loci targeted for "shallow-scale" phylogenetic/phylogeographic studies [79], and microsatellites [80] or SNPs [81] for extremely shallow phylogeographic histories [82]. The microsatellite or SNP data should be useful for genotyping individuals for SDL studies of very young species [48].

More efficient and less costly whole-genome sequencing is becoming available on a regular basis, a trend that started with the first-generation technology (Sanger capillary-sequencing), followed by the second-generation (i.e., SOLiD 454, Illumina, Solexa, etc; [83]), and continuing today with the recently introduced third-generation 'nanopore' sequencing [84,85]. A significant by-product of these single-molecule sequencing methods is their ability to automatically resolve the allelic phases of heterozygotes, in contrast to the time-consuming phase estimation and/or cloning required after direct dideoxy-sequencing [86]. In addition, the uniform sampling of hundreds of loci across the genome can help identifying "outlier" loci via genome scans, which can represent candidate genes with fitness value, subject to selection and linked to processes such as ecological speciation [87].

### 2.2.3. Advantages of Multi-Species Coalescent-Based Methods (MSCM)

*Model-based.*–Because these SDL are based in the multi-species coalescent model, the likelihood of the data can be evaluated to find maximum-likelihood and posterior probability estimates of parameters and testing alternative SDL models under different criteria (e.g., likelihood-ratio test, Akaike information criterion, Bayes factors [46,88]). More importantly, these methods implement SDL in a hypothesis-testing framework, and taking into account uncertainty due to genetic processes and insufficient sampling [89,90]. In addition, coalescent simulations generated under a null hypothesis of no-speciation and the alternative hypothesis of speciation can be used for evaluating the performance of these methods based on estimations of inferential errors (type I and II errors, see [91]). For example, the accuracy of three coalescent-based SDL (SpeDeSTEM, BP&P, and ABC) has been compared using simulations under a model of speciation for variable sampling densities and parameter values to estimate type II error (i.e., failing to reject no-speciation when it is false) across a range of conditions [66,69,92]. When there is no migration, SpeDeSTEM can delimit species that have diverged as recently as $0.5N_e$ generations ago using only 5 loci and 5 alleles per species [66] while BP&P could detect speciation at shorter divergence times ($0.4N_e$ generations ago) with the same sampling design [67,92]. In agreement with these results, a comparison under identical simulation conditions showed that BPP outperformed SpeDeSTEM (and also ABC) when speciation takes place with or without gene flow [69]. In spite of these simulations covering different speciation scenarios, sampling designs, and SDLs, the practical question of the appropriate balance between number of loci and number of alleles sequenced has not been explicitly explored until now. Below, in the last section of this chapter, we performed some simulations for a preliminary evaluation of the relative benefits of sampling more alleles vs. loci for accurate species delimitation.

*Neutral loci.*–These markers should be insensitive to 'phenotypic plasticity', the phenotypic response to environmental variation that is not genetically-based (in contrast to adaptive variation), which could bias morphological-based taxonomy. Environmental variation in different parts of the range can lead to a plastic phenotypic response, which can be revealed and distinguished from local adaptation via reciprocal transplant or 'common garden' experiments [93]. In these cases, morphological variation as a result of this plastic response could be used as a criterion to delimit species, while neutral markers would indicate that there is no genetic differentiation [94,95]. In contrast, in cases of morphologically-cryptic species due to for example to niche conservatism [96], genetic divergence and lineage sorting is expected to occur in neutral markers due to independent evolution in isolation, and those markers with higher mutation rates and smaller effective population size (e.g. mtDNA) should be ideal for species delimitation [97,98]. Moreover, it has been suggested that neutral loci will also differ in their usefulness for species delimitation since those with higher rates of intra-specific gene flow will be less sensitive to the effect of inter-specific introgression [99]. However, the mitochondrial locus does not always meet assumptions of neutrality [100], and it frequently introgresses across species boundaries [101], so in our view it should be used to identify "candidate species" [102], which can then be verified with independent lines of evidence [103].

*Repeatability.*–The results of a SDL analysis can be replicated exactly when using the same data and the same analytical methods, which eliminates much of the subjectivity and/or investigator bias for/against certain kinds of data (morphology vs. molecular, etc.). Because these methods rely on explicit predictions about genealogical patterns under alternative models of lineage divergence, it is possible to carry out species delimitation in a more objective and bias-free fashion compared to diagnosability-based SDL methods [90]. In addition, because inferences are dependent upon a specific sampling design and the method used, one can make explicit statements about how robust a given species delimitation method is to variation in these parameters, and to violations of the method's assumptions.

*Universality.*–The same SDL method and the same kind of data (i.e, DNA sequences or gene trees from homologous regions of the genome) can be used for SDL across different taxa, making these approaches comparable across all parts of the Tree of Life, as long as the assumptions of the method are reasonable for the taxon under study (see below). Another advantage associated with the use of neutral markers in coalescent-based SDLs is related to the standard criterion used for assigning species status across a variety of taxa when using the same markers and analyses [90], assuming that these markers offer similar resolving power. This is a desirable property for a SDL method since a uniform criterion implies that the species level could be compared readily among different higher-level taxa, thereby allowing meaningful analyses of species diversity among communities typical of ecological studies [91].

### 2.2.4. Disadvantages of MSCM

Many of the advantages listed above also impose some limitations of MSCM and other SDL methods for different reasons. First, these are *model-based* methods, and any violations of assumptions of the standard coalescent are expected to introduce inference errors. For instance, and most relevant to the SDL problem, while the standard coalescent assumes panmixis within populations, it is clear that in most natural populations, there is almost always some degree of population structure (i.e, demes connected by limited gene flow). In fact, a recent study using the Brownie's SDL method found that more dense sampling increased the chances of detecting population structure, supporting more species boundaries, and consequently, inflating estimates of the number of species [104]. Thus, MSCM could be more prone to split a single real species into multiple lineages due to intra-specific population structure alone, increasing type I error (i.e., rejecting a true hypothesis of a single species), and leading to 'taxonomic inflation' [91]. Fortunately, some flexible MSCM methods allow incorporating population structure within species via coalescent simulation of island, stepping-stone, and other potential models, and subsequent comparison of SDL hypotheses with ABC approaches [69].

Another frequent assumption of most MSCM is that species have diverged from a common ancestral species without gene flow even though speciation with gene flow seems to be rather common in nature, especially in cases of ecological speciation [22,95,105]. While these methods ignore the effects of gene flow, simulation testing has shown that some of them are relatively robust to low levels of gene flow [66,92], and that its impact on delimitation accuracy is ameliorated when gene flow is explicitly incorporated in the speciation model [69]. This result supports the suggestion that, in order to distinguish between species- and population-level differentiation, it is necessary to jointly consider the two components of the divergence process: time since splitting and gene flow after divergence [59].

Second, *sampling effort* is well known to strongly impact coalescent-based and other SDL methods. A number of studies evaluating the accuracy of several MSCM methods suggest that limited sampling of loci and sequences will decrease the probability of detecting speciation when this hypothesis is correct [66,69,92] and consequently, increasing type II error [91]. In addition, these simulation results also support the intuitive idea that the problem of insufficient sampling becomes more serious when SDL is more difficult: shorter divergence times, larger population size, and increasing inter-specific gene flow. However, more simulations are necessary to evaluate the appropriate balance between sampling intensity and design (e.g., geographic vs. genealogical dimensions, [91]) for different parameter configurations, in a power analysis context to provide further guidance to empirical studies [106]. In addition to limited geographic sampling, the collected sequence data also impose a limit to the amount of genetic data available for analysis. In the next section, we explore how accuracy in species delimitation responds to variable sampling of loci and alleles for a fixed sequencing effort.

Third, coalescent-based SDL approaches assume *selective neutrality* of gene regions used, but divergent selection on ecological traits, across habitats or along an environmental gradient, can lead to local adaptation and correlated reproductive isolation in a process of ecological speciation [95,107]. Phenotypic divergence can be so fast that mutation rates could produce little or no differentiation at all in neutral markers used in SDL approaches. Only those "outlier loci" under selection revealed by genomic scans, which are potentially associated with the selected traits, would be appropriate markers under these scenarios [87].

Fourth, as in other methods *data conflict* may be evident when multiple data sets are used. These SDL methods are not expected to resolve the discordance among different kinds of data sets (i.e., morphological, behavioral, ecological, molecular, etc.) since they typically use sequence data or gene trees from presumably neutral loci. However, Bayesian approaches have the potential of incorporating previous information about species limits derived from non-molecular data into prior distributions of genetic-based analyses [67].

Fifth, there may be conflicts with *traditional taxonomic practices*. The discovery of new cryptic species with coalescent-based SDL in a statistical framework, is still insufficient for formal taxonomic descriptions, since nomenclatural rules still require traditional morphology-based diagnoses [108,109]. While these methods will help diagnosing new cryptic diversity, many taxonomists will be reluctant to formally describe new species based on molecular-data alone, which ultimately will further expand the 'taxonomy-phylogeny' gap [91]. While the description of cryptic species is complicated by the lack of morphological diagnostic characters, another difficulty relies in the inability of MSCM to assign newly collected specimens to species (i.e., taxonomic determination) unless new analyses are carried out to re-evaluate species limits.

## 3. Future directions

*Statistical testing of SDL.*–The ongoing surge in the new generation of SDL methods will probably encourage many taxonomists to apply these methods empirically, especially for recently evolved, cryptic taxa that cannot be delimited with other data. The ability to frame species limits as statistical hypotheses that can be tested objectively with multi-locus and multi-species analyses make these new SDL methods very appealing for empirical systematists in the context of an 'integrative taxonomy' [4,110,111]. In addition to empirical application to real data sets, we also expect that more simulation studies will be carried out to compare the performance of different data sets, under different methods/assumptions, and for variable sampling designs, using statistical power analyses. Previous studies have compared methods for a limited set of parameter conditions (e.g., usually population size has been assumed to remain constant) and have examined the effect of increased sampling effort for loci or sequences separately. However, performance of these SDL methods has not been evaluated for a variable sampling design and a fixed sampling effort; in other words, what should be the optimal balance between number of loci and number of sequences when the total number of sequenced base pairs (bp) is the same?

In order to provide a preliminary evaluation of the impact of sampling design on performance of new SDM, we simulated coalescent genealogies with the program ms [112] and sequence data with the program Seq-Gen [113] for a speciation model between species A and B for three increasing divergence times: 0.25, 0.5, and $1N_e$ (Figure 2A). We assumed a constant $\theta$ per site = 0.01, 500 bp per locus, and ~50 variable sites per locus. For each divergence time, we simulated 5 combinations of number of loci (1, 2, 4, 10, and 20) and number of sequences per species (1, 2, 5, 10, and 20) while keeping the total sequencing effort constant (20 sequences per species). We simulated 100 replicates for each sampling treatment which were analyzed with BP&P to calculate the mean speciation probability between species A and B across replicates, which represents the accuracy of the method (i.e., the probability of detecting speciation when it is the true hypothesis). We also simulated a no-speciation model where sequences from species A and B were collapsed into a single lineage, and repeated the same sampling and analytical procedure to examine the performance of the method based on a plot of true positive and false positives rates (i.e., ROC plot; [114]).

The results show that under the conditions examined, more sequences per species is better than more loci at least in the range of 1-20 loci and sequences per species (Figure 2B). The ROC plots for the 5 sampling treatments at a divergence time of $0.5N_e$ show that performance is higher (i.e., area under ROC curve is larger) when sampling 20 sequences for 1 locus or 10 sequences for 2 loci, but performance gradually decreased with more loci and fewer sequences (Figure 2C). These results are congruent with the impact of sampling design on the accuracy of species-tree methods (STM) at shallow divergence times [115,116], which is an expected outcome because both STM and SDL methods share the same basic multispecies coalescent model [67,117]. However, our results are contingent upon the conditions simulated, in particular the assumptions of panmixia within species, and a constant $\theta$ across the species tree. This second assumption is a critical parameter of coalescent models, which can be estimated more accurately with a larger sample of loci [118]. Our attempt with this simulation example was to show how we can evaluate the performance of a SDL method under a variety of sampling conditions based on a power analysis, and that this same approach can be applied for comparisons across different SDL methods and more complex speciation scenarios than those that have been examined so far.

*Population and species delimitation.*–The application of coalescent-based SDM, which can delimit species at very shallow levels of divergence [66,69,92] should reduce the 'taxonomy-phylogeny' gap and help decrease the type I error of biological-species criteria that often fail to detect species, when reproductive isolation is not yet complete [91]. Thus, coalescent-based SDL methods will probably help to delimit entities, name taxonomic units, and give appropriate conservation priority to the increasing amounts of cryptic diversity being discovered in nature [91]. On the other hand, MLCM should be used with caution to avoid confusing species-level divergence with intra-specific population structure and therefore, over-splitting lineages, with serious consequences for conservation science since limited resources would be potentially wasted due to bad taxonomy [91].

A potential protocol for an informed species delimitation approach that takes into account population structure, could consist of first applying a clustering/population aggregation method to identify the smaller clusters of individuals under a population genetics criterion based on genotype or allele frequency data ('e.g., Structure 48, 58, 60). Subsequently, a SDL method can be applied to test if these clusters also represent independent evolutionary lineages based on the pattern of allele coalescence in gene genealogies (e.g. BP&P). Because initial population divergence starts with differentiation in allele frequencies and secondly, with random lineage sorting and mutation that further differentiates lineages during speciation [59], population genetics approaches are expected to detect lineages earlier than SDL approaches. For example, an empirical analysis of West African forest geckos (*Hemidactylus fasciatus*) found ~10 populations with Structure, which were considered as 'candidate' species in a subsequent BPP analysis that collapsed them into 4 species [119]. This two-stage approach would provide a consistent and standard criterion for distinguishing between population- and species-level divergence, a threshold that has been difficult to resolve with genetic parameters measuring amounts of evolutionary differentiation [59].

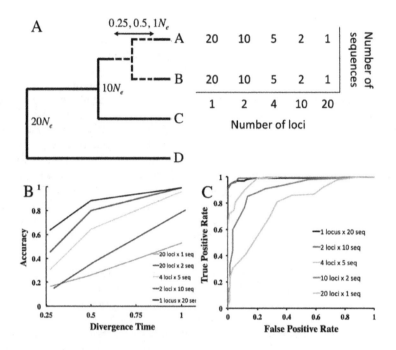

**Figure 2.** Simulation-based testing of the accuracy of BP&P to detect speciation between species A and B using five alternative sampling designs with the same sequencing effort and at three increasing divergence times (0.25, 0.5, and 1Ne) (A). Plot of accuracy and divergence time for each sampling design (B).ROC plot for each sampling design when divergence time = 0.5Ne (C).

*The next generation of SDL methods.*–We have emphasized that species delimitation should take into account the speciation processes that have shaped the patterns of trait divergence in genetic, morphological, and ecological data [89]. In a process-oriented classification of modes of speciation, we can distinguish between 'passive' modes driven by random divergence associated with the classic allopatric models, and the 'adaptive' modes of speciation. The formulation of a null hypothesis of speciation due to stochastic forces (i.e., 'passive divergence' or 'drift-only' model) should facilitate testing this mode of speciation, because rejecting this hypothesis is probably easier than demonstrating 'adaptive' speciation due to deterministic processes [120]. In nature, both speciation models appear to interact and work in concert during diversification of closely related lineages [121,122]. Adaptive speciation in turn can be subdivided into 'ecological' speciation, reproductive isolation due to disruptive natural selection operating on ecological traits [95], and speciation due to sexual selection that results in divergent mating preferences and assortative mating [123]. In theory, both kinds of selection seem to be necessary to drive speciation to completion [124], and limited empirical data supports the role of this interaction during diversification [125]. Due to this variety in speciation processes, we should expect different patterns of trait divergence, and consequently, different kinds of data would be more appropriate for species delimitation under each speciation scenario. Therefore, relying on any single kind of trait could potentially miss the detection of a speciation event, for example using exclusively morphological data will fail to recognize cryptic species. Similarly, if we use only the typical neutral genetic markers of phylogeography and population genetics, we could miss many instances of ecological speciation that takes place in contemporary time scales [126], and/or without divergence in neutral loci [127].

## 4. Conclusion

There is an ongoing genomics revolution for the study of adaptation in ecological and evolutionary non-model organisms derived from (NGS) technologies [76,128]. Decreasing sequencing costs and new protocols for discovering and screening thousands of markers scattered throughout the genome [79], is now allowing application of population genomics approaches to identifying the candidate loci underlying adaptive traits with ecological significance [87]. In fact, recent studies have found genomic regions and/or specific loci related to repeated local adaptation, population divergence, and reproductive isolation between ecotypes in different habitats or hosts [129,130]. We anticipate that these 'speciation genomics' approaches will become more common in non-model organisms and will provide a basis for species delimitation in scenarios of adaptive speciation SDL methods, complementing current SDL methods. Moreover, this plurality of criteria for species delimitation based on multiple kinds of traits is consistent with the GLC of species that views these organismal traits as evolving in different temporal order depending on how speciation has actually taken place [9,12]. In addition, it is also compatible with the more recent 'differential fitness' concept, which is based on those organismal features of one species that have negative fitness effects in other species and cannot be exchanged upon contact [17].

## Acknowledgments

AC acknowledges a postdoctoral fellowship from CONICET (Argentina). For financial support we thank thank NSF awards OISE 0530267 and AToL 0334966 to JWS, as well as BYU graduate research and graduate mentoring awards, and student research awards from the Society of Systematic Biologists and the Society for the Study of Amphibians and Reptiles, to AC. We both also received support from the BYU Dept. of Biology and the Bean Life Science Museum.

# Author details

Arley Camargo[1] and Jack Jr. Sites[2*]

*Address all correspondence to: jack_sites@byu.edu

1 Unidad de Diversidad, Sistemática y Evolución, Centro Nacional Patagónico, Consejo Nacional de Investigaciones Científicas y Técnicas, Puerto Madryn, Chubut, Argentina

2 Department of Biology and Bean Life Science Museum, Brigham Young University, Provo, Utah, USA

# References

[1] Sites JW, Jr., Marshall JC. Delimiting species: a Renaissance issue in systematic biology. Trends in Ecology and Evolution 2003;18:462-420.

[2] Wiens JJ. Species delimitation: new approaches for discovering diversity. Systematic Biology 2007;55: 875-878.

[3] Agapow PM, Bininda-Edmonds ORP, Crandall KA, Gittleman JL, Mace GM, Marshall JC, Purvis A. The impact of species concept on biodiversity studies. Quarterly Reviews in Biology 2004;79:161-179.

[4] Padial JM, Castroviejo-Fisher S, Köhler J, Vilá C, Chaparro JC, De la Riva I. Deciphering the products of evolution at the species level: the need for an integrative taxonomy. ZoologicaScripta 2009;38:431-447.

[5] Frost DE, Kluge AG. A consideration of epistemology in systematic biology, with special reference to species. Cladistics 1994;10:259-294.

[6] Mayden RL. A hierarchy of species concepts: the denoument in the saga of the species of the species problem. In: Claridge, M.F., H.A. Dawah, and M.R. Wilson (eds.) Species: The Units of Biodiversity. Chapman and Hall; 1997. p381-424.

[7] deQueiroz K. The general lineage concept of species, species criteria, and the process of speciation. In: DJ Howard, Berlocher SH (eds.) Endless Forms: Species and Speciation. Oxford University Press;1998. p.57-75.

[8] Simpson, G.G. 1961. Principles of Animal Taxonomy. Columbia Univ. Press, 247 pp.

[9] deQueiroz K. Ernst Mayr and the modern concept of species. Proceedings of the National Academy of Sciences of the USA 2005;102:6600-6607.

[10] Coyne JA, Orr HA. Speciation. Sunderland, MA: Sinauer Associates; 2004.

[11] Price TD. Speciation in Birds. Greenwood Village, CO: Roberts & Co.; 2008.

[12]  deQueiroz K. A unified concept of species and its consequences for the future of taxonomy. Proceedings of the California Academy of Sciences 2005;56:195-215.

[13]  Sites JW, Jr., Marshall JC. Operational criteria for delimiting species. Annual Review of Ecology, Evolution, and Systematics 2004;35:199-227.

[14]  Pons J, Barraclough TG, Gomez-Zurita J, Cardoso A, Duran DP, Hazell S, Kamoun S, Sumlin WD, and Vogler AP. Sequence-based species delimitation for the DNA taxonomy of undescribed insects. Systematic Biology 2006;55:595-609.

[15]  Marshall JC, Arévalo E, Benavides E, Sites JL, Sites, Jr. JW. Delimiting species: comparing methods for Mendelian loci using lizards of the *Sceloporus grammicus* complex (Phrynosomatidae). Evolution2006; 60:1050-1065.

[16]  de Queiroz K. Species concepts and species delimitation. Systematic Biology 2007;56:879-886.

[17]  Hausdorf B. Progress toward a general species concept. Evolution 2011;65:923-931.

[18]  Hey J, Waples RS, Arnold ML, Butlin RK, Harrison RG. Understanding and confronting species uncertainty in biology and conservation. Trends in Ecology and Evolution 2003;18:597-603.

[19]  Wu C-I. The genic view of the process of speciation. Journal of Evolutionary Biology 2001;14:851-856.

[20]  Wu C-I, Ting CT. Genes and speciation. Nature Reviews Genetics 2004;5:114-122.

[21]  Butlin R, The Marie Curie SPECIATION Network. What do we need to know about speciation? Trends in Ecology and Evolution 2011;27:27-39.

[22]  Nosil P. Speciation with gene flow could be common. Molecular Ecology 2008;17:2103-2106.

[23]  Nosil P, Feder JL. Genomic divergence during speciation: causes and consequences. Philosophical Transactions of the Royal Society, Series B 2012;367:332-342.

[24]  Nosil P, Crespi BJ, Sandoval CP. Host-plant adaptation drives the parallel evolution of reproductive isolation. Nature 2002;417:440-443.

[25]  Rundle HD, Nagel L, Boughman JW, Schluter D. Natural selection and parallel speciation in sympatric sticklebacks. Science 2000;287:306-308.

[26]  Soltis DE, Soltis PS. Polyploidy: recurrent formation and genome evolution. Trends in Ecology and Evolution 1999;14:348-352.

[27]  Soltis PS, Soltis DE. The role of genetic and genomic attributes in the success of polyploids. Proceedings of the National Academy of Sciences of the USA 2000;97:7051-7057.

[28]   Fontaneto D, Herniou EA, Boschetti C, Caprioli M, Melone G, Ricci C, Barraclough
       TG. Independently evolving species in asexual bdelloid rotifers. PLoS Biology
       2007;5:914-921.

[29]   Edwards SV. Is a new and general theory of molecular systematics emerging? Evolu-
       tion 2009;63:1-19.

[30]   Padial JM, Miralles A, De la Riva I, Vences M. The integrative future of taxonomy.
       Frontiers in Zoology 2010;7:16.

[31]   Highton RR. Biochemical evolution in the slimy salamanders of the *Plethodonglutino-
       sus* complex in the eastern United States. Part 1. Geographic protein variation. Illinois
       Biological Monographs 1989;57: 1-78.

[32]   Highton RR. Taxonomic treatment of genetically differentiated populations. Herpe-
       tologica 1990;46:114-121.

[33]   Puorto G, Salomão MG, Theakston RDG, Thorpe RS. Combining mitochondrial DNA
       sequences and morphological data to infer species boundaries: phylogeography of
       lanceheadpitvipers in the Brazilian Atlantic forest, and the status of *Bothrops pradoi*
       (Squamata: Serpentes: Viperidae). Journal of Evolutionary Biology 2001;14:527-538.

[34]   Good DA, Wake DB. Geographic variation and speciation in the torrent salamanders
       of the genus *Rhyacotriton* (Caudata: Rhyacotritonidae). University of California Publi-
       cations in Zoology 1992;126:1-91.

[35]   Porter AH. Testing nominal species boundaries using gene flow statistics: taxonomy
       of two hybridizing admiral butterflies (*Limenitis*: Nymphalidae). Systematic Zoology
       1990;39:131-147.

[36]   Davis JI, Nixon KC. Populations, genetic variation, and the delimitation of phyloge-
       netic species. Systematic Biology 1992;41:421-435.

[37]   Doyle J. The irrelevance of allele tree topologies for species delimitation, and a non-
       topological alternative. Systematic Botany 1995;20:574-588.

[38]   Mallet J. A species definition for the modern synthesis. Trends in Ecology and Evolu-
       tion 1995;10:294-299.

[39]   Brooks DR, McLennan DA. Species: turning a conundrum into a research program.
       Journal of Nematology 1999;31:117-133.

[40]   Brower AVZ. Delimitation of phylogenetic species with DNA sequences: a critique of
       Davis and Nixon's population aggregation analysis. Systematic Biology
       1999;48:199-213.

[41]   Wiens JJ, Penkrot TA. Species delimitation in systematics: inferring diagnostic differ-
       ences between species. Proceedings of the Royal Society of London, Series B
       2002;267:631-636.

[42] Baum DA, Shaw KL. Genealogical perspectives on the species problem. In: Hoch PC, Stephenson AG (eds.) Molecular and experimental approaches to plant biosystematics. St. Louis: Missouri Botanical Garden; 1995. p.289-303.

[43] Templeton AR, Routman E, Phillips CA. Separating population structure from history: a cladistic analysis of the geographical distribution of mitochondrial DNA haplotype in the tiger salamander, *Ambystoma tigrinum*. Genetics 1995;140:767-782.

[44] Templeton AR. Using phylogeographic analyses of gene trees to test species status and boundaries. Molecular Ecology 2001;10:779-791.

[45] Dettman JR, Jacobson DJ, Taylor JW. A multilocus genealogical approach to phylogenetic species recognition in the model eukaryote *Neurospora*. Evolution 2003;57:2703-2720.

[46] Dettman JR, Jacobson DJ, Turner E, Pringle A, Taylor JW. Reproductive isolation and phylogenetic divergence in *Neurospora*: comparing methods of species recognition in a model eukaryote. Evolution 2003;57: 2721-2741.

[47] Knowles LL, Carstens BC. Delimiting species without monophyletic gene trees. Systematic Biology 2007;56:887-895.

[48] Shaffer HB, Thomson RC. Delimiting species in recent radiations. Systematic Biology 2007;56:896-906.

[49] Raxworthy C, Ingram CM, Rabibisoa N, Pearson RG. Applications of ecological niche modeling for species delimitation: A review and empirical evaluating using day geckos (*Phelsuma*) from Madagascar. Systematic Biology 2007;56:907-923.

[50] Rissler LJ, Apodaca JJ. Adding more ecology into species delimitation: Ecological niche models and phylogeography help define cryptic species in the black salamander (*Aneides flavipunctatus*). Systematic Biology 2007;56:924-942.

[51] Leaché AD, Koo MS, Spencer CL, Papenfuss TJ, Fisher RN, McGuire JA. Quantifying ecological, morphological, and genetic variation to delimit species in the coast horned lizard species complex (*Phrynosoma*). Proceedings of the National Academy of Sciences of the USA 2009;106:12418-12423.

[52] Bond JE, Stockman AK. An integrative method for delimiting cohesion species: Finding the population-species interface in a group of California trapdoor spiders with extreme genetic divergence and geographic structuring. Systematic Biology 2008;57:628-646.

[53] Templeton AR. Species and speciation: Geography, population structure, ecology, and gene trees. In: (Howard DJ, Berlocher SH (eds.) Endless Forms: Species and Speciation. New York, NY:Oxford University Press;1989. p.32-41,

[54] Cummings MP, Neel MC, Shaw KL. A genealogical approach to quantifying lineage divergence. Evolution 2008;62:2411-2422.

[55] Weisrock DW, Rasoloarison RM, Fiorentino I, Ralison JM, Goodman SM, Kappeler PM, Yoder AD. Delimiting Species without Nuclear Monophyly in Madagascar's Mouse Lemurs. PLoS ONE 2010;5: e9883. doi:10.1371/journal.pone.0009883

[56] Puillandre N, Modica MV, Zhang Y, Sirovich L, Boisselier MC, Cruaud C, Holford M, Samadi S. Large-scale species delimitation method for hyperdiverse groups. Molecular Ecology 2012;21:2671-2691.

[57] Puillandre N, Lambert A, Brouillet S, Achaz G. ABGD, Automatic Barcode Gap Discovery for primary species delimitation. Molecular Ecology 2012;21(8):1864-1877.

[58] Hausdorf B, Hennig C. Species delimitation using dominant and codominantmultilocus markers. Systematic Biology 2010;59:491-503.

[59] Hey J, Pinho C. Population genetics and objectivity in species diagnosis. Evolution 2012;66:1413-1429.

[60] Guillot G, Renaud S, Ledevin R, Michaux J, Claude J. A unifying model for the analysis of phenotypic, genetic, and geographic data. Systematic Biology 2012; doi:10.1093/sysbio/sys038.

[61] Maddison WP. Gene trees in species trees. Systematic Biology 1997;46:523-536

[62] Rannala B, Yang Z. Bayes estimation of species divergence times and ancestral population sizes using DNA sequences from multiple loci. Genetics 2003;164:1645-1656.

[63] Degnan JH, Rosenberg NA. Gene tree discordance, phylogenetic inference and the multispecies coalescent. Trends in Ecology and Evolution 2009;24:332-340.

[64] Kubatko LS, Carstens BC, Knowles LL. STEM: species tree estimation using maximum likelihood for gene trees under coalescence. Bioinformatics 2009;25:971-973.

[65] Liu L. BEST: Bayesian estimation of species trees under the coalescent model. Bioinformatics 2008;24:2542-2543.

[66] Ence DD, Carstens BC. SpedeSTEM: a rapid and accurate method for species delimitation. Molecular Ecology Resources 2011;11:473-480.

[67] Yang Z, Rannala B. Bayesian species delimitation using multilocus sequence data. Proceedings of the National Academy of Sciences of the USA 2010;107:9264-9269.

[68] O'Meara BC. New heursitic methods for joint species delimitation and species tree inference. Systematic Biology 2010;59:59-73.

[69] Camargo A, Morando M, Avila LJ, Sites, Jr. JW. Species delimitation with ABC and other coalescent-based methods: a test of accuracy with simulations and an empirical example with lizards of the *Liolaemus darwinii* complex (Squamata: Liolaemidae). Evolution 2012;doi:66:2834-2849

[70] Zapata F, Jiménez I. Species delimitation: inferring gaps in morphology across geography. Systematic Biology 2012;61:179-194.

[71] Thomson RC, Wang IJ, Johnson JR. Genome-enabled development of DNA markers for ecology, evolution and conservation. Molecular Ecology 2010;19:2184,àí2195.

[72] Townsend TM, Alegre RE, Kelley ST, Wiens JJ, Reeder TW. Rapid development of multiple nuclear loci for phylogenetic analysis using genomic resources: an example from squamate reptiles. Molecular Phylogenetics and Evolution 2008;47:129-142.

[73] Portik DM, Wood Jr., PL, Grismer JL, Stanley EL, Jackman TR. Identification of 104 rapidly-evolving nuclear protein-coding markers for amplification across scaled reptiles using genomic resources. Conservation Genetics Resources 2012;4:1-10.

[74] Holsinger KE. Next generation population genetics and phylogeography. Molecular Ecology 2010;19:2361-2363.

[75] Rokas A, Abbot P. Harnessing genomics for evolutionary insights. Trends in Ecology and Evolution 2009;24:192-200.

[76] Tautz D, Ellegren H, Weigel D. Next generation molecular ecology. Trends in Ecology and Evolution 2010;19(Suppl. 1):1-3.

[77] Alföldi J, Di Palma F, Grabherr M, Williams C, Kong L, Mauceli E, Russell P, Lowe CB, Glor RE, Jaffe JD, Ray DA, Boissinot S, Shedlock AM, Botka C, Castoe TA, Colbourne JK, Fujita MK, Godinez-Moreno R, ten Hallers BF, Haussler D, Heger A, Heiman, D Janes DE, Johnson J, de Jong PJ, Koriabine MY, Lara M, Novick PA, Organ CL, Peach SE, Poe S, Pollock DD, de Queiroz K, Sanger T, Searle S, Smith JD, Smith Z, Swofford R, Turner-Maier J, Wade J, Young S, Zadissa A, Edwards SV, Glenn TC, Schneider CJ, Losos JB, Lander ES, Breen M, Ponting CP, Lindblad-Toh K. The genome of the green anole lizard and a comparative analysis with birds and mammals. Nature 2011;477:587-591.

[78] Faircloth BC, McCormack JE, Crawford NG, Harvey MG, Brumfield RT, Glenn TC. Ultraconserved elements anchor thousands of genetic markers spanning multiple evolutionary timescales. Systematic Biology 2012;61:717-726.

[79] Lemmon AR, Emme S, Lemmon EM. Anchored hybrid enrichment for massively high-throughput phylogenomics. Systematic Biology 2012;61:727-744.

[80] Castoe TA, Poole AW, Gu W, Jason de Koning AP, Daza JM, Smith EN, Pollock DD. Rapid identification of thousands of copperhead snake (*Agkistrodon contortrix*) microsatellite loci from modest amounts of 454 shotgun genome sequence. Molecular Ecology Resources 2010;10:341-347.

[81] Baird NA, Etter PD, Atwood TS, Currey MC, Shiver AL, Lewis ZA, Selker EU, Cresko WA, Johnson EA. Rapid SNP discovery and genetic mapping using sequenced RAD markers. PLoS ONE 2008;3(10):e3376./journal.pone.0003376.

[82] Emerson KJ, Merz CR, Catchen JM, Hohenlohe PA, Cresko WA, Bradshaw WE, Holzapfel CM. Resolving post-glacial phylogeography using high-throughput sequencing. Proceedings of the National Academy of Sciences of the USA 2010;107:16196-16200.

[83]  Metzker ML. Sequencing technologies the next generation. Nature Reviews Genetics 2010;11:31-46.

[84]  Schneider GF, Dekker C. DNA sequencing with nanopores. Nature Biotechnology 2012;30: 326-328.

[85]  Pennisi E. Search for Pore-fection. Science 2012;336:534-537.

[86]  Brito PH, Edwards SV. Multilocus phylogeography and phylogenetics using sequence-based markers. Genetica 2009;135:439-455.

[87]  Nosil P, Funk DJ, Ortiz-Barrientos D. Divergent selection and heterogeneous genomic divergence. Molecular Ecology 2009;18:385-402.

[88]  Carstens BC, Dewey TA. Species delimitation using a combined coalescent and information theoretic approach: An example from North American *Myotis* bats. Systematic Biology 2010;59:400-414.

[89]  Camargo A, Sinervo B, Sites, Jr., JW. Lizards as model organisms for linking phylogeographic and speciation studies. Molecular Ecology 2010;19:3250-3270.

[90]  Fujita MK, Leaché AD, Burbrink FT, McGuire JA, Moritz C. Coalescent-based species delimitation in an integrative taxonomy. Trends in Ecology and Evolution 2012;27:480-488.

[91]  Bernardo J. A critical appraisal of the meaning and diagnosability of cryptic evolutionary diversity, and its implications for conservation in the face of climate change. In: Hodkinson TR, Jones MB, Waldren S, Parnell JAN (eds.) Climate Change, Ecology and Systematics. Cambridge University Press:The Systematics Association; 2011.p380-438.

[92]  Zhang C, Zhang D-X, Zhu T, Yang Z. Evaluation of a Bayesian coalescent method of species delimitation. Systematic Biology 2011;60:747-761.

[93]  Kawecki TJ, Ebert D. Conceptual issues in local adaptation. Ecology Letters 2004;7:1225-1241.

[94]  Schluter D. Ecology and the origin of species. Trends in Ecology and Evolution 2001;16:372-380

[95]  Schluter D. Evidence for ecological speciation and its alternative. Science 2009;323:737-741.

[96]  Wiens JJ, Ackerly DD, Allen AP, Anacker BL, Buckley LB, Cornell HV, Damschen EI, Davies TJ, Grytnes J-A, Harrison SP, Hawkins BA, Holt RD, McCain CM, Stephens PR. Niche conservatism as an emerging principle in ecology and conservation biology. Ecology Letters 2010;13:1310-1324.

[97]  Avise JC. Molecular markers, natural history, and evolution. Second edition. Sinauer Associates; 2004.

[98]   Zink RM, Barrowclough GF. Mitochondrial DNA under siege in avian phylogeography. Molecular Ecology 2008;17:2107-2121.

[99]   Petit RJ, Excoffier L. Gene flow and species delimitation. Trends in Ecology and Evolution 2009;24:386-393.

[100]  Galter N, Nabholz B, Glémin S, Hurst GDG. Mitochondrial DNA as a marker of molecular diversity. Molecular Ecology 2009;18:4541-4550.

[101]  Funk DJ, Omland KE. Species-level paraphyly and polyphyly: frequency, causes, and consequences, with insights from animal mitochondrial DNA. Annual Review of Ecology, Evolution, and Systematics 2003;34:397-423.

[102]  Morando M, Avila LJ, Sites, Jr. JW. Sampling strategies for delimiting species: genes, individuals and populations in the *Liolaemus elongatus-kriegi* complex (Squamata: Liolaeminidae) in Andean Patagonian South America. Systematic Biology 2003.52:159-185.

[103]  Funk WC, Caminer M, Ron SR. High levels of cryptic species diversity uncovered in Amazonian frogs. Proceedings of the Royal Society, Series B, 2012;279:1806-1814.

[104]  Niemiller ML, Near TJ, Fitzpatrick BM. Delimiting species using multilocus data: diagnosing cryptic diversity in the southern cavefish, *Typhlichthys subterraneus* (Teleostei: Amblyopsidae). Evolution 2012;66:846-866.

[105]  Agrawal AF, Feder JL, Nosil P. Ecological divergence and the origins of intrinsic postmating isolation with gene flow. International Journal of Ecology, 2011;doi: 10.1155/2011/435357

[106]  Case LD, Ambrosius WT. Power and sample size. Methods in Molecular Biology 2007;404:377-408.

[107]  Rundle HD, Nosil P. Ecological speciation. Ecology Letters 2005;8:336-352.

[108]  Bauer AM, Parham JF, Brown RM, Stuart BL, Grismer L, Papenfuss TJ, Böhme W, Savage JM, Carranza S, Grismer JL, Wagner P, Schmitz A, Ananjeva NB, Inger RF. Availability of new Bayesian-delimited gecko names and the importance of character-based species descriptions. Proceedings of the Royal Society Series B 2011;278:490-492.

[109]  Fujita MK, Leaché AD. A coalescent perspective on delimiting and naming species: a reply to Bauer et al. Proceedings of the Royal Society Series B 2011;278:493-495.

[110]  Schlick-Steiner BC, Steiner FM, Seifert B, Stauffer C, Christian E, Crozier RH. Integrative taxonomy: a multisource approach to exploring biodiversity. Annual Review of Entomology 2010;55:421-438.

[111]  Yeates DK, Seago S, Nelson L, Cameron SL, Joseph L, Trueman JWH. Integrative taxonomy, or iterative taxonomy? Systematic Entomology 2011;36:209-217.

[112] Hudson RR. Generating samples under a Wright-Fisher neutral model. Bioinformatics 2002;18:337-338.

[113] Rambaut A, Grassly NC. Seq-Gen: An application for the Monte Carlo simulation of DNA sequence evolution along phylogenetic trees. Computer Applications in the Biosciences 1997;13:235-238.

[114] Fielding AH, Bell JF. A review of methods for the assessment of prediction errors in conservation presence/absence models. Environmental Conservation 1997;24:38-49.

[115] McCormack JE, Huang H, Knowles LL. Maximum likelihood estimates of species trees: how accuracy of phylogenetic inference depends upon the divergence history and sampling design. Systematic Biology 2009;58:501-508.

[116] Knowles LL. Sampling strategies for species tree estiation. In: Knowles LL, Kubatko LS (eds.) Species Trees: Practical and Theoretical Aspects. Wiley-Blackwell;2010. p163-174.

[117] Knowles LL, Kubatko LS. Estimating species trees: An introduction to concepts and models. In: Knowles LL, Kubatko LS (eds.) Species Trees: Practical and Theoretical Aspects. Wiley-Blackwell;2010. p.1-14.

[118] Felsenstein J. Accuracy of coalescent likelihood estimates: do we need more sites, more sequences, or more loci? Molecular Biology ad Evolution 2006;23:691-700.

[119] Leaché AD, Fujita M. Bayesian species delimitation in West African forest geckos (*Hemidactylus fasciatus*). Proceedings of the Royal Society Series B: Biological Sciences 2010;277:3071-3077.

[120] Futuyma DJ. Progress on the origin of species. PLoS Biology 2005;3:e62.

[121] Thorpe RS, Surget-Groba Y, Johansson H. Genetic test for ecological and allopatric speciation in anoles on an island archipelago. PLoS Genetics 2010;6:e1000929.

[122] Pereira RJ, Wake DB. Genetic leakage after adaptive divergence and nonadaptive divergence in the *Ensatina eschscholtzii* ring species. Evolution 2011;63:2288-2301.

[123] Weissing FJ, Edelaar P, van Doorn GS. Adaptive speciation theory: a conceptual review. Behavioral Ecology and Sociobiology 2011;65:461-480.

[124] Van Doorn GS, Edelaar P, Weissing FJ. On the origin of species by natural and sexual selection. Science 2009;326:1704-1707.

[125] Maan ME, Seehausen O. Ecology, sexual selection, and speciation. Ecology Letters 2011,14:591-602.

[126] Hendry AP, Nosil P, Rieseberg LH. The speed of ecological speciation. Functional Ecology 2007;21:455-464.

[127] Thibert-Plante X, Hendry AP. When can ecological speciation be detected with neutral loci? Molecular Ecology 2010;19:2301-2314.

[128] Stapley J, Reger J, Feulner PGD, Smadja C, Galindo J, Ekblom R, Bennison C, Ball AD, Beckerman AP, Slate J. Adaptation genomics: the next generation. Trends in Ecology and Evolution 2010;25:705-712.

[129] Roesti M, Hendry AP, Salzburger W, Berner D. Genome divergence during evolutionary diversification as revealed in replicate lake-stream stickleback population pairs. Molecular Ecology 2012;21:2852-2862.

[130] Smadja CM, Canbäck B, Vitalis R, Gautier M, Ferrari J, Zhou J-J, Butlin RK. Large-scale candidate gene scan reveals the role of chemoreceptor genes in host plant specialization and speciation in the pea aphid. Evolution, 2012;doi: 66:2723-2738

# Conspecific Recognition Systems and the Rehabilitation of the Biological Species Concept in Ornithology

V. S. Friedmann

Additional information is available at the end of the chapter

## 1. Introduction

The species problem, as it has been discussed by zoologists for over a hundred years, is formed by two aspects. On the one hand, there is an objective controversy, a kind of a gap, between species as they actually exist in nature and the 'classifier's species' identified by a group-expert taxonomist; on the other hand, there is everyone's natural aspiration to identify the species in accordance with nature, to purposefully narrow this gap, and ideally, to close it completely. [15, 9, 43]. One of the creators of the modern evolutionary synthesis, N. V. Timofeeff-Ressovsky had good grounds for considering the notion of discreteness of life to be the crucial part of biological thinking, i.e. the idea of particulate essence of biodiversity. In other words, life-forms don't flow into one another smoothly and graduattly, but rather are divided into units, partly totally discreet, partly connected with transitional forms — the ones that haven't separated completely.

Being included into such a group as species proves to be important for the specimens it comprises. It is even more important than the individual survival and reproduction, primarily because an individual can't improve its adjustment alone, but only in association with the others within a species-specific pattern of relations (to the territory, as well as social, habitat and other relations) in the superindividual population system [55, 57-58, 10]. To this end, specimens competing with each other for a territory and a mate invest their energy into the long-term reproduction of the specific pattern of spacial-ethological structure of the population, and simultaneously try to occupy the best position in this structure of relations within one of the population groups. It can be a dominant status in the community structured by the hierarchy (based on the agonistic dominance), taking of the best territories for territorial species (where their 'quality' is defined by the position in the anisotropic space of the group and by an earlier time of acquisition), attraction of a greater number of 'better quality' partners, and so on [58, 11-12].

Anyway, achieving superiority over the others (distribution in space or in mosaic habitats, and other homologous processes) through the competitive social communication within the species-specific relations allows the individual to exploit those relations more effectively as a resource of even more superiority and/or maximising the individual's final adjustment.

The latter can be understood differently: as an advantageous (in comparison with other specimens that have chosen alternative strategies) replication of the individual's progenies or genes. In any case, being bound by specific relations to the others within the superindividual population system and a more effective maintenance of these relations, both are crucial for the individual's successful struggle for survival [10-11]. And consequently, for the distribution of life strategies carried by the successful individuals. At present this 'morphological' understanding of a population is just beginning to form; the view of population as an integral system characterised by a pattern of relations that are consistently reproducible in the sequence of generations in the stochastics of collisions and interactions of specimens [10].

Accordingly, separate individuals with their activeness, adjustment and destiny here are only a means to identify the most effective set of life strategies and configuration pattern of spatial-ethological structure of the system in the direction of its most sustainable / long-term reproduction (given the local features of external 'perturbations' of the ecological environment and the demographic 'tension' within the system).

We are mentioning this understanding of population for two reasons. First, it surprisingly consorts with Vavilov's understanding of species (an essential part of the biological species concept) as a system of populations that interact in a particular area. And inter alia regulate redistribution of individuals in the mosaic habitats to maintain integrity in the system and maximize its reproductive output (despite the demographic randomness and environmental instability).

Secondly, it follows that the key moment of an elementary evolutionary phenomenon - the isolated process of selective changes in biological form, which make up everything, including the formation of species, is not in itself the victory of the individual over another in the struggle for existence. This inclusion (or exclusion, fragile, unstable inclusion, acception on the "worst position") in the form specific to spatial-ethological structure of populations. Only this makes it possible to maximize the fitness of individuals in competition with each other, but only within the processes of social communication, structured in accordance with the specific "rules of the game", where "errors" or "fraud" exclude the specimens from the system [11, 19].

This means that the 'friend-foe' recognition within the corresponding interactions, especially when the individual is included in the system of relations inside the species population, is important for maintaining the integrity in the species ("Vavilov's species", which is the most consistent with the "species in nature", if we adhere to the population approach versus the typological one). Further on we will discuss what means of intersystem regulation are used to perform this task; while here we shall note that this very same 'friend-foe' recognition will naturally maintain the isolation of the species where the nature "puts it to a test". For example, in the areas of secondary contact with hybridization, where the isolating mechanisms are not yet developed [49] and/or existing obstacles for cross-breeding are insufficient, but are not perfected by selection [53], nevertheless the isolation of forms remains without a 'blur' [1, 7, 24, 50].

## 2. Population as a system and «the species problem»

This results in our hypothesis that the integrity of population subunits within the species and its isolation from close forms outside are supported by the same in-system regulation mechanism related to the directional selection of the specimens between groups by the potentialities of development of certain behaviors, or by the selection of one of the alternative strategies in a particular environment. As the author is an ornithologist, the examples and argumentation are taken mainly from the ornithological material, but in such a way that they could be used as the basis for general theoretical propositions. The more so because a significant difference between the taxon of birds and other macrotaxa is that the description of its structure of biodiversity (in the interpretation of A.A.Nazarenko [42]) is overall complete [18, 40]. Already, the changes in the numbers of bird species are more due to the change of the prevailing concepts of species and/or concepts of speciation in the scientific community, rather than to the discovery of new forms [1, 32, 65].

Directional migration of non-resident specimens between groups, which connect the network of settlements scattered or isolated from each other on the "islands" of habitats, have been described recently and are just beginning to be studied [10-12]. It becomes clear that they are capable of directional redistribution of individuals from "reserve"[1] to restore species population that has declined locally due to a depression of internal nature, extermination or a natural disaster. These mechanisms work quite accurately both in respect of the population size, correlated with the capacity of "vacant" habitats, and of the spatial-ethological structure of newly created settlements, correlated with species-specific "ideal model" of the latter. The most important work here was done by N.A.Shchipanov [57-58] on the functional structure of the populations of small mammals and Y.P.Altukhov [2] on the intra- and interpopulation regulation of the optimal level of gene frequencies through differential dispersion of different genotypes.

On the other hand, it has been shown (mainly for birds and mammals, but also for juvenile freshwater fish — [48]) that the flow of the dispersion of individuals within a population, whether the dispersion of the young or the resettlement of the adult

---

1 the composition of the latter is extremely heterogeneous (e.g., [57-58]). On the one hand, these are all kinds of "losers" in the competition, who have lost their territory and/or a partner because of the inefficiency of behavior, the wrong choice of life strategy, etc., or non-territorial vagrant individuals that had never had either. On the other hand — they are the deported from areas where the species habitat had been destroyed by a local disaster, such as fire, floods, deforestation, etc. On the third — they are, on the contrary, the "best" specimens, the winners of the competition in their own group, relocating to another, "better" one to maximize their reproductive success (on these directed migrations see [11]).

a.   are directed and not random;

b.   are predetermined not so much by "guide lines" of environmental or landscape charac-
     ter, or gradients of habitat change within the area, as by the anisotropy in the position
     of different groups in the "network" of the species' settlements in the corresponding
     area, the presence of the "best" (stable / large) and "worst" (unstable, more temporary
     and small) groups, between which the streams of migrants are moving. From the first to
     the second evict mostly unsuccessful individuals displaced from the "best" neighbor-
     hoods, characterized by higher density and more intense competition for territory and /
     or a partner; in the opposite direction — their antagonists that increase their reproduc-
     tive success by resettlement [20, 36, 12];

c.   the sustainable reproduction of populations of different levels, from local communities
     (demes) to geographic populations and, most likely, subspecies too is more dependent
     on constant influx of immigrants from the outside than on the efficiency of the resi-
     dents' breeding [11, 59]. In any case, it has been shown for different species that the re-
     duction of the flow while keeping the average (or even higher) level of breeding success
     of residents leads to a directed decline in population size and invariably leads to extinc-
     tion sooner or later, unless the flow is restored. Therefore, species are so vulnerable to
     fragmentation of the natural landscape, the isolation of settlements in the emerging "is-
     lands" [59, 57]. Particularly vulnerable are those with underdeveloped "restoration" sub-
     system of the population system, a "reserve" of non-resident specimens is small and
     inactive, and the residents adhere to the K-strategy, are firmly tied to the social part-
     ners, territories and/or habitats [57, 11].

In other words, connectivity of territorial divisions "within" the system, its ability to "manage" the movements and interactions of individuals in a wide range is more important for
population viability than breeding efficiency in local groups. Therefore, factors, processes
and mechanisms that improve intrasystemic regulation, increasing the stability and direction of migration of individuals between groups, "in terms of" natural selection have an advantage over local adaptations;

d.   the settlement of individuals within an area by no means eliminates morphobiological
     differences between populations or lead them to a certain common denominator. On
     the contrary, it *enhances them*, because happens asymmetrically and includes a "sorting"
     of specimens by the potentialities of different behavioral / life strategies that exist with-
     in the area [10, 12].

So it is quite a common situation when a more than significant intraspecific differentiation
of forms (based only on their degree of morpho-ecological divergence they could be considered a separate species) is achieved not through isolation, but while maintaining sustainable
exchange of specimens with the rest of populations of the species. An example is a resident
endemic form of the Albion Mountains crossbill in southern Idaho, which coevolved together with the lodgepole pine *Pinus contorta lafifolia*. Analysis of amplified fragment length
polymorphism (method AFLP) showed a divergence of about 5% of loci, despite the existence of a stable gene exchange with other forms of this species. A similar pattern is shown

for nine more forms in a group of common crossbills *Loxia curvirostra* in North America, complex for a taxonomer as they have significantly diverged morphologically, and in the vocalization type [46]. Examples of this are plentiful ([14] and others).

It is now clear that the morphological, ecological, and genetic divergence of intraspecific forms, including reaching the level of "good species" does not require isolation and disjunction of the range, but is achievable also when a binding flow of genes is present. The movements of individuals controlling it are not homogenizing, but differentiating for the subunits of a population system [10, 12].

Conclusions **a—d** bring us back to the subject of the work - the system of 'friend-foe' recognitionas a key to the renaissance of the biological species concept. Based on the systemic understanding of the population and the "morphological approach" to the population structure and dynamics, we can rectify the main disadvantage of using the biological species concept at the peak of its popularity in zoology in 1960-1970's, associated with the inconsistency in the conduct of its main principles — "species differ not by differences [in character] but by isolation [of population systems of different forms from each other]". See [21, 34, 52].

The more so because among biologists, at least among evolutionists, there aren't any discrepancies in relation to "what is population is". In evolutionary biology, population is the only subject of evolution, and its specimens are real carriers both of "standard development" programs and evolutionary innovations" [42: *181*]. considering the current dissonance of ideas about what species actually is (even in theory), this uniformity in respect of population approach may be a common basis for the definition of "species in nature", and a directed movement towards them from the "taxonomer's species ".

Unfortunately, routinization in the practical application of the concept has led to the fact that the required 'isolation' is understood solely as a reproductional discreteness of form — sterility or other inferiority of hybrids in respect of fertility/survival, bringing to life the "selection against hybrids". If at the beginning of hybridization they do not exist, selection, aimed at preserving the locally co-adapted gene complexes, "perfects" the isolating mechanisms, which promotes the establishment of the required discreteness [34, 44, 52]. We shall see that the "conservation" and "perfection" must be taken into question.


## 3. Completion of speciation "natural experiments"

First, they are the different scale disjunction of areas, the number associated with depression in population, man-made, climatic and other environmental changes, ranging from short to encompassing historical time scales. Second, they are the hybridization in the zones of secondary contact with similar forms, long-term and regular enough, that there arise fully hybridogeneous populations or at least containing a significant fraction of hybrids. The change in population of a species in response to the former shows integrity and connectivity of territorial elements of its population system, the latter — its isolation from the population systems of other species with which this one hybridizes. In fact, "isolation" in Ernst Mayr's understanding is one aspect of the modern population concept that assumes a well-defined system[2] characterized by a specific pattern of structure and the ability to consistently repro-duce the species-specific morphotype and a pattern of spatial-ethological structure of the species population despite various "tensions" inside and "disturbances" from the outside. Even as powerful as a steady, long-term gene flow across the hybrid zone.

Therefore, the "morphological approach" is as productive in relation to population system of the species, as to the species morphotype. On the basis of it, the isolation of the species level is understood as the ability of the population system to stably reproduce the specific type (which includes not only the biology, but any specific pattern of ecological, behavioral traits, etc.), and the specificity of the population structure in spite of the two main types of "distur-bances" mentioned above. Then the results of the two "natural experiments" provide an ob-jective and complete test for the evaluation of "speciesness" of this form.

It follows that the "nature-friendly" species concept, aimed at defining the "species in na-ture", must follow the events and situations in which individuals of the two different forms (for which a taxonomist doubts whether they have reached the species level of divergence or not) manifest their own specific isolation separating themselves from the "foes". Or, con-versely, "refute" it, merging with them in a single group or forming a series of transitional forms. This, of course, is not a rare event and individual "mistakes", but the bulk of popula-tion processes purposefully unfolding before the researcher.

Fortunately, the natural processes of areas dynamics — environmental, demographic, etc. — always create both types of "test cases". And they need to be used to test and improve the species concepts we use, as well as to refute the hypothesis about the status of these forms (for any taxonomer's propositions are hypotheses, [47]).

The biological species concept has not been able to explain some of the results of both "natu-ral experiments". Among them:

a.  persistance of the isolation of forms in the presence of a long-standing and stable gene flow between them, so that gene pools are long and well blended. Given that the im-provement of the isolation mechanisms by selection is not observed either, hybridiza-tion in contact zones takes place unobstructed, without any assortative mating at all, or with an insufficient and a constant level of assortativeness [43, 50, 51]. These are classic examples of zones of secondary contact with hybridization: gray and black crows in Eu-rope and Asia; three forms of northern flickers in North America (*Colaptes (auratus?) auratus, cafer* and *chrysoides*) and others [9]. Actually, such cases prevail, while the "rein-forcement" (discrimination of forms in the area of contact) and "character displacement" postulated by the theory have to sought after almost with a magnifying glass [50, 53, 62], and usually the alternate explanation turns out to be more convincing.

2 instead of the previous understanding of it as a sample. This change in understanding well reflects a shift from the first meaning of the word population — "number, quantity" to the second, "people". For biological species, this means the transition from analysis of a sample of individuals living in a certain area, and characterized by the average statisti-cal values of different characteristics to the structure of relations — social, territorial, habitat and others. Relations are implemented in a particular area ("characteristic area of detection" of the group), the size of which is specifically associ-ated with both the construction of relations in a population system, and the level and nature of the unstability of the environment in which the "construction" is reproduced in the area.

**b.** the indeterminacy of the species/subspecies status of well-differentiated forms from allopatric isolates of varying age created by different-scale disjunctions in the area. The classic example is the Pyrenean (2) and the Far East (6) subspecies of blue magpie *Cyanopica cyanus* (based on the phylogenetic species concept the former is distinguished as a separate species *C. cooki*) and other cases of the same kind.

Alas, the biological species concept in the form that prevailed in the 1950−1970s rather than trying to resolve these "difficult cases" was distancing itself from them, attempting to put them down to some other cases or to ignore them completely [8, 54]. Both attitudes were especially common among the supporters of the biological species concept, where by the 1980s it had finally become apparent that the isolating mechanisms are not "perfected" by selection, at least in the specially studied secondary contact hybridization zones of "well-differentiated subspecies" and other forms of birds [44] and other vertebrates [37].

A nonoperative concept that claims to be universal, in cases which, according to its own postulates, are included in its "domain" and "range of values", significantly contributed to the fact that it was shelved and superseded by competitors. Among them, in the West prevailed the phylogenetic and evolutionary concepts of species (with some others that emphasize different special cases of these two, [25, 47]), in our country it was the morphological (typological) concept [54].

On the other hand, it has discredited the very idea of the possibility of developing a universal species concept; an idea arose to replace it with a kind of **convention of species**, worked out for reasons of comprehension and usability, and then it could be possible, instead of "fruitless theoretical debates", to switch to an in-depth analysis of the interesting special cases [42].

We must admit that it is bad, bad as any routinization of a theory. The situation must be improved. What can be the essence of the improvement? First of all, it must be understood that the contradiction between the universality of the species as a category and the characteristic aspects of the notion of species in the different groups of biota (caused by the unique tradition of identification of species in each of them) is only seeming.

Once we assume after E. Mayr that "species differ not by character, but by isolation", everything falls into place. The basis of allocation of all the groups in the biota is the same *principle* (the isolation of population systems, forming a kind of "natural body" of the species, from similar systems, which form other forms of the same rank). Indeed, this principle in different groups of biota is implemented differently by taxonomers, based on different features, evidentially relating to the isolation in this particular group; the taxonomers attract amateur and personal knowledge, but to the extent that they follow just it, they identify universals, comparable on the same basis, as "species".

Another thing is that in many groups because of their poor state of exploration, high incompleteness even of the usual inventorization of forms, these principles still can not be applied, population studies of related species just haven't started or are extremely difficult. So, for example, in the studies of macrogroups of biota the cumulative curve of the number of described genera, families, and other taxa in the late twentieth century, leveled off, while the curve of the number of species isn't even close, *except for the birds* [18, 40]. By the way, it follows that it is the ornithological material that is the most relevant for the discussion of the species concept and, more broadly, the "species problem" as such.

"Species in nature" are "Vavilov's species", population systems, effectively supporting in its specific area the integrity "inside" and the isolation from "outside". If the biological concept had closely followed his principles, it would have allocated species according to this very persistence, that is, naturally. The confronting pile of "morphological" concepts that distinguish species according to character, whether the character of the organisms themselves (species morphotype separated by the hiatus from similar morphotypes of other species) or synapomorphies that mark individual "branches" in the trees (built by phylogenetic algorithms based on various data sources), is defective if only because it does not allow either to draw or to apply such a universal principle.

## 4. Ways to overcome the internal contradictions of the biological concept

The biological species concept grew out of the polytypic approach to the description of species by German ornithologists Hartert and Stresemann [21]. According to it, the variation of individuals within the area was considered important, the selection of well-differentiated geographic populations with their taxonomic designation as "subspecies", etc. Hence different species are definable by one specimen, but subspecies — only in a series and, as morphological characters were "involved" in the analysis of intraspecific variation and/or isolation of intraspecific forms, the main criterion of the species yurned out to be the reproductive isolation — as a "circle of races or forms" — from the others, with which it can be confused.

All isolation means **heterophobia** (term by G.A. Zavarzin [64]). It is the ability in some way to recognize 'friend' from 'foe', and reproduce only with 'friends', without "making mistakes". Or, if an error is made anyway, the ability to effectively correct it — not allowing the products of hybridization, specimens with hybrid phenotype, if they turn out viable and fertile, to become agents of implementing this error further. Which also requires recognition only at other times of the life cycle — not at the meeting of sexual partners and early courtship, but at the inclusion/non-inclusion of the individuals in the aggregation — population groups with their specific territorial / social structure (mating spots, colonies, settlements and other units). In general, with birds and mammals, solitary individuals settling outside groups, implement only the first percents of reproductive potential, with the increased risk of death [10-11].

Hence the main features of the biological species concept, as they have come to us in Russian translation by E. Mayr [34]:

- **the population-based approach instead of the typological**. The answer to the question whether the forms have reached the species level of divergence or not requires an analysis of the interaction of populations in nature (for example, the dynamics of the different phenotypes in the zone of secondary contact and hybridization), but not comparisons of hiatuses by the characters of museum specimens. To some extent, this is exaggeration, but without significant loss of meaning: instead of "morphology" any other indications can be used, for example, the genetic distance.

As whith the analysis of any population processes and the following elementary evolutionary phenomena, we can't dispense with an analysis of selective pressures and selection processes. The latter, in general, can be both "for" and "against" the deepening of the launched hybridization.

- **species are not determined by differences, but by isolation**. The "differences" were understood as a sustained hiatus of character, "isolation" — as inability to cross with a similar form in the zone of secondary contact. "Species is a population with a closed genome" (*Table 1*), "protected" against the penetration of foreign genes and phenes through hybridization because of the ability of individuals in the area of secondary contact to "avoid mistakes" too often, so as not to produce hybrids and backcrosses. Or, if the "error" do occur, the products of hybridization show a significant reduction in the viability and/or fertility, even to complete sterility.

Therefore, an important part in the concept structure is designated to "isolating mechanisms" that ensure the latter. They are either formed during the period of separate living of forms, and in a zone of secondary contact with hybridization they only appear to stop it (if the latter does not work immediately, they are amplified by the selection to the desired level, [7, 53]). Or they don't exist until the contact and are created by selection (primarily sexual, changing the song, courtship demonstrations, etc. signs directing pair formation), "out of nothing right in the hybrid zone (*Table 1A*). Therefore, such importance is attributed to the idea of "isolating mechanisms to improve the selection", and reinforcement + character displacement as "traces" of effective "perfection".

Above, I argued the merits of the biological species concept over its opponents so hard, it's time to ask — why then, in spite of them, since 1980-90's has it been in oblivion, and the phylogenetic concept has triumphed? Why have population studies of isolation of forms in nature been substituted by the "technique" of phyloclad discernment? [1].

I think this is a direct consequence of the fact that the biological species concept, as it became popular among the masses of field naturalists in the 1960-70's underwent considerable simplification, routinization. Routinization means degradation of concept to the most obvious illustration of it, comes from the principle of economy in the form in which it can not be endorsed — the desire to save mental effort [30].

Alas, for the naturalist it is always "more fun to watch than to think", so that such bias is our professional risk. Routinisation allows fielders to explore more and more situations without wasting any time to discuss the most difficult — the domain and range of the concept. And the understanding of the biological species concept, accepted at the peak of its popularity in the 1950s and 70s, helped that a lot.

| A. BCOS, TRADITIONAL UNDERSTANDING | B. PCOS | C. BCOS, OUR UNDERSTANDING |
|---|---|---|
| **1. Theoretical content (ideology)** | | |
| - **The population approach instead of the typological.** - Species is a population system, separate from the other population systems to the same extent to which it is integral and integrated inside | - Typological approach while ignoring the population in the form of "dendrogram thinking" in which any taxa are not perceived as real natural objects (groups of populations), but as a minimum units of phylogeny. | - The population approach instead of the typological. Species is a population system, separate from the other population systems to the same extent to which it is integral and integrated inside |
| - **Species are determined not by differences but by isolation.** Species actually exist in nature and are distinguished by isolation; its absence leads to considering a form to be subspecies, however much it has diverged from the original. | - **This option is generally not included in the ontological description of the species.** Species is a population, characterized by a unique combination of features and distinguished from other such forms by arbitrarily insignificant features and their combination. | - **Species are not determined by differences and isolation.** Species actually exist in nature and are distinguished by isolation; its absence leads to considering a form to be subspecies, however much it has diverged from the original. |
| **2. Suggested ways of implementation (ontology)** | | |
| - Putting isolation down to inability of crossing; species is a population with a closed genome. - Inability of crossing arises due to low frequency of "recognition errors" during the choice of sexual partners and mating with them. Its reduction is usually associated with the failure of such matings, stopped by selection due to full or partial sterility (often in combination with reduced viability) of hybrids, but can occur at full fertility / viability of the latter. Then the selection improves the isolating mechanisms producing pre-mating obstacles to crossing de novo for the sake of saving the co-adapted gene complexes from the damaging effects of hybridization. | The current dominance of cladistic methods leads to a revival of the typology, as apomorphy of each clade corresponds to a feature used in the logical divisions within the construction of classifications of descending series. | In the area of secondary contact and hybridization — isolation of the species level specifies is dictated by the effective 'friend-foe' recognition at the moment when individuals of different phenotypes are included in the spatial-ethological structure of populations of some form; For allopatric isolates the isolation of species level is set by the comparison of the degree of divergence of the DNA-genealogies between the isolates populations with the degree of divergence of similar pedigrees of close forms with the same type of area, but having retained the intermediate populations, which have disappeared between the isolates in question. |

| *A*. BCOS, TRADITIONAL UNDERSTANDING | *B*. PCOS | *C*. BCOS, OUR UNDERSTANDING |
|---|---|---|
| **3. Application to real populations** | | |
| - The degree of similarity and difference between the populations is not critical.<br>- cases of reinforcement and character displacement are essential for the proof of the status of the species<br>- Species can be mono-, para- and polyphyletic, i.e. be comprised by many non-identical populations (polytypic species). Species is polymorphous and politypic to the extent that is necessary for survival in fluctuating environments, it is always ready to gemmate a population, "groping" for new niches.<br>- taxonomic "halftone" are allowed: subspecies, semispecies etc.<br>- "Evolutionary event" and speciation are separated in time: "speciation" is the moment of acquisition by population of reproductive isolation mechanisms<br>- Secondary contacts and hybrid zones between them act as a powerful tool for evaluating the taxonomic rank of these populations. | - Species is only a monophyletic population (monotypic species). The "subspecies" category cannot exist in principle.<br>- "Evolutionary event" and speciation happen simultaneously.<br>- The phenomenon of secondary contact zones and hybridization is completely devoid of any heuristic value as the genetic exchange between non-identical populations is, by definition, interspecific hybridization | - Population is not a sample, but a heterophobic system, the structure of which is subject to "morphological method" analysis.<br>- The same regulatory mechanisms *maintain the integrity* of the population system of the species from the "inside" and its isolation "outside", in the secondary contacts with another form<br>- 'Friend' recognition and 'foe' rejection is not so much on the level of marital interactions of specimens as in case of including the descendants of the former in the population structure.<br>- Individuals very often "err", producing hybrids and backcrosses. Whether they will be the agents to enhance hybridization or not, whether the isolation of the forms continues or not is determined by their increased non-inclusion compared with "pure" specimens in the population system of both forms<br>- Even in the course of introgressive hybridization, population systems of forms are significantly selective to an influx of foreign genes and phenes, which is why a hybrid phenotype "gets completed" only in the hybrid zone<br>- Inflow of foreign genes into the population system, in principle, does not threaten its isolation and species status of the form, if only the "border" remains semipermeable, heterophobia is maintained, and the system successfully selects and filters the gene flow. |

**Table 1.** Comparison of the biological species concept with the phylogenetic (currently the most popular of the "character"-based concepts). After [42], with changes.

Firstly, under the "population" of a species, *without any kind of discussion*, researchers have come to understand, "not people, but population", not a system of relationships, a specific pattern of structure that is to be remodelled, but simply individuals who have fallen in the researcher's "box" here and now. According to R.L. Smith (1990, cited from [58], "Most of the populations have no limits, other than those made up by the ecologists themselves". Now we know that the population is structured in a species-specific manner even in a homogene-ous environment, and even in it different populations indicate the presence of boundaries, connected with the internal "structure" of the system [12, 58]. Naturally, such "populations" as instant "frames" of the real structure of settlement types, different quality and detail will be uncomparable between different authors working in connection with different tasks.

Second, the "isolation" of forms *again without discussion* was understood as the inability to cross at the *specimen level*: they "do not make mistakes", and do not produce hybrids, or the hybrids are fully/partly sterile. See a mention of this as common place among various au-thors, who have investigated the problem of the taxonomic status and hybridized and hy-bridogeneous forms of birds from different positions [15, 25, 44, 54]. In other words, there is a narrowing of the original meaning of the concept to the emasculation of the original theo-retical content which went unnoticed, also largely for its creator. This was noted by P. Beur-ton in his work on the evolution of biological species concept; Mayr, in response to it, in general agreed with him [21].

The solution of another difficult problem — defining the status of allopatric isolates — was simply postponed. This supported the false belief of the opponents of the biological species concept that the latter does not have the means to breach the subject [42, 54].

Thus, it is the "internal" deformation of the biological species concept that has created its "difficulties", which it then could not resolve, though in principle they are quite solvable. First, its practical application was immediately complicated, because questions arose:

a.   What is the minimum percenatge of hybrids in the zone of secondary contact between two forms still indicates the mainenance of their isolation, and what does not, and how do we determine this threshold in a biologically meaningful way?

b.   what to do with allopatric isolates — how much should they diverge so as to be consid-ered "isolated"? Especially considering that birds and other classes of vertebrates don't have a strict correspondence between the ranks — subspecies, species, subgenus, genus, family, etc. — and the levels of divergence, morphological or genetic. See [8, 14, 15, 44].

Second, in the routinized form, the biological species concept could not adequately inter-pret a whole range of empirical evidence, which is why the latter seemed to be "objec-tions" against her. The most important of them are the following: individuals are very "often wrong", producing hybrids; and precisely in the situation of the secondary contact of similar forms the "errors" are not a rare event, they depend at least on the same factors as the "exact" recognition, which means that in certain circumstances, they can build up and thicken directionally [8, 44, 51].

Christopher Randler's survey [51] of cases of interspecific hybridization and hybrid zones among birds, testing theories of possible causes of the "recognition errors" shows that the above is a general rule. Such errors even among good species are not rare events, but a regular phenomenon (marked between 850 species, it is 10% of the world fauna). Moreover, the specific circumstances of "meeting" of the two forms on the edge of the area rather promote hybridization than hinder it. This decrease by 1-2 orders of both the number of forms in the contact area, compared with the level in similar habitats of continuous area nearby. Lack of partners of the same species here and the subsequent deprivation increases the motivation of the birds to mate with everything that somehow seems to be "legitimate" objects of court-ship. More frequent depressions of numbers, leading to a particular rarity of one of the spe-cies, the "disturbances" of the environment, more common there than in the center of the area (the frequently happening "atypical" course of spring, shifting phenodates, etc.) — all this greatly increases the frequency of the "errors" individuals make, even in spite of the pre-zygotic isolation under normal conditions.

Therefore, individuals do not "know" that they belong to different species, and so very often interbreed with closely related forms, which is greatly facilitated by shifting of the area in space and of the number in time. This puts an end to the idea of perfecting the isolating mechanisms in an already started hybridization to prevent destruction of co-adapted gene complexes of both forms. Apparently, there are only those obstacles to the crossing which are a by-product of a separate development of forms at the stage of isolates, and the com-pleteness of their formation only shows in the zone of secondary contact. As well as infertili-ty in hybridization, the degree of which increases along a parabola in accordance with the model by Dobzhansky-Moeller over the time of the individual form developement [35].

## 5. Reinforcement and character displacement in nature is a rarity, not the ordinary means of preserving the species level of isolation

The great importance of E.N.Panov's book [44] "Natural hybridization and ethological isola-tion in birds " — as well as the subsequent works by Y.Haffer [15] and T. Price [50] is in the demonstration of the above on a vast factual material. Although the book by E.N.Panov is not without drawbacks associated with biased expectations of the outcome[3], both of the above theses, contrary to the traditional understanding of the biological species concept, were proved at least for birds.

That is independently confirmed by research in genetics of speciation, whereby hybrid zones are rather a channel for the flow of genes between closely related forms than an obsta-cle to it, and often the flow continues until the unification of both gene pools [56]. Further, the time of formation of the inability to cross for an "average" bird species greatly exceeds

---

3 The author defended the view that interspecific hybridization among birds is an important way of formation of new specific and species-like rank forms. And he included into the appendix list of conditions of hybridization everything that even remotely looked like interspecific hybridization, even cases that he could not have failed to know were not examples of it. For instance, he considered a carcass, acquired in 1929 in Europe and kept in the Berlin Museum to be a hybrid of the black-headed gull Larus ichthyaetus and the brown-headed gull L.brunnicephalus, although it is known that this was a specimen of a new kind — a relic gull L.relictus, discovered only in 1969 in Kazakhstan and then un-known to zoologists. Or an even more curious interpretation of the same kind — a crested shelduck Tadorna cristata in his list was named a hybrid (of widgeon and shelduck), although the view that it is not a separate species, but a prod-uct of hybridization, had been refuted in the early twentieth century.

the time of speciation [49]. Consequently, a large proportion of cases of differentiation at the level of forms effectively begins with the total gene pool and goes to the end (separation of the species level), without separation and divergence of the latter, rather the genetic differentiation is the result of isolation. Finally, differentiation at the level of forms appears *earlier* than at the level of genes [41, 49], and remains stable even after as a result of secondary contact with hybridization their gene pools are re-united.

"At the dawn of youth" of the biological species concept, many people thought that in response to the secondary contact and hybridization, selection in the population itself will perfect the insulating mechanisms, "protecting" the isolation from the introgression. Accordingly, a wide distribution of so-called reinforcement + character displacement was expected, when close forms are quite similar to each other in the allopatric zone, but in the contact zone and overlapping areas are the more dissimilar behaviorally, morphologically, coloration and so on, the longer the contact.

Initially the "displacement characteristics" as an adaptive process, specifically directed at the protection of isolation of the species in hybridization conditions with a similar form, was though to include three selective processes:

1.  selection against hybrids, reducing fertility and/or adaptation of the latter, "works" for the production of postzygotic isolation;

2.  selection, increasing prezygotic precopulation barriers to interbreeding, improves the "mutual recognition" of species, including through the "discrepancies" of character responsible for it, thus reducing the likelihood of individuals' "errors" leading to hybridization;

3.  the directed selection establishment of *de novo* initially absent isolating mechanisms that prevent malicious for both forms destructions of locally co-adapted gene complexes.

It is important that Mayr the and first generation of researchers of the 'new synthesis' considered case 3) to be crucial, since its discovery in nature would allow to consider the isolation on the species level to be a particular case of adaptation, developed essentially in the same way as other types of useful adaptations of a species to the environment.

Then, for about 50 years, all these three cases have been persistently searched for in nature, especially since it was assumed that they would be found often. Alas! They are rare, and to understand the discovered ones other schemes are more convincing than the Dobzhansky-Mayr explanation. At the same time, according to the theory, situations 1)—3) should prevail in all the "difficult cases" of incomplete speciation, secondary contact with hybridization of closely related forms, etc. Much (by 2—3 orders of magnitude) more common is a couple of cases not covered by the theory, and even directly contradicting it.

If during isolation the contact forms accumulated considerable incompatibility, hybrids are completely or partially sterile, sometimes also with reduced viability [50-51]. However, their products are in the contact zone is ongoing and the expected "plugging of leaks" is not happening — not due to the reinforcement or character displacement. Neither the reinforcement nor the character displacement are observed in situations where the gene flow between the forms should be stopped, based on Dobzhansky's ideas.

But in cases where both reinforcement and character displacement are surely present [7, 53], an effective reduction in the intensity of hybridization in time has never been shown — for instance, for such number of generations, which is sufficient for a basic evolutionary phenomenon. These phenomena do exist, but the termination of hybridization of closely related forms, "flashing" on the frontier of one of them settling in the area of the other, is stopped in other ways, by 'friend/foe' recognition of the population systems of both.

In other words, now we can openly say that the expectations of the classics were not confirmed, and the isolation of the species level is maintained otherwise. It is important that it is the problems with the search for "perfection of isolating mechanisms" in nature contributed to the loss of interest in the biological species concept in the early 1980s. That is, the decline of interest began even before the rise of phylogeography, "dendrogram thinking" and the phylogenetic species concept, which are usually stated as reasons for it [1].

Situations 1) and 2) provided by Dobzhansky's model, happen in nature, but much more seldom than the theory. In addition, the vast majority of examples of reinforcement, presented in the surveys [7, 53] etc., are much better explained not by Dobzhansky's model, but by two other methods.

The first is: the observed reinforcement nor the character displacement in the contact zone is the consequence and the effect of trends of variation connected to the adaptation of each form to the environmental conditions of its main area (for example, climatic gradients, changes in vegetation structure in the populated habitats, etc.).

This is just becasue these trends are opposite to each other, among other things, because the "splitting off" of child forms of the original is often associated with the adaptation to the territories with conditions alternative to those in the historical center of the area. Since both forms before the secondary contact spread from the refuges towards each other, it is clear that associated with them difference in morphology, behavior, etc., to be found there will be the greatest. Therefore, the reinforcement is often asymmetrical, is seen only in one of the forms or is stronger in one form than the other.

The second alternative explanation is as follows. The observer, looking for confirmation of the idea of "perfection of isolating mechanisms", sees it in something that in fact is a fixation of the beginning of population systems "splitting" due to the mechanism we proposed for the 'friend/foe' recognition. This impression is created due to the presence of reinforcement or the character displacement and defectiveness of hybrids between two forms. In this case, hybridization does not ever stop, the "leak" is not plugged, and yet the "cost of eliminating" inferior hybrids is "bearable" for both forms, the "reproductive self-destruction" doesn't happen. A good example here is the relationship of benthic and limnetic forms of three-spined stickleback *Gasterosteus aculeatus* in the post-glacial lakes of North America, described in [53], as well as the above mentioned "splitting" of population systems among birds.

50 years of research into speciation in nature showed that the divergence of forms (including when it is distinct populations within the maternal species) and the formation of isolation at

the species level — are processes that take place in different moments, "protected" by different mechanisms, etc. Accordingly, the required reinforcement in practice is rare and almost always can be explained otherwise. But the "perfection of isolating mechanisms" does not happen at all.

In other words, a survey of modern data on this issue confirms the old idea of Darwin (1939, cited from [44]) that the forms' inability to cross (as well as infertility in hybridization) is not a selective advantage and is not accumulated by selection directly. The latter occurs indirectly, through a period of isolation of the separated forms [53, 56]. Consequently, the separation of species is a non-linear function of the time they stay in isolation.

Much better than with the idea of character reinforcement, all of the above is consistent

a.   with the Dobzhansky — Moeller model of formation of hybrid incompatibility

b.   with our model maintaining the isolation at the species level, according to which the main obstacle to the crossing is not created in the time of formation of the mixed pairs and the production of hybrids, but at the "integration" of hybrids and backcrosses in the population groups of each form [9].

Mechanisms postulated by b) assist the mechanisms postulated by a) and vice versa. According to the Dobzhansky — Moeller model, the gradual accumulation of innovations in the genes of the "friend" form produces the side effect of quadratically increasing reproductive incompatibility with "foe" forms of the same kind, because selection can not "test" their compatibility with innovations that accumulate in homologous genes of the "foe" gene pool. Therefore b) will necessarily lead to the fact that despite the formal presence of intense hybridization, "clean" specimens of both forms live and breed mainly in the environment of other "clean" ones, which actually form the groups.

While hybrids, even numerically dominant in the hybrid zone, *do not form* their own groups, but *one by one join* the settlements of some form or reproduce outside the settlements — with predictably poor results. Thanks to that, the possibility to test new fixed mutations for "compatibility" with the innovations in the gene pool of another form drops dramatically before *any meaningful cesure of hybridization*, genetic incompatibility of forms increases, which, with the continued production of hybrids and backcrosses, reduces the stability of their behavior and the ability to create their own stable groups, even with the numerical dominance. The circle closes.

If b) is true, then there is a synergy between the behavioral (or ecological) mechanisms to maintain isolation and the genetic incompatibility between the two forms due to which cases of contact either "merge" population systems into one, or allow preservation of an isolation for an unlimited time, "confirming" the achievement of the species level. But if "the traditional form" of the biological species concept is true (where the main point of 'friend/foe' recognition depends on whether mixed couples are formed or not), such synergy will not be observed. If so, the two sides of the overall process of speciation — precopulation and postzygotic isolation mechanisms (*species recognition* and *hybrids incompatibility*) are mutually exclusive, and not connected by positive correlation.

This result is quite verifiable. Indeed, among birds are observed

- either (for the youngest species) isolation, created by b) type mechanisms and sustainable differentiation of forms in the absence of genetic incompatibility, and often with a completely common gene pool, "mixed" by introgressive hybridization,

- or (for the "good species") a positive correlation between the degree of *species recogntition* and *hybrids imcompatibility* in some forms (the former reflects the development of the pre-copulation, the latter — postcopulation barriers to crossing) [7, 43, 49]. The traditional view of the speciation involves the *regression of the process* — first the accumulation of genetic changes that lead to inconsistencies, in the period of independent development, then during the secondary contact with hybridization — a rapid development of behavioral and morphological adaptations such as "mating plumage" within the framework of perfection of isolating mechanisms [7, 52].

In reality, with the birds and other vertebrates, we are seeing the opposite — differentiation of forms at the species level often begins and ends before gaining the inability to cross, it is possible with a fully common gene pool. Conversely, the effective separation of population systems of two forms by the b) mechanism accelerates the slow and gradual development of the inability to cross, lagging far behind the pace of speciation [41, 49].

In cases of real character reinforcement, selection against hybrids or selection towards better recognition in the formation of pairs, which increases the contacting forms isolation from each other, produces, as a rule, an external biocenotic agent. It can be insect pollinators with their different reaction to the flowers of different colors, different species-models to simulate, which the developing forms of mimicing species begin to imitate, carnivorous bats, catching tree frogs, focusing on the mating male calls, etc. See [35, 53].

In a general case there might not be a biocenotic agent, and most likely there won't be one; most predators, pollinators, etc. are not so specialized to detect the differences in the signals of the incipient species. At the same time, the process of speciation in all species concepts, including the original understanding of Darwin [37] is thought of as proceeding spontaneously, affected only by the internal forces.

## 6. The "Renaissance" of the biological species concept: Possible approaches

Appealing to the data on hybridization as to objections, the proponents of alternative concepts (especially typological, phylogenetic and evolutionary) by the early 2000's actually excluded the biological concept from the list of discussed, especially in the West [1-2, 65].

However, the obvious disadvantages of alternative concepts are demanding the restoration of the latter on more grounds than ever, which allows to incorporate the objections and use them for the development of the concept. It is possible to put forward the following theses for its recovery, returning to the original understanding of the two key points related to population-based approach and isolation as a criterion of species-level differentiation of forms.

- Population is not a sample, but a heterophobic system, the structure of which is to be analysed by the "morphological method".

- The same regulatory mechanisms *maintain the integrity* of the population system of the species "inside" and *its isolation from "outside"*, during the secondary contact with another form

- The viability of populations "sitting" on a "center-periphery" gradient of the species area is much more defined by the inflow of non-residents than by the local reproduction and local adaptations.

- Resettlement of individuals leads to differentiation of populations, and not to "blurring the differences", because individuals are sorted according to behavioral potentiality.

- 'Friend' recognition and 'foe' rejection is not so much on the level of mating interactions of individuals as when the descendants of the former are integrated in the population structure.

- Individuals very often "err", producing hybrids and backcrosses. Whether they become agents of hybridization enhancement or not, whether the isolation of the forms is continued or not, is determined by the worst position on which they are included in comparison with "pure" individuals in the population system of both forms

- Even with introgressive hybridization, population systems of forms are substantially selective to an influx of foreign genes and phenes, which is why a hybrid phenotype is "put together" only in the hybrid zone

- Like the presence of a foreign currency in the pocket does not make us closer to the psychological makeup of its residents, so the inflow of foreign genes in the population system, in principle, does not threaten its isolation and the species status of the form. If only the "border" is semipermeable, heterophobia is maintained, and the system successfully selects and filters the gene flow, that is, it is taken under control. Overall, current evidence supports the conclusion by E.N.Panov [44] on the creative role of hybridization processes in the zones of secondary contact of the "separated" initially allopatric species. But they did not "create" new hybridogeneous forms that can exist along with the parent ones, but provides the parent forms with the "necessary" genes and phenes to adapt either to a hybrid zone habitats or other conditions of the area, without disturbing the original isolation of both.

Thus, according to mtDNA, common mallards and spotbills are "like crows" [28], gene pools have joined together, but the forms remain isolated. This situation is different from hybridization with American black duck *Anas rubripes* in the U.S., where there is a classical Mayr's "merging of forms" [34]. The same situation as for common mallards and spotbills is viewed for six sympatric species of ducks, as the mallard, gadwall, pintail, wigeon, teal and tufted duck, despite the fact that the last three species are far enough from the first three. Gene exchange through hybridization (preserving fertility and viability of hybrids) between these species has been so great that the authors compare it with the "horizontal transfer" of genes in bacteria. However, the "blurring" of species boundaries hasn't happened even here [24].

An interesting point is that, unlike the mallard, forms such as *Anas zonorhyncha* and *Anas rubripes* are "female-coloured" and don't have bright males significantly different from females in coloration, "braids" of decorating feathers, etc. Furthermore, in all experiments, where females of "female-coloured" forms had to choose different males, the colorful, heavy and strong mallard drake is invariably beyond competition. Females prefer him, rejecting the "humble" males of their own species. Therefore, it was believed that in nature the hybridization is always asymmetrical: male mallards breed with females of monomorphic species. However, nature reveals no less powerful gene flow associated with the hybridization in the opposite direction, which is easily detected by mtDNA analysis, when the hybrid population is dominated by haplotypes not of monomorphic species, but of mallard [28]. Apparently, this is a consequence of interactions related to the "dark side" like forced copulation after pairing.

In summary, the proposed improvement of the biological species concept can be illustrated by the addition of the third column of *Table 1*, opposing the biological species concept (in the traditional sense) to the phylogenetic.

Indeed, the main idea that is firmly confirmed from the original concept — is the discreteness of species, and the reality of their existence in the wild. Including the situations of secondary contact of forms. Hybrid zone is either restricted by narrow limits and constant over time despite the long-standing hybridization (but is not reduced, because there is no "perfection of the isolating mechanisms"). Or, if the form does not reach the level of species, introgression between them in the hybrid zone deepens, the latter "blurs". Eventually, a new hybridogeneous population emerges, joining the previously independent forms A and B into a single population system, in fact another subspecies among subspecies, with restoration of variability trends to a smooth cline. Such, for instance, is the situation in the zone of contact of European and Siberian chiffchaff *Phylloscopus collybita abietinus* and *Ph.c.tristis* in the Cis-Ural region and in the Urals. As a result of the unobstructed deepening of hybridization, the contact zone "blurred" over a vast territory occupied by hybridogeneous form *fulvescens*, in fact, yet another in a series of other subspecies. This "natural experiment" shows that both chiffchaffs, no matter how much they separated, still fall short of the level of species: their secondary contact restored the unity of the population system [31].

Here we propose the idea of "own-alien" recognition following the inclusion of phenotypic variant individuals into the spatial and ethological structure of both populations, differing in their phenoforms. This mechanism could explain the phenomenon of populations isolation without obvious hybridization obstacles. In case of close similarity of biological forms hybrid offsprings are to be characterized with normal survival and reproduction abilities. In the other case these properties of hybrids should be significantly reduced. Therefore in the latter variant there is no hybridogenic population forming. By the way, isolated crossbreeding cases occur more of less frequently. In that follows, some hybridization obstacles probably could arise and stop crossbreeding. On the contrary, without such preventing mechanisms, hybridization frequency remains stable.

There is an prominent example of rapid settlement of Syrian woodpecker *Dendrocopos syria-cus* in the "traditional" into the natural habitat of great spotted woodpecker *D.major*. The first species was previously resident of Balkans and Asia Minor. Since 1930s it moved to north and east: in 1994 it was found the Uzhgorod; during next 50 years it was spread in Ukraine; in 90s it reached Voronezh and Volgograd regions and further "ended the circle" in Ciscaucasia. During first years of this travel the ourbursts of crossbreeding were detected (I have also seen several mixed pairs), while later the hybridization reduced [6]. There were a lot of causes of such effect: the allobiotope establishments due to Syrian woodpecker pref-erence to gardens and parks during dissemination; the destabilization of hybrids behavior including those with normal reproductive possibilities. Such birds were characterized with disorders in the search of potential partner. In individual cases the reaction of hybrids to specific stimuli varied a lot including even more strong response than for "pure" ones. However, such response was both less stable and less specific. Consequently, the offsprings of mixed pairs were in disadvantage to pure ones during partner searching and/or territo-ries occupation especially in the higher probability of occasional migration [6, 8].

Another interesting observation was described for two chipmunk species from the Rockies: *Tamias ruficaudus* and *T.amoenus*. These chipminks vary significantly basing on their bacu-lum morphology. Previously baculum differences were considered as a guarantee from hy-bridization. Surprisingly, the frequent traces of alien mtDNA introgression were revealed for these non-sister species. Three described evolutionary events probably were associated with recent asymmetric mtDNA introgression in morphologically distinct secondary contact zones. Additionally, the traces of ancient hybridization events leading to alien mtDNA fixa-tion were also detected. By the way, such events were characterized with the unchanged phenotype and remaining "own" nuclear DNA [13].

In is well-known that in the case of inter-species hybridization of birds most of mixed pairs are unstable with the duration of "staying together" varying from hours to several days. It is true for forms differing in their signaling repertoires. But the remaining mixed pairs could be extremely stable for the whole season and even during several years. The fraction of such stable "misalliances" is higher if signaling languages are more similar. Note that this effect is independent on the descendants survival and even on the fertilization of eggs [8, 44]. For example, in the narrow contact zone of two shrikes, *Lanius collurio* and *L.cristatus* in the Western Siberia, mixed pairs were described. Most of such "marriages" are quickly ended due to different signaling repertoires. But the rare cases of stable mixed *Lanius* pairs are characterized with viable and fertile offsprings [26].

In such cases *one-type* signals provide recognition of aliens and fast decay of a part of "misal-liances" after initial contacts of excited individuals. The *other-type* signals (or another signal-ing regime) support the "survived" pairs stability. As expected, these "*other*" signals just minimally diverge between closely related taxa. For example, for two South African turtle-dove species, the *Streptopelia vinacea* and *S.capensis,* the narrow hybridization zone in Ugan-da was described. This region population contains the stable fraction of viable and fertile hybrids. Both birds are capable of producing homologous signals: "cooing at roost" and "cooing in bow". The stereotypic usage of these signals is important for the protection of the territory and/or for attracting potential partners theres. However, the reproductive behavior and corresponding signals vary for these birds.

Consequently, the first "territorial" signals, but not the second "reproductive" signals lead to preferencial choice of "owns", not "aliens" in the allopatric areas – but not in the hybrid zone [17, 22-23]. Respectively, the first shout type provide heterophobia of both populations and the second one consolidates pairs formed after "filtering" at the first step. Hybrids are characterized with disordered "mixed" in varying proportions signals, as it was shown basing on the acoustic comparative analysis. Unlike parents, the hybrids are unable to distinguish "owns" and "aliens": there response to the "cooing on the roost" is independent from the similarity of this signal acoustic characteristics to their own. Saying more, their response is *worse* in comparison to "pure" individuals and "own" signals. Consequently, hybrids are defective in territorial seizure and potential female partners attraction [17, 22-23]. While speaking on the parental species, let us note that the response to the "cooing on the roost" of *S.vinaceae* is stronger than to those of *S.capensis* in allopatric populations. This asymmetry persists in hybrid offspring, thus the hybridization is also asymmetric: the *capensis* area is enlarging into the *vinacea* area, but not vice versa. Despite the remaining isolation of these forms, the *vinacea* alleles are frequently added into the gene pool of *capensis*. Nevertheless, outside the hybridization zone both morphology and behavior of *capensis* is unchanged from the normal features of this species [17, 22-23].

Traditionally the classics (and Mayr, especially) proposed the populational approach, not the typological. But this idea contained the time bomb exploded in 70-80s: there are some forms. Are they differentiated enough to be considered as species? It's impossible to answer this question without typological approach [29]. However the typological procedures should be *discussed and taken under control* initially and not passed over silence: thus, the populationism could use them. The "natural typology" of systems of populations may be developed based on the described above morphological approach to their structure and dynamics. This way allows us to reconstruct the "natural formation" with it's inner regulators providing the ethological and populational structure despite the environmental instability outside and the stochastic demography inside the system [10]. The reason is that is any ideology lacks the indispensable components and methods, they in any case "pierce their way" – but probably in a spontaneous barbaric manner.

Indeed, the application of *such* a typology for the analysis of intraspecific variation and interspecific hiatuses could be helpful for solving the following old problem of biological concepts of species: the time-stable hybrid zones of two forms. Such zones remains narrow without tendencies to grow in spite of the long-standing period of hybridization (hundreds and thousands of years) and the overall viability and fertility of hybrid offsping (leading to common gene pools for contacting species).

The number of such forms could be different and greater than two. The common area of the "race circle" contains numerous their subtypes, located serially just like beads on a string and hybridizing in pairs and/or totally. The example of such "race circle" is known for golden woodpecker *Colaptes auratus s.l. – auratus s.str., cafer* and *chrysoides* in North America [38-39, 60]. Another case was revealed for black wheatear *Oenanthe picata – picata, capistrata* and *opistoleuca* [45]. By the way, the classic case of stable narrow hybridization zones between black and gray crows could be included here also.

Here we have presented just a few examples among numerous ones [see reviews: 43]. All of them illustrate the possibility of common gene pool emergence and phenotypes mixing as a result of long-term crossbreeding. The molecular genetic evidences of this phenomenon were published, for example, for black and gray crows [16, 27]. Several markers between neighboring crow groups in the hybridization zone demonstrate the similar level of differences as those for black crows of Eastern and Western Siberia or for gray and black crows. Furthermore, the form B phenes are spread far to the habitat of the form A up to the third form intergradation [see fig. 81 in 45].

What one can see in such a situation? The commonality of gene pools and long history of hybridization makes in theoretically possible that in the *area of the form A* the significant fraction (up to 5-15%) of *hybrid A-B phenotype* could exist. Such phenotype may consist of numerous XYZ features characterized by opposite values in A and B forms and combined – in hybrids. However, this mixed phenotype occurs only in *hybrid zone*! On the contrary, such "alien" phenes in form A and B areas could be found only one by one and not in complex. These findings are the more interesting the more prominent phenotypically distinct features could be found for hybrids. Such phenoforms were even described as species [44].

Let us look at the example: the above mentioned golden woodpeckers are characterized with the hybrid phenotype A-B, which additionally to A/B contact zone arises in the hybridization boundary with form C (as well as the hybrid A-C phenotype) [38-39, 60]. This remarkable fact tells us that the B-form signs successfully crossed the total form A habitat; but the hybrid A-B phenotype couldn't be generated (in contrast to "pure" B forms) in other place than the hybridization zone – even *other* than A-B. In the inner form A area these phenes could be found only separately, not together. However, the argued molecular genetic basics without enhancement of isolation mechanisms are insufficient for explaining listed observations.

Are there any missed factors of natural selection? We propose the impact of ontogenesis here. The population structure in the "inner" habitat of species is strongly regulated and stabilized using various mechanisms. For example, the "incorrect" behavior of individual leads to it's "culling". The normal course of ontogenesis under such conditions excludes the "alien" phenes. In extreme cases, they could be expressed only as separate signs. Thus, the final morphotype is formed under the pronounced regulatory pressure selecting mostly "own" but not "alien" phenes [9].

The narrow zone at the area borders of contacting species does not allow such "morphoselective" mechanisms to work and to switch on the recognition of "owns" and "aliens". Therefore, the hybrid morphotypes occur only in such crossbreeding loci. The population "soup" already contains all gene- and phene components, and the only thing you need to collect the hybrid puzzle is the shutdown of listed regulators. At the narrow strip of hybridization zone the structure of relations, behavioral traits and other features are disordered and the overall population is destabilized.

It's no wonder that population sizes in *the same loci* of hybrid zone are up to two orders smaller than in the *nearest* allopatric loci. Therefore such "demographic" decrease is unrelated to such expected factors: extreme environmental conditions, bad quality of habitats etc. The real reason is the lack of stable networks of groups and migrating individuals between them: population structure should be formed only through cooperation and integration. The boundaries of species areas lack such systemic interconnections or, probably, they are extremely destabilized [10-11]. It is important to note that the abilities to distinguish and sort "owns" and "aliens" are better at the level of population system, not at the individual level. These effects are prominent both for case of population-forming behavior and relations and in the case of morphotype-forming ontogenesis.

Therefore, both contacting species are able to persist as somehow isolated forms in spite of their long-term history of introgressive hybridization [4, 5]. Such a phenomenon include various features: morphological, behavioral, environmental, biotopic specialization etc. Indeed, recent studies of gene flux through hybridization zone demonstrate that both forms are *highly selective* to the choice of particular genomic elements to be introduced into the hybrid gene pools [reviewed in 56].

Mitochondrial DNA is prone to integration into other species genome. Thus, those groups where males were heterogametic, could be considered using Haldane rule: mtDNA is female-inherited and, additionally, females are highly fertile. Surprisingly, the similar effects were described for BIRDS – known to be female-heterogametic. In several cases it was shown that the "alien" mtDNA have replaced "own" mtDNA totally (for *Vermivora pinus* and *V.chrysoptera*). However, the "alien" mtDNA pervasion is significantly asymmetric illustrating the existence of selection "filters" for gene exchange. The pronounced asymmetry of mtDNA invasion was revealed in 50 among 80 cases [63].

As for nuclear DNA, it's recombinational hotspots are known to be mostly introduced into alien genomes. These loci are short and their shortness is also associated with high recombination frequency. The Dobzhansky-Muller hypothesis fits well these data: the genomic incompability between species is considered to be associated with a few genes, prone to recombination and providing better adaptation of their carriers. Other loci, even fast evolving but slowly recombining – such as pericentromeric heterochromatin – are rarely adopted [56]. Moreover, pericentromeric regions are connected with building the barrier for interspecies hybridization.

Such mechanisms probably regulate individual development of forms under the pressure of scud of the invasive alien genes. All these pathways lead to preferences of "pure" morphoforms in each area and complicate the development of mixed phenotypes in other places than contacting area [9].

Thus in the common area of both forms "alien" phenes could be realized only partially, as a traces in addition to "normal" phenotypes: even in the case of significant fraction of genetic hybrids and enhanced introgression.

In the recent review [56] it was noted that hybrid zones were not *barriers* but *channels* for gene exchange. Various factors of natural selection don't stop the introgression – just make it asymmetric. The examples were shown for *Cyanoliseus patagonus* [33], *Manacus vitellinus* [61], *Foudia* [62] and other species. In case of reaching the higher level of isolation these form can form new species through the forming of regulatory mechanisms for such gene exchange. The "maturity" of such regulatory mechanism(s) gives us an indication of the real species differentiation level (including numerous populational, morphological, molecular-genetic and other characteristics).

Genetic processes only couldn't lead to proper isolation without further selective pathways: lacking these selectors all specific morphotypes should be connected in the continuum. But mostly we reveal the opposite effects confirming therefore the presence of evolutionary selective mechanism preferring distinct morphotypes. This mechanism doesn't deal with such features as crossbreeding barriers, ethological signals, behavioral species recognition, hybrids incompability etc. My idea is that THIS mechanism works well only at *population* level, not individuals. And thus the offsping of proper of mistaked breeding are integrated into the whole population structure.

The secondary contact zone with frequent crossbreeding is characterized with stable colonies of A and B forms separately. While the hybrids and back-crosses are able only to accompany poure "citizens" and can't form the populational units. That's why the "improper" individuals inclusion into population and their reproduction probability is low. Thus, the individual-level breeding mistakes are corrected at this population level. Hybrids and back-crosses are fertile enough but they are "deprived of their rights" to deepen the reproductive influence. Different hybrid variants are sorted between A-form and B-form habitats exposing to stabilizing selection (with various selective morphological, biotopic and behavioral parameters). The "improper" phenotypically hybrid individuals are unable to integrate to normal population "etiquette" – during their lifespan and/or reproduction. The behavioural standard should be involved significantly in such culling mechanisms: the mixed signaling of hybrids leads to frequent mistakes in contrast to "pure" ones [9].

The heterophobic events at the boundary of two populations could be demonstrated using the well-known sponges experiment. They were grinded extremely to single cells and then mixed. The components of this "cocktail" reassociate with their specific cellular partners, not aliens. Therefore, the key stage of breaking or stabilization of the isolationing barrier between populations is the *inclusion* of the offspring into the population structure, but not their fertility of other features. The mechanism of such "culling" is the *communication*. The proper generation of this mechanism stabilize the populations and their isolation even in the presence of frequent crossbreeding with fertile hybrid offspring. It seems that our ideas work well also for variable systems with subspecies or races characterized with differences in sizes, melanization, coloring etc. Therefore there could be distinct *unbiased* signs of *species*, not of *subspecies*. The absence of "culling" mechanism in studied populations let us consider them as single species despite any measured morphological variations.

## 7. The problem of allopatry

The most complicated case of our analysis is the situation with allopatric isolates. How one can determine when the level of biological divergence is high enough to consider the iso- lates as distinct species? Noteworthy, the baseline of divergence is always present for such objects, including cases of "well-differentiated subspecies".

The answer is simple, indeed. Using the proposed morphological approach to populations analysis we can describe:

- the *species* = the Vavilov's species definition as a system of populations interacting inside the area and differentiating at the boundaries at the center- periphery gradient basing on the flux of migrating individuals;

- the *species* remains intact until it's interior populations are integrated through the stable exchange of individuals. Such traffic don't smear the differences between subpopulations, on the contrary, it enchances the differentiation due to selection of "proper" individuals [10-11]. In the case of stable reduction of the population size and/or ecological changes leading to area fragmentation the several, mostly peripheric loci, are prone to further dif- ferentiation even up to the level of new species. How can we detect it if the area is already fragmented and the secondary hybridization is impossible?

- Before the separation of "far populations" to the isolate(s) the species area is gradiented in the center-periphery direction. Most event of differentiation are distributed along this ax- is. The social stress and the competition are more important in the center characterized with dense population. As for environmental stress, it's pressure is stronger at the periph- ery. Thus, individuals are sorted geographically according their abilities to overcome the first – or the second stresses (Fig.1).

This mechanism provide further differentiation between contacting populations not *despite,* but *through* the stable migration amongst, and the variants of patientness-competitiveness of life strategies could be distributed in a bell-shaped curve for both populations (Fig.1).

The specific systemic pattern the the key *regulator* and *"controlling force"*, and the individuals migration, communication etc. form the *controlled* response. I propose that the whole mor- phoecological differentiation between isolates was present at the stages of their presence in- side the integral population system of the same species. The period of isolated provide the grow of separation only.

Thus, the metrics of the specific differences for allopatric isolates could be easily proposed (for example, for the case of blue magpies from Pyrenees and the Far East. The molecular- genetic divergence values between them should be compared to those between related spe- cies with the similar area(s) which are still connected.

The *Pica pica* magpie with it's European and Far East races is perfectly suited for this pur- pose; also one can study such Corvidae species as *Cyanopica cyana* and it's Far East *pastinator* form as well as *Corvus monedula* and *C.monedula dauricus* [16, 25, 27]. Transpaleoarctic Corvidae species studied by Kryukov's group could be considered as a good data for the case of conclude the state of blue magpies (basing on the *C.corax, Perisoreus infaustus, Nucifraga caryocatactes* etc). Other allopatric situations could be analyzed by similar way.

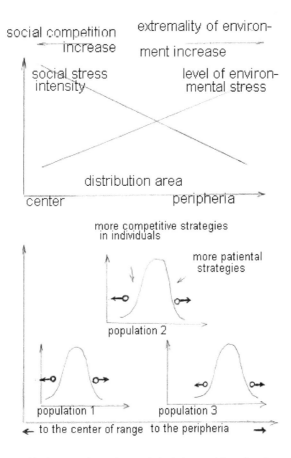

**Figure 1.** Population system of the form according to the "morphological approach". gradients "center - periphery" of the range (top), sorted by species potencies (degree of competition and patience strategies) along these lines (bottom).

Finally, these data demonstrates that various *Corvidae* are able to form divergent Eastern and Western subtypes with varying level of differences. This level is high and equal for magpies and blue magpies but higher in comparison to rooks or jackdaws. Thus, the blue magpies are not unique and these populations are still the same species: the difference between Eastern and Western populations is quite similar to those for other magpie population system characterized by continuous area.

## 8. Conclusion

The biological species concept is defined *biologically* according to our aim to divide the total biodiversity into *real* classes. The supra-individual systems – populations and species – are characterized by the presence of regulators stabilizing their integrity and isolation from other similar systems. Various mechanisms are involved it the process: gene pool(s), migration, communication, ontogenesis etc [see 3]. This "natural control pathway" works like the restorer of ancient painings.

Let us overcome the last complication: the species are present in all taxa. But only the well-studied phylas give us enough data to distinguish species correctly. For others we know only several individuals and nothing about their populations.

Thus, the specie should be defined basing on the presence of the precise hiatus, using morphological conception and typological approach. It is convenient but depends on the stability of the hiatus – and this stability is unknown.

If new data destroy the previously defined classification we're to start the work again. We'd prefer to use more "natural" classifiers stable to such new findings. But there are still two poles basing on the prevalent role of the typological (morphological species concept) or populationist approaches (biological species concept). All other concept are based also on the same features while the influences of these two approaches are combined in different proportions.

The species problem is associated with difficulties in reaching the proper compromise here – and remaining the usability of the proposed classifiers.

## Acknowledgements

I am grateful to Igor Yakovlevitch Pavlinov for the invitation to talk about species problem in Zoological Museum; it helped me to bring order to my thinking and this article preparation. I am also thankful to A.A.Makhrov, S.V.Polevova and A.A.Pozdnyakov for their remarks and discussions. I'd like to thank VIGG researchers M.V.Friedmann and N.Yu.Oparina for their help and consultations on the genetic mechanisms of hybridization barriers and other discussions.

## Author details

V. S. Friedmann

Moscow State University, Faculty of Biology, Moscow, Russia

# References

[1] Abramson N.I. Molecular markers, phylogeography and search for the criteria for delimiting species// Trudy ZIN. 2009; 313 (1) Suppl. 185-198. [in Russian, English summary].

[2] Altukhov Yu.P. Genetic processes in populations. Moscow: Akademkniga; 2003. [in Russian, English summary].

[3] Artamonova V.S., Makhrov A.A. Genetical systems as regulators for adaptation and speciation processes (toward the system theory of microevolution)// Modern problems of biological evolution. Proceedings of the Int. Conference on the 100th anniversary of the State Darwin Museum. 2008. 381-401. [in Russian].

[4] Bell D.A. Genetic differentiation, geographic variation and hybridization in gulls of the Larus glaucescens-occidentalis complex // Condor. 1996; 98(3) 527-546.

[5] Bell D.A. Hybridization and reproductive performance in gulls of the Larus glaucescens-occidentalis complex// Condor. 1997; 99 585-594.

[6] Butyev V.T., Friedmann V.S. Great spotted woodpecker. Syrian Woodpecker. Middle spotted woodpecker. White-backed woodpecker. Three-toed woodpecker // Birds of Russia and neighboring regions. From Owls to Woodpeckers. Moscow: KMK Scientific Press; 2005. 325-398. [in Russian, English summary].

[7] Edwards S.V., Kingan S.B., Calkins J.D., Balakrishnan Ch.N., Jennings W.B. Swanson W.J., Sorenson M.D. Speciation in birds: Genes, geography, and sexual selection// PNAS. 2007; 102 6550-6557.

[8] Friedmann V.S. The communication in the mixed pairs between different species of woodpeckers: some ethological barriers to mating // Zh. Obshchei biologii. 1993; 54 (1) 294-310. [in Russian, English summary].

[9] Friedmann V.S. 'Friend-foe' recognition systems and the Renaissance of the biological species concept// Modern problems of evolution. XXIth reading by A.Lyubishchev. Ulyanovsk; 2007 201-215. http://www.evolbiol.ru/fridman.doc [in Russian, English summary].

[10] Friedmann V.S. Communication signals in vertebrates: from stimulus to symbols. Part 2., chapter 2.5. Population as a complex system and communication tasks. M.: URSS; 2012 160-190. [in Russian, English summary].

[11] Friedmann V.S., Eremkin G.S. Urbanization in "wild" bird species in context of urbolandscape evolution. Moscow: URSS; 2009.

[12] Goltsman D.W. Differentiation by dispersal // Nature. 2005; 433 (6) 23-24.

[13] Good J.M., Demboski J.R., Nagorsen D.W., Sullivan J. Phylogeography and introgressive hybridization: chipmunks (Genus Tamias) in the northern Rocky Mountains // Evolution (USA). 2003; 57 (8) 1900-1916.

[14]  Grant P.B., Grant R.B. Speciation and Hybridization in Island Birds [and Discussion] // Phil. Trans. R. Soc. Lond. B 1996; 351 765-772.

[15]  Haffer J. Parapatrische Vogelarten der paläarktischen Region// Journal of Ornithology. 1989; 130 (4) 475-512.

[16]  Haring E., Gamauf A., Kryukov A. Phylogeographic patterns in widespread corvid birds // Molecular Phylogenetics and Evolution. 2007; 45 840–862.

[17]  den Hartog P.M., de Kort S.R., ten Cate C. Hybrid vocalizations are effective within, but not outside, an avian hybrid zone// Behav. Ecol. Vol. 2007; 18 608-614.

[18]  Joppa L.N., Roberts D.L., Pimm S.L. The population ecology and social behaviour of taxonomists // Trends of Ecology and Evolution. 2011; 26 (11) 551–553.

[19]  Kerimov A.B., Ivankina E.V. The proportion of current and final indicators of reproductive success of individuals in the suburban population of the Great Tit, Parus major L. // Trudy Zvenigorodskoi biologicheskoi stantsii. 2005; 4 221-231. [in Russian, English summary].

[20]  Kharitonov S.P. Waterbird colony structure: system approach. // Ornithology. Moscow; MSU Publ. 1998; 27 C.26-37. [in Russian, English summary].

[21]  Kolchinski E.I. Ernst Mayr and modern evolutionary synthesis. Moscow: KMK Publ.; 2006. [in Russian].

[22]  de Kort S.R., den Hartog P.M., ten Cate C. Vocal Signals, Isolation and Hybridization in the Vinaceous Dove (Streptopelia vinacea) and the Ring-Necked Dove (S. capicola)// Behavioral Ecology and Sociobiology. 2002a; 51 (4) 378-385

[23]  de Kort S.R., den Hartog P.M., ten Cate C. Diverge or merge? The effect of sympatric occurrence on the territorial vocalizations of the vinaceous dove Streptopelia vinacea and the ring-necked dove S. capicola // J. of Avian Biology. 2002b; 33 150-158.

[24]  Kraus R.H.S., Kerstens H.H.D., van Hooft P., Megens H.-J., Elmberg J., et al. Widespread horizontal genomic exchange does not erode species barriers among sympatric ducks// BMC Evolutionary Biology. 2012; 12 45 http://www.biomedcentral.com/1471-2148/12/45

[25]  Kryukov A.P. The modern species concepts and the role of Russian biologists in their development // Problemy evolutsii. Vladivostok: Dalnauka. 2003; 5 31-39. [in Russian, English summary]

[26]  Kryukov A.P., Gureev S.P. New data on interrelations of red-backed and brown shrikes (Lanius collurio and L. cristatus, Aves) in zone of sympatry // Zoologichesky zhurnal. 1997; 76 (10) 1193-1201. [in Russian, English summary]

[27]  Kryukov A., Suzuki H., Haring E. The Phylogeny and evolutionary history of corvids // Natural history of crows. Sapporo. 2010; 3-19.

[28]  Kulikova I.V., Zhuravlev Yu.N. Molecular-Genetic Studies of Anatinae Interspecies Hybridization from the Example of Anas platyrhynchos Susraspecific Complex // Uspekhi sovremennoi biologii. 2009; 129 (2) 162-163. [in Russian, English summary].

[29]  Lyubarsky G.Yu. Archetype, Style, and Rank in Biological Systematics. Moscow.: KMK Scientific Press; 1996 [in Russian, English summary].

[30]  Lyubishev A.A. The problem of whether// Problems of form, systematics and evolution of organisms. Moscow: Nauka; 1982 [in Russian, English summary].

[31]  Marova I.M., Fedorov V.V., Shipilinna D.A., Alexeyev V.N. Genetic differentiation and vocal in hybrid zones songbirds: Siberian and European Chiffchaff (Phylloscopus (collybita) tristis и Ph. (c.) abietinus) in the southern Urals // Doklady Akademii Nauk. 2009; 427 (6): 848-850. [in Russian].

[32]  Martin G. Birds in double trouble// Nature. 1996; 380 666-667.

[33]  Masello J.F., Quillfeldt P., Munimanda G.K., Klauke N., Segelbacher G., Schaefer H.M., Failla M., Cortés M., Moodley Y. The high Andes, gene flow and a stable hybrid zone shape the genetic structure of a wide-ranging South American parrot // Frontiers in Zoology. 2011; 8 1-16. http://www.frontiersinzoology.com/content/8/1/16

[34]  Mayr E. Zoological species and evolution. Moscow: Mir.; 1968 [in Russian, English summary].

[35]  Matute D.R., Butler I.A., Turissini D.A., Coyne J.A. A Test of the Snowball Theory for the Rate of Evolution of Hybrid Incompatibilities // Science. 2010; 329 1518–1521.

[36]  McDonald D.B. Microsatellite DNA evidence for gene flow in neotropical lek-mating long-tailed manakin // Condor. 2003; 105 580–586.

[37]  Mina M. V. Evolution of Species Concept from Darwin to Our Days: Progress or Wandering? // Charles Darwin and modern biology. Proceedings of the International Academic Conference 21–23 September, 2009. Saint-Petersburg: Nestor-Historia; 2010 228-236. [in Russian, English summary].

[38]  Moore W.S. Random mating in the northern flicker hybrid zone: implications for the evolution of bright and contrasting plumage patterns in birds// Evolution 1987; 41 539-546.

[39]  Moore W.S., Graham J.H., Jef T. P. Mitochondrial DNA Variation in the Northern Flicker (Colaptes auratus, Aves)// Mol.Biol. Evol. 1991; 8 (3) 327-344.

[40]  Mora C., Tittensor D.P., Adl S., Simpson A.G.B., Worm B. How many species are there on Earth and in the Ocean? // PLoS Biology. 2011; 9 (8) 1-8. http://www.plosbiology.org/article/info:doi/10.1371/journal.pbio.1001127

[41]  Morrow E.H., Pitcher T.E., Arnqvist G. No evidence that sexual selection is an 'engine of speciation' in birds // Ecology Letters. 2003; 6 228-234.

[42] Nazarenko A.A. Is unified conseption of species in ornithology possible or not? (Opinion of practicing taxonomist) // Zhurnal obshchei biologii. 2001; 62 (2) 180-186. [in Russian, English summary].

[43] Newton I. The speciation and biogeography of birds. Academic Press, Cambridge; 2003

[44] Panov E.N. Natural hybridization and ethological isolation in birds. M.: Nauka; 1989 [in Russian, English summary].

[45] Panov E.N. Wheatears of Palearctic: ecology, behaviour and evolution. M.: KMK Scientific Press; 1999 [in Russian, English summary].

[46] Parchman T.L., Benkman C.W. Britch S.C. Patterns of genetic variation in the adaptive radiation of New World crossbills (Aves: Loxia) // Molecular Ecology. 2006; 15 1873-1887

[47] Pavlinov I.Ya. The species problem in biology – another look // Trudy ZIN. 2009; 313 (1) Suppl. 250-271. [in Russian, English summary].

[48] Pavlov D.S., Lupandin A.I., Kostin V.V. Mechanisms downstream migration of young river fish. M.: Nauka; 2007. [in Russian, English summary].

[49] Price T.D., Bouvier M.M. The evolution of $F_1$ postzygotic incompatibilities in birds // Evolution. 2002; 56 (10) 2083-2089.

[50] Price T.D. Speciation in Birds. Roberts & Company Publishers, Greenwood Village, Colorado; 2008

[51] Randler C. Avian hybridization, mixed pairing and female choice // Animal Behaviour. 2002; 63 (1) 103-119

[52] Rasnitsyn A.P. On the species and speciation // Problems of evolution. Vol.IV. Vorontsov N.N. (ed.). Nauka Press, Novosibirsk. 1975; 221-230. [in Russian, English summary].

[53] Servedio M.R., Noor M. A.F. The role of reinforcement in speciation// Annu. Rev. Ecol. Evol. Syst. 2003; 34 339–64

[54] Severtsov A.S. The modern species concepts // Bull. MOIP. Biological Series. 1988; 93 (6) 3-14. [in Russian, English summary].

[55] Severtsov A.S. Possible microevolutionary consequences of the hierarchical organizations in vertebrates populations // Zoologichesky zhurnal. 2003; 83 (2) 109-118. [in Russian, English summary].

[56] Shamoni M., Barbash D.A. The Genetics of Hybrid Incompatibilities // Annu.Rev.Genet. 2011; 45 331–355

[57] Shchipanov N.A. Aspects of sustainable persistence of small mammals population stability //Uspekhi sovremennoi biologii. 2000; 120 (1) 73-87. [in Russian, English summary].

[58]  Shchipanov N.A. Population as a unit of species existence. Small mammals // Ento-
      mological Review. 2003; 83 142-160.

[59]  Shilova S.A. Population Ecology As a Basis for Small Mammal Population Control
      M.: Nauka; 1993. [in Russian, English summary].

[60]  Short L.L. Hybridization in the flickers (Colaptes) of North America// Bull. Am. Mu-
      seum Nat. Hist. 1965; 129 307-428.

[61]  Uy J.A.C., Stein A.C. Variable visual habitats may influence the spread of colourful
      plumage across an avian hybrid zone // J. Evol. Biol. 2007; 20 (5) 1847-1858.

[62]  Warren B.H., Bermingham E., Bourgeois Y., Estep L. K., Prys-Jones R. P., Strasberg
      D., Thébaud C. Hybridization and barriers to gene flow in an island bird radiation //
      Evolution (USA). 2012; 66 1490-1505.

[63]  Wirtz P. Mother species - father species: Unidirectional hybridisation in animals with
      female choice // Animal Behaviour. 1999; 58 1-12

[64]  Zavarsin G.A. Biodiversity as part of the Biosphere-Geosphere System of order out of
      chaos // The methodology of biology: new ideas. Synergetics. Semiotics. Coevolution.
      Moscow: Editorial URSS. 2001; 151-176. [in Russian, English summary].

[65]  Zink R. The role of subspecies in obscuring avian biological diversity and misleading
      conservation policy// Proc. R. Soc. Lond. B. 2004; 271 561–564

# Historical Issue

# Chapter 10

# Darwin's Species Concept Revisited

David N. Stamos

Additional information is available at the end of the chapter

## 1. Introduction

What did Darwin *really* mean by "species"? There are different ways of answering this question, different ways of doing history, and they need not converge on the same answer.

One way is to look at Darwin's definitions of the term "species" and to use a comparative method analogous to the comparative method in biology, comparing and contrasting Darwin's definitions with those of his contemporaries, with his predecessors, and with definitions today. This is the method favoured typically by philosophers of biology and biologists who are concerned with history, and it might be called the *pure history of ideas approach*.

One problem with this approach is that it assumes that Darwin meant what he said in his definitions. But why would one assume that? Darwin was a great scientist, to be sure, as well as a wonderful husband and father and an upstanding member of society, but he was also a human being and human beings do not always mean what they say, either intentionally or unintentionally. Why should a scientist be exempt from this general truth about humans, especially when there already exist examples of dissimulation, even radical dissimulation, among leading intellectuals? (I shall turn to some famous cases at the end of this chapter.) To think otherwise of Darwin, to place him on a pedestal above human nature, is to descend to the level of religious iconography.

Another way to approach what Darwin really meant by "species" is to deal with Darwin's writings in the context of his time, not just the scientific context but the personal, social, cultural, economic, and political contexts as well. As Bernard Lightman [1] puts it in his Introduction to an anthology devoted to contextualist analyses of Victorian science, "The hallmark of contextualist studies is their emphasis on the way scientific ideas are embedded in material culture such that there are no insides or outsides of science." This, he says, "allowed historians to avoid the false analytical distinction between science and society (or base and superstructure), dissolve the categories external and internal, and begin to transcend the science/society dualism" (p. 7). For at least the past two decades now, this is the method of doing history of science that seems most favored by professional historians of science, epitomized by the books by Desmond and Moore [2, 3], and may be called the *pure contextualist approach.*

This approach suffers from a number of difficulties, however, not the least of which is that, as Lightman [1] points out, these historians need not always agree on the same matter because "there are many different kinds of contexts" (p. 7). A much greater difficulty, arguably the biggest, concerns the background assumptions of the approach, such as what Lightman calls "the false analytical distinction between science and society." That the distinction is a false one might be the conclusion of a massive process of induction, and hence synthetic, even an inference to the best explanation. But if historians of this persuasion are not open to the possibility that a scientist can have an idea or theory that is not the product of his or her time, then the approach is dogmatic and this way of doing history takes on the character of an ideology. There is also the problem of a self-referential paradox, in that one has to wonder how these historians have managed to transcend their cultural milieu and arrive at objective, accurate causal explanations but not natural scientists. Surely a more balanced approach would be to allow that many scientific ideas and practices are to some extent products of their environment but not necessarily all. So then we are back to what Darwin really meant by "species," and whether what he meant was a product of his time or something unique, something that one would not have been able to predict (let alone retrodict).

What I have called the *pure contextualist approach* to history of science is not the only way of doing contextualist history. Another kind of contextualist history involves a careful consideration of the writings of an author within the author's environmental context *so as better to understand what the author meant.* There is no assumption here that the ideas of the author were purely the product of their context so that the author could not have had ideas that were contrary to his or her environment. There is no assumption even that the author could not have had ideas that were ahead of his or her own time. Richard Ashcraft [4] uses this approach (though not purely) to better understand Locke's political philosophy as given in his *Two Treatises of Government*, in contrast to the approach to Locke typically found in philosophy, which considers his writings almost in a Platonic heaven and examines them for validity and soundness. As Ashcraft puts it, his approach is "to raise questions concerning the meaning of a particular political theory that are referable to the actor's social life-world, the nature of the intended audience, and the purposes for which the political theory were formulated" (p. 6). Hence his focus is on the radical political movement of which Locke was a part, headed by Shaftesbury, involving former Levellers, secret codes, and plots to kill the King such as the Rye House Plot, rather than on the rise of capitalism, mercantilism, the scientific revolution, and so on.

But this approach is not necessarily preferable when it comes to Darwin on "species." Darwin was the head of a scientific revolution, to be sure, which today goes by his name, but one cannot assume that he wrote in a specific manner that only his followers would have fully understood. He might have, but the parallel with Locke is highly strained at best. In the case of Darwin, unlike with Locke, we have to be open to the possibility that he sometimes wrote in a manner that was aimed at manipulating his followers along with the rest of his readers in ways they might not have recognized. But how could one ever hope to find that out if true?

This brings us to the method of doing history that seems to me perfectly suited to the present purpose. It is taken from philosophy of language, from Ludwig Wittgenstein to be specific, and is employed in a manner which he might not have agreed with or even imagined but which follows from his theory of meaning. In short (and I shall have more to say about his theory in the third section), what I shall do in this chapter is go beyond Darwin's *definitions* of "species" in his writings and pay careful attention to his *use*. In the famous words of Wittgenstein which I here paraphrase, when it comes to what a word means in a language community *don't ask for the meaning, ask for the use*. The same principle can also be applied to the writings of a single author, not only to a particular work but to the entire corpus, viewed in much the same way as a palaeontologist looks at strata. In the case of Darwin on "species," I hope to show in the present chapter the enormous power of this method and some of its fruits. The method, moreover, should not be viewed as incompatible with history of ideas approaches, or with contextualist approaches either, but in the case of Darwin it should be seen that the approach is best suited as the preliminary one to any further investigations.

## 2. Survey of the current state of scholarship

Before we begin by applying a Wittgensteinian approach to Darwin on "species," it is important to take a brief survey of the current state of scholarship on Darwin's species concept. What is utterly remarkable is the near total state of confusion. After all the years since Darwin first published *On the Origin of Species* [5], one would naturally have expected better.

What is perhaps most surprising is the paucity of attention that contextualist historians of science have given to Darwin's species concept, in stark contrast to the amount of effort they have given to other key concepts and theories of his, notably natural selection, divergence, adaptation, and the Tree of Life. For example, in his book devoted to the development of Darwin's theory of evolution by natural selection from 1838–1859, Dov Ospovat [6] makes much of the influence of British natural theology and biological theories of progress, including the influence of particular naturalists such as Owen and von Baer, but he nowhere deals with Darwin's species concept. Granted, the work was seminal, and this important young historian could not be expected to do everything in his book (especially since he did not live long enough to see it in print). As Adrian Desmond points out in his Foreword to the book, Ospovat "wanted to make Darwin less of a seer, standing out of time, and more a man of his day" (p. ix), with the consequence that "Dov set many hares running, and we all chased them" (p. xi). It is therefore all the more surprising that none of the contextualists seem to have bothered to chase Darwin's "species" hare, or to even have noticed that it needed (or was worth) catching. Desmond and Moore's *Darwin* [2], as a prime example, "seeks to portray the scientific expert as a product of his time" (p. xviii), and they "want to understand how his theories and strategies were embedded in a reforming Whig society" (p. xix), but one looks in vain in their book for a contextualist analysis of what Darwin meant by "species." The same holds true for their latest book on Darwin [3], which they claim "is the untold story of how Darwin's abhorrence of slavery led to our modern understanding of evolution" (p. xxi). This is especially surprising given that Darwin categorized all the humans races as members of the same species.

When we turn to philosophers of biology and biologists devoted to a historical understanding of Darwin, we find an entirely opposite situation, marked by lots of publications on Darwin's species concept and lots of different answers. In order to indicate the current state of scholarship on this topic, what may be called *the problem situation*, I want to confine myself to a summary of five publications from the past few years. The methodology they share is basically the history of ideas approach, but clearly something else is going on, much of it arguably not history in any sense of the term.

We begin with a book by the philosopher John Wilkins [7], which is devoted entirely to a detailed examination of the history of the concept of biological species. His book explicitly is not an example of contextualist history of science but instead is written in the tradition of history of ideas, and hence is "a conceptual history" (p. ix). His purpose, moreover, is not simply to do history for the sake of history but to make a contribution to the present debate on "species" in biology and philosophy of biology. In the very least, he says, "Knowing the past may also help scientists to avoid repeating it unnecessarily" (p. 7). He recognizes that "Darwin acts as a focal point" (p. 5) in the history of the debate, not only because Darwin did have something to say about the nature of species but because the *Origin* "changed *every* scientist's way of looking at species thereafter" (p. 130). Accordingly Wilkins devotes an entire chapter to "Darwin and the Darwinians." Here, through roughly twenty pages of analysis of Darwin's writings from his mature period, we find that Darwin "presents species as real," that he thought that species as *"groupings"* of organisms exist in nature and are not necessarily arbitrary like constellations, but that he did not think of species "as a formal and fixed rank" (p. 144). The claim seems to be the same as that developed by John Beatty [8], who in this respect followed the original claim of Michael Ghiselin [9], which is that Darwin was a species taxa realist but not a species category realist and accordingly did not have a species concept. Wilkins [7] tells us that "Darwin was not a cladist, but he was pretty close to it" (p. 153), that Darwin was "not a nominalist but a pluralist with regard to what makes a species distinct" (p. 158)—elsewhere he states that "he was a pluralist as to the degree of difference between, and causes of, species" (p. 230)—and that "in the end, Darwin proposed a 'snowflake' theory of species: all members are alike in some ways, but they are also unique individuals" (p. 158). In the last few pages of his book, in his summary, Wilkins states that Darwin's work was "perhaps the first and most complete attempt to deal with the implications of the transmutation of species" (p. 231) and he seems to add that Darwin got it basically right. This is because "the standard stories and assumptions from the architects of the modern synthesis are often simply incorrect" (p. 233), such that "We might stop trying to overgeneralize species concept(ion)s or species mechanisms to all species" (p. 234).

Second in our list is a chapter on Darwin on "species" by the philosopher Phillip Sloan [10]. In his chapter Sloan is mainly concerned with historical influences which he believes impacted on Darwin's species concept. Linnaeus crystallized what Sloan calls the "species$_L$" approach to species, which was logical and classificatory, based on necessary and sufficient criteria for class membership, while Buffon crystallized the "species$_H$" approach, which was historical and ontological. Immanuel Kant attempted to systematize the distinction, Charles Lyell framed the question about the reality of species in terms of species$_L$, whereas Johann Jacob Bernhardi framed the question about the reality of species in terms of species$_H$. Darwin, it is then argued by Sloan, employed and "synthesized" both traditions in *Natural Selection* and then in *Origin*, the species$_L$ tradition in characterizing species as "tolerably well-defined" at "any one period" and the species$_H$ tradition when he "interprets species and varieties as genealogical lineages that display varying degrees of historical relationship" (p. 79).

Third on our list is a paper by the biologists Mark Ellis and Paul Wolf [11]. The paper is devoted to the proper teaching of the concept of species to biology students, which they consider fundamental to the understanding of evolution. Much of their paper is devoted to the history of the concept of species, with Darwin as the hero of the story: "Charles Darwin would finally provide the keys to understanding and accepting the reality of evolving species, breaking the millennial stranglehold" (p. 92). But no actual species concept is explicitly attributed to Darwin. The species concept that they themselves subscribe to, however, and which they imply Darwin would have accepted had he known of it, is the ecumenical species concept developed by the biologist R.L. Mayden [12], his "overarching nonoperational species concept" as they [11] put it (p. 94), which takes most modern competing species concepts as operational only and incorporates them into a hierarchical concept. The proper concept of species, in short, according to these two authors, is that of "organisms in one or more populations that together form a cohesive, reproductive unit—a separate lineage on its own evolutionary trajectory" (p. 90). Mayden himself [12] places the diachronic evolutionary species concept of G.G. Simpson on top of the hierarchy, as the "primary concept" (p. 419), with "all of the other concepts," reckoned as being at least 21 concepts, serving as secondary, operational concepts "at some level" (p. 417).

In a paper devoted to examining why Darwin's view on species was rejected by most twentieth-century biologists, the biologist James Mallet [13] spends some effort in examining those passages in the *Origin* which Ernst Mayr especially (and the many who followed him) made much of, passages which Mallet maintains were either misunderstood in themselves or because they were not taken in their textual context. According to Mallet, in short, Darwin was not confused about species nor was he a species nominalist. Instead, "Darwin had a good idea of what he was talking about after all" (p. 498), which was that species taxa are delimited by "gaps in morphology in nature" (p. 499). "He believed," adds Mallet, "that species existed, but that they were not 'fundamentally' different from varieties, and that they did not have essences." Hence varieties for Darwin were "incipient species" and in explaining the origin of varieties and species Darwin also explained why "borderline cases" are to be expected. In all of this, Mallet adds still further, "we appear to be returning to more

Darwinian views on species, and to a fuller appreciation of what Darwin meant" (p. 498). This is because "Today, genetic markers are available and widely used to delimit species, for example using assignment tests: genetics has replaced a Darwinian reliance on morphology for detecting gaps between species" (p. 497). Again: "Darwin's view of species as clusters of similar individuals separated by gaps remains relevant today. Species are multilocus genotypic clusters that retain identity when in sympatry with close relatives" (p. 502). Darwin, then, according to Mallet, was not only right to focus on morphological gaps in delimiting real species in nature but should be viewed as the father of the genetic marker approach to species and speciation.

Finally, in a paper devoted to examining "Darwin's solution to the species problem," the philosopher Marc Ereshefsky [14] explicitly follows Ghiselin [9] and Beatty [8] in interpreting Darwin as a species taxa but not species category realist: Darwin's "skepticism of the species category did not extend to taxa, and in particular those taxa called 'species'" (p. 409). Ereshefsky also follows Beatty in holding that Darwin maintained the reference of the species taxa so designated by his fellow naturalists and that he did this for "pragmatic reasons" (p. 421). But Ereshefsky goes beyond Ghiselin and Beatty when he argues that Darwin was basically right. According to Ereshefsky, different modern species concepts pick out different real taxa and modern biology requires a pluralism of species concepts in order to adequately capture the diversity of life. Darwin, says Ereshefsky, rejected the reality of the species category "based on his skepticism of the species/variety distinction," whereas modern biology "implies that Darwin's skepticism of the species category is correct" (p. 425) because it has uncovered a heterogeneous lot of real taxa in the world called "species" (reproductively isolated, phylogenetic, ecological, asexual, etc.), taxa that are equally meaningful in terms of information, prediction, and explanation. Given that this is so, given that disagreement over the definition of "species" has actually increased in biology, and given the enormous impracticality of removing the term "species" from the biological lexicon, Ereshefsky concludes that "Current biological theory confirms Darwin's solution to the species problem" (p. 410).

## 3. Problems with the current state of scholarship

One thing that is remarkable about the five recent publications summarized above is how different their understandings are of what Darwin meant by "species." After decades of research and publications on what Darwin meant by one of the central terms in what is now called "the Darwinian revolution," there is still no consensus on what Darwin meant by "species," and little of what might be called progress on the matter. To be sure, few if any accuse Darwin any longer of being an outright species taxa nominalist, but there is no consensus on where Darwin stood concerning the binary categories that virtually define the modern species problem: category monism or pluralism, speciation monism or pluralism, taxa as primarily horizontal or vertical, taxa as process or pattern/product entities, taxa as classes or individuals (or something else), taxa as monophyletic or polyphyletic.

There is something further that is striking about the publications examined in the previous section. With the exception of Sloan, each of the other authors remind me of a phenomenon Albert Schweitzer observed over a hundred years ago in his *The Quest of the Historical Jesus*, first published in German in 1906. Examining many biographies of Jesus, Schweitzer found that each author of a *Life* tended to infuse into the teachings of Jesus his or her own theological beliefs, liberal or otherwise, instead of remaining neutral and objective. As Schweitzer put it in his autobiography [15], speaking of the individuals in the field collectively, "we managed to interpret Jesus' teaching as if it were in agreement with our own worldview" (p. 55). Here surely is a research project for a bright young graduate student, the quest of Darwin's species concept from 1859 to the present.

The fundamental point is that if we really want to get to the history of the matter, if we really believe it is possible to reconstruct to a significant degree what was going on inside Darwin's head when he wrote about species (taxa and category), the first thing we have to do is to completely put aside our preferred solutions to the species problem. In short, what we have to do is make a serious effort to avoid the evil that historians calls *presentism*, which is reading the present (whether personal or collective) into the past, in this case reading into Darwin's writings what we think species are.

The second thing we have to do is to look beyond Darwin's various definitions of "species." Looking to the *Origin* [5] alone, it is very easy to conclude that Darwin was a species nominalist, that he thought that the species category (as it is now called) is completely manmade and that the designation of organisms into species taxa is likewise manmade. In the final chapter, for example, he says, "systematists will have only to decide (not that this will be easy) whether any form be sufficiently constant and distinct from other forms, to be capable of definition; and if definable, whether the differences be sufficiently important to deserve a specific name" (p. 484). Again, "In short, we shall have to treat species in the same manner as those naturalists treat genera, who admit that genera are merely artificial combinations made for convenience.... we shall at least be freed from the vain search for the undiscovered and undiscoverable essence of the term species" (p. 485). Or one could go outside the *Origin* and focus on its predecessor, Darwin's unfinished *Natural Selection* [16], as the historian Gordon McOuat [17] does, which he says provides Darwin's "only one clear definition of species in any of his work" (p. 4n10), which is: "In the following pages I mean by species, those collections of individuals, which have commonly been so designated by naturalists" (p. 98).

To say that we should "look beyond" Darwin's definitions is not to say that we should ignore them. But it is to say that we have to look deeper. At this point a famous injunction of Wittgenstein comes strongly to mind, as found in his *Philosophical Investigations* [18]. I am not a "Wittgensteinian," a philosopher who believes that philosophy should be confined to conceptual elucidation, but Wittgenstein described part of the elephant of meaning when we wrote that "For a *large* class of cases... the meaning of a word is its use in the language" (§43). Hence his repeated injunction when trying to figure out the meaning of a word: "don't think, but look!," "*look and see*" (§66). This is the motto that should guide research into Darwin's species concept, given the collective muddle of conflicting claims over what Darwin meant by "species." In other words, one should be willing to "do a Wittgenstein" on Darwin.

What this means is that we have to look beyond and get over what Darwin *says* about species, whether about the term or about species taxa in general. That information might prove to be important at some point (I believe it is important), but it would be none but a superficial analysis to leave it at that. To really be able to see whether Darwin had a species concept, and what it was if he had one, we have to turn mainly to his *use*, to what he *does* with particular species designations. This will be no easy matter, as the writings Darwin left us are enormous. But it is what we must do if we want to determine whether Darwin actually employed a species *concept* throughout his writings. To have a concept, of course, does not necessarily mean that one has a definition (although one might). But if one repeatedly and consistently applies a term in a variety of contexts, then it is legitimate to suppose that a concept is in operation, with rules for its use, involving a meaning. It might be a "family resemblance" concept, as Wittgenstein claimed is the case for the term "game," or it might be a concept where every example of its use has features "common to them all." The only way to know is to "look and see."

In the case of Darwin on "species," many have thought and continue to think that in his early period he employed something like the modern biological species concept, which is based on reproductive isolation. But that is not the period in his life that should mainly concern us. Instead, if we want to speak of "Darwin's species concept," we should want to focus on his *mature* period as an evolutionist, which should probably include his barnacle work given its role in the redevelopment of his theory of evolution and that it immediately preceded his work on *Natural Selection*, which he intended to be his scientific case for evolution to the world.

The catalyst for my own research on this topic [19, 20] was John Beatty's often-cited paper [8]. At the core of his thesis is the claim that

His species concept was therefore interestingly minimal: species were, for Darwin, just what expert naturalists called "species." By trying to talk about the same things that his contemporaries were talking about, he hoped his language would conform satisfactorily enough for him to communicate his position to them. [p. 266]

In all of this Beatty was taking the thesis of Michael Ghiselin [9] a step further, who argued, again, that Darwin was a species taxa realist but not a species category realist. What is remarkable, as I point out in my book, is how many biologists, philosophers of biology, and even historians of biology followed Beatty in his central claim. No one, as far as I could find, ever thought to test Beatty's thesis empirically. Scientists routinely think of ways to test novel and even established theories, but this habit of thinking is typically missing in the humanities.

The point of it all is that Beatty's theory can indeed be tested. And the way to test is to "do a Wittgenstein" on Darwin's writings, focusing on his mature period. What this means is that we need to go through Darwin's writings much as a paleontologist goes through strata, looking for patterns that perhaps everyone else has missed. In the case of Darwin this means going through not only his books, such as *Origin* and *The Descent of Man*, but also his articles, letters, and marginalia. This is no small task, given the enormous amount of writings that Darwin left to us, but the effort is richly rewarding. What follows is a representative sample of my own efforts in this regard, some from the larger sample in my book but noticeably improved, and some entirely new.

In looking at these examples, it should become apparent that Darwin did not simply follow the species designations of his fellow naturalists. But what is especially significant is why Darwin deviated when he deviated from the species designations of his fellow naturalists. This is in fact the key to unlocking the door to what Darwin really meant by "species." In what follows, what is required on the part of the reader to fully grasp this key and to see what is behind the door is not only a considerable degree of patience, but also a strong ability to perceive a recurring pattern, what may rightly be called a kind of *scientific imagination*.


## 4. Evidence from Darwin's barnacle work

As a preliminary, we need first to juxtapose two claims made by Darwin in the *Origin* [5]. The first is that, "To sum up, I believe that species come to be tolerably well-defined objects, and do not at any one period present an inextricable chaos of varying and intermediate links" (p. 177). The second is that, "Lastly, looking not at any one time, but to all time, if my theory be true, numberless intermediate varieties, linking most closely all the species of the same group together, must assuredly have existed" (p. 179). From these passages alone, it should be clear that Darwin, to use modern terminology, thought of species taxa as real and that their reality is primarily horizontal or synchronic. This is not to say that he excluded the extension of the reality of species taxa over long stretches of time. What he has to say about "living fossils" (pp. 107, 486), along with the fact that he capitalizes species E and F in his one and only diagram in the *Origin*, species "either unaltered or altered only in a slight degree" (p. 124), indicates that he allowed a vertical or diachronic reality to species taxa. But the reality of species taxa was for him still primarily horizontal or synchronic—evidenced alone by the fact that he twice calls species $F^{14}$ a "new species" (pp. 117, 124).

The above should make it clear that Darwin was not a cladist, or a proto-cladist, or that he anticipated the theory of punctuated equilibria. Nevertheless, what Darwin often said about the "natural system" of classification in his scheme as being primarily genealogical, that "all true classification is genealogical; that community of descent is the hidden bond which naturalists have been unconsciously seeking" (p. 420), has led many to believe otherwise, to believe that Darwin was a proto-cladist, taking genealogy entirely as the basis of classification [21, p. 356], and that he believed species to be spatially limited and temporally extended individuals [9, p. 85]. What Darwin actually did with species designations, however, especially when he went against his fellow naturalists, should make it clear enough that he thought of species taxa as primarily horizontal entities and that he did not employ a concept of monophyly, in whatever way the latter is defined today, as part of the ontology of species taxa.

But first, did Darwin really believe that at any one time most species are "good species"? To answer this question we need to turn to Darwin's most important contribution to taxonomy, namely, his eight grueling years of work on barnacles from 1846–1854, mainly for which he received the Royal Medal of the Royal Society in 1853. What struck Darwin more than anything else right from the start was the amount of variation in barnacles. As he puts it in his volume on the Balanidae [22], "it is hopeless to find in any species, *which has a wide range, and of which numerous specimens from different districts* are presented for examination, any one part or organ—which from differing in the different species is fitted for offering specific characters—absolutely invariable in form or structure" (p. 155). And yet, when we turn to Darwin's pages devoted to the genus *Balanus*, which for Darwin was the largest genus of barnacles and which he considered to have "an especial amount of variation" (p. 156)—and which accordingly caused him the greatest frustration—we find that of the 45 species he describes, most of the species, once they are "disarticulated," are described as good species: as "quite distinct species," as "well-marked species," as "well-defined species," as "strongly characterized species," and so on. At this point he was almost near the end of his eight years of taxonomic work on barnacles, and Darwin never did formal taxonomy again, so that when he says in the *Origin* that "species come to be tolerably well-defined objects, and do not at any one period present an inextricable chaos of varying and intermediate links," we need to take him at his word.

Darwin, of course, was not with his barnacle work trying to communicate his theory of evolution by natural selection to his fellow naturalists. He had not yet made public his evolutionism, and was still working on preparing his case, which had to be as empirical as possible to be accepted as scientific. So there is no point asking whether he was following his fellow naturalists here in what they *called* "species." What needs to be asked at this point, instead, is on what basis Darwin made his species. Barnacle taxonomy prior to Darwin was an enormous mess, not only because the "book species" and even "book genera" were given different names by different taxonomists, the problem of synonymy, but because barnacle taxonomy had been done based only on external characters. As Darwin puts it in one of his letters at the time, "not one naturalist has ever taken the trouble to open the shell of any species to describe it scientifically" [23, p. 207].

What we need to do, then, is to look and see how Darwin made his "book species" of barnacles. What he did for barnacle taxonomy was entirely new: he focused mainly on the anatomy of the organisms. He found much variability, to be sure, but again, even in the highly variable *Balanus*, Darwin concluded that most organisms divided into good species, as noted above.

But what made a "good species" *good*? If we are to find an origin for Darwin's mature species concept, Darwin's only major taxonomic work would seem the natural place to start. Interestingly, in the same volume in which he deals with *Balanus*, Darwin [22] states quite clearly that "In determining what forms to call varieties, I have followed one common rule: namely, the discovery of such closely allied, intermediate forms, that the application of a specific name to any one step in the series, was obviously impossible; or, when such intermediate forms have not actually been found, the knowledge that the differences of structure in question were such as, in *several allied forms*, certainly arose from variation" (p. 156). In other words, Darwin was much more of a lumper than a splitter. And yet in the largest and most variable genus, *Balanus*, Darwin found that most species were good species.

But there is more to the matter. It wasn't simply continuity in traits between barnacle specimens punctuated by gaps that defined species taxa for Darwin. There were "specific or diagnostic characters" that Darwin used to distinguish one barnacle species from another (p. 1). The amount of detail in Darwin's anatomical work is staggering, and it must be kept in mind that the specific functions of many of the internal parts that he studied were either unknown or imperfectly known. Yet Darwin marched on. Confining ourselves to *Balanus*, an instructive example is that of *B. improvisus*, a species hitherto unrecognized by Darwin's fellow naturalists, having been confused with either *B. crenatus* or *B. balanoides* (p. 252), the latter two usually confused together as well (p. 261). Darwin recognizes *improvisus* as a distinct species, not only because of differences in many fine points of anatomy, such as in teeth and thoracic cirri, but mainly because it is adapted to both freshwater and saltwater, and saltwater is a "deadly poison" to most species of the genus (p. 253). *B. balanoides* is indeed not adapted to salt water (p. 272), though with *crenatus* Darwin doesn't say either way. *B crenatus*, however, unlike the other two, has a "great geographical range" (p. 264), from the tropics to the arctic. *B. improvisus* extends from the tropics only northward to Britain. *B. balanoides* is a "tidal species," confined to shallow water apparently for the need of air (p. 272), and is found only along the coasts of the north Atlantic. In all of this, it becomes apparent that the "specific or diagnostic characters" that Darwin is aiming at are ultimately adaptive ones.

This becomes more evident when we look at his designation of varieties for each of the above three species. Only some longitudinal lines distinguish *var. assimilis* from *B. improvisus*, and "it is impossible to consider so trifling a character as specific" (p. 252). *B. crenatus* is not designated any varieties, which is surprising given its great range. *B. balanoides* is given one variety, which Darwin simply calls "the remarkable variety (*a*)," and it is interesting to see why it is not designated as a species by Darwin. He gives four reasons, the two most important of which are: (i) "all the characters by which this variety differs from the common *B. balanoides*, are those which are variable in the latter," and (ii) there are specimens from Ayrshire, located on the shores of south-west Scotland, such that "it was impossible to decide whether to rank the Ayrshire specimen under *var*. (*a*) or under the common form, so that I was compelled to give up *var*. (*a*) as a species" (pp. 270–271). This last reason speaks to Darwin's "one common rule" above, while the former speaks to the lack of a distinguishing adaptation.

The period from 1846, when Darwin began his work on barnacles, to 1856, when he began his big book on species, *Natural Selection*, was one during which Darwin's theory of evolution underwent substantial change. His barnacle work, of course, gave him much of the empirical data that he needed to make his case for evolution by natural selection, specifically, plenty of individual variation. But it also contributed to the development of his "principle of divergence," his principle of the ecological division of labor, whether it was developed during a period of reflection shortly after his barnacle work [6, pp. 170–184] or was brewing during his barnacle work [24, pp. 101–108]. A further issue concerns Ospovat's [6] claim that even to the end of his barnacle work Darwin retained the belief in perfect adaptation, inherited from British natural theology, which he combined with intermittent selection, and only changed it shortly thereafter to relative or imperfect adaptation combined with the constancy of natural selection and the principle of divergence, all of which makes its first appearance in *Natural Selection* [pp. 2–3, 205–207]. Throughout all of this the temptation is to think that Darwin's species concept, assuming for the moment that he had one, might also have undergone a change. I have argued nevertheless that in Darwin's barnacle work it is already possible to discern a species concept that distinguishes species primarily by different adaptations. What I shall now argue is that, when we turn to Darwin's *Origin* and beyond, this picture becomes much clearer, such that it becomes possible to speak confidently of "Darwin's mature species concept." (We could also begin with *Natural Selection*, but the nature of the *Origin* as an "Abstract" of the former makes it the more desirable choice.)

## 5. Evidence from the *Origin* and beyond

Beginning with the *Origin* [5], then, what proves especially instructive is when we look at the particular cases where Darwin went *against* the species designations of his fellow naturalists, the cases that empirically falsify Beatty's central claim. The striking fact is not only that there are plenty of examples, but that in example after example there is a recurring pattern, which not only indicates that a species concept on Darwin's part was in operation, and a consistent one at that, but that it is possible to actually reconstruct his species concept, especially if we pay close attention to the details and do some cross-referencing.

A good example to begin with is Darwin on primroses and cowslips. In the *Origin* Darwin tells us that primroses and cowslips are "united by many intermediate links" (p. 50) and so are "generally acknowledged to be merely varieties" (p. 485), which Darwin on the latter page states as a matter of principle: "for differences, however slight, between any two forms, if not blended by intermediate gradations, are looked at by most naturalists as sufficient to raise both forms to the rank of species." But intermediate links was not the fundamental linking principle. Instead it was the creationist principle of common descent, which in the minds of many naturalists allowed for the production of varieties by secondary laws. As Darwin puts it, returning specifically to the case of primroses and cowslips, "every naturalist has indeed brought descent into his classification.... He who believes that the cowslip is descended from the primrose, or conversely, ranks them together as a single species, and gives a single definition" (p. 424). Darwin himself believed that primroses and cowslips "descend from common parents" (p. 50), by which he meant a common parental species, but nevertheless he did not follow his fellow naturalists and call primroses and cowslips merely varieties. He tells us that "it is very doubtful whether these [intermediate] links are hybrids." But more importantly, he tells us that "These plants differ considerably in appearance; they have a different flavor and emit a different odour; they flower at slightly different periods; they grow in somewhat different stations; they ascend mountains to different heights; they have different geographical ranges; and lastly, according to numerous experiments made during several years by that most careful observer Gärtner, they can only be crossed with much difficulty. We could hardly wish for better evidence of the two forms being specifically distinct" (pp. 49–50).

The last reason in the list above was apparently thrown in by Darwin for rhetorical effect. For a start, Darwin argues later in the *Origin*, against the majority of his fellow naturalists including Kölreuter and Gärtner, that "neither sterility nor fertility affords any clear distinction between species and varieties" (p. 248). Secondly, unlike the other differences between primroses and cowslips that Darwin lists above, sterility between species was not believed by Darwin to be a product of natural selection, only a byproduct, and so not an adaptation. As he puts it in the *Origin*, the sterility (of whatever degree) "of first crosses and of hybrids... is not a special endowment, but is incidental on slowly acquired modifications, more especially in the reproductive systems of the forms which are crossed" (p. 272). This was to remain Darwin's view, with a brief possible exception occurring in late 1862, during which he toyed with the idea that sterility between species might possibly be selected [25, pp. 700–711]. Third and finally, as we examine more cases of where Darwin went against his fellow naturalists in calling a form a "species" or not, it will become apparent that he was amazingly consistent—either explicitly or implicitly—in appealing only to adaptations produced by natural selection.

Darwin, of course, knew of what today are called "sibling species," species so outwardly identical that biologists confused them as one species until innate reproductive isolation was discovered between them. As he puts it in the *Origin* [5], "many species there are, which, though resembling each other most closely, are utterly sterile when intercrossed" (p. 268). But it should not be assumed that reproductive isolation was part of his own species concept. A case in point is willow wrens, which Darwin refers to in a number of places in his writings. As he puts it in *Natural Selection* [16], they are "so close that the most experienced ornithologists can hardly distinguish them," especially given that they "inhabit the same country [? county]," and yet they are "undoubted species." But Darwin never mentions reproductive isolation in any sense, only adaptive characters, such as differences in "their voice, & the materials with which they line their nests" (p. 99).

Returning to the case of primroses and cowslips, what needs to be noticed is that the rest of Darwin's reasons for distinguishing them as two species speak to adaptation by natural selection, especially his phrase "different stations," which is borrowed from the distinction between "stations" and "habitations" emphasized by Charles Lyell [26, p. 69], and which today would be translated as "niches" and "habitats." That Darwin does not explicitly appeal to differences in adaptation between primroses and cowslips in the passage we examined above is possibly because his chapter on natural selection was not to come for another two chapters. But that Darwin's real focus was exclusively on adaptations produced by natural selection will become abundantly evident as we look at more examples.

We are not quite finished, however, with primroses and cowslips. At the end of the *Origin* [5] Darwin repeats that in classifying primroses and cowslips as two species instead of one he is going against his fellow naturalists. He says, "It is quite possible that forms now generally acknowledged to be merely varieties may hereafter be thought worthy of specific names, as with the primrose and cowslip; and in this case scientific and common language will come into accordance" (p. 485). In using the word "generally" Darwin might possibly have misunderstood, misremembered, or even misrepresented botanists in the classification of primroses and cowslips. As Darwin's friend the botanist H.C. Watson [27] pointed out many years earlier, Linnaeus had indeed classified primroses as one species with cowslips as a "subordinate" variety, but, says Watson, "This view will scarce find favour in the eyes of those botanists who labour under the 'species-splitting' monomania" (p. 219). Linneaus himself had complained of the mania for splitting among his fellow botanists, the making of species out of mere varieties, as had Lamarck after him and not only Watson but J.D. Hooker in Darwin's day. But even worse, in a companion paper published in the same issue, which Darwin refers to in his correspondence and in *Natural Selection*, Watson [28] states that the "prevailing opinion" (p. 145), the "majority" view among naturalists (p. 161), is that primroses and cowslips are "truly distinct species." Then again, Darwin might simply have been confused about the situation, since in the *Origin* [5] he actually misnames the common primrose and the common cowslip "Primulaveris and elatior" (p. 49), a point corrected by Watson in a letter to Darwin shortly after the publication of the *Origin* [29, p. 408]—it should have been *Primula vulgaris* and *Primula veris*, a matter of naming he had previously gotten right in *Natural Selection* [16, p. 128], in which he concludes that "common practice & common language is right in giving to the primrose & cowslip distinct names" (p. 133). But no matter. In all of this the point still stands: Darwin in the *Origin*, if we take him at his word, believed that he was going *against* the classification of primroses and cowslips by his fellow naturalists, and it is important to see why.

Darwin's reasons for going against his fellow naturalists (assuming such) are remarkably similar to his position on the races of man in *The Descent of Man* [30]. The big debate, of course, was over whether the races of man are subspecific (monogenism) or specific (polygenism). Darwin defended monogenism, the view that the human races (whatever their number) are all of one species, and it is important to see why. Darwin readily acknowledges, as he puts it in his concluding chapter, that some of the races of man, "for instance the Negro and European, are so distinct that, if specimens had been brought to a naturalist without any further information, they would undoubtedly have been considered by him as good and true species" (p. 388). And yet in his chapter on the races of man, Darwin argues that "Those naturalists, on the other hand, who admit the principle of evolution,... will feel no doubt that all the races of man are descended from a single primitive stock" (p. 229). This was because of "a close agreement in numerous small details," all suggesting descent "from a common stock" (p. 233). Darwin takes it, however, that "the most weighty of all the arguments against treating the races of man as distinct species, is that they graduate into each other, independently in many cases, as far as we can judge, of their having intercrossed" (p. 226).

As we have seen, Darwin also argued that primroses and cowslips descended from a common stock and had between them numerous intermediate links, which he did not think were the result of intercrossing, and yet he went against his fellow naturalists (or so in the

*Origin* he apparently thought) and classified primroses and cowslips as separate species. But when it comes to the races of man he does not classify them as separate species, and once again it is quite informative to see why.

It was not because he thought the human races are interfertile. Darwin was well aware that there were naturalists who claimed empirically that some of the human races had a low degree of fertility between them, and that the children of mixed parentage were either low in fertility or low in vitality. He rejected the evidence as being "almost valueless" (p. 220). But what is important is his claim that even if he was proved right, even if "it should hereafter be proved that all the races of men were perfectly fertile together," nevertheless "fertility and sterility are not safe criterions of specific distinctness" (p. 222), a claim, as we have seen above, that he had made earlier in the *Origin*.

Darwin's reasons were different, even though he recognized that geographically there is a "considerable amount of divergence of character in the several races" (p. 234). For a start, he claims that "as far as we are enabled to judge (although always liable to error on this head) not one of the external differences between the races of man are of any direct or special service to him" (pp. 248–249). In other words, Darwin does not believe that the external differences used to distinguish human races, such as skin color, hair, and the shape of the nose, are adaptations, and so he does not believe that they were produced by natural selection. Instead he thinks that they were produced by "Sexual Selection" (pp. 249–250), and as a simple matter of varying concepts of beauty between the races (pp. 338–384). Consequently he does not think of external racial differences as adaptations at all.

And what of internal characters, particularly social, moral, and intellectual faculties such as reason and language? Here again, although Darwin does not refrain from ranking human races as higher and lower, in the very least as more civilized and less civilized, with the superstition and cannibalism of the Fuegians among the lowest, he nowhere argues that the differences are a matter of different adaptations, but instead treats them only as a matter of degree. As Darwin puts it in a characteristic passage, "The American aborigines, Negroes and Europeans differ as much from each other in mind as any three races can be named; yet I was incessantly struck, whilst living with the Fuegians on board the 'Beagle,' with the many little traits of character, shewing how similar their minds were to ours; and so it was with a full-blooded negro with whom I happened once to be intimate" (p. 232). And it is surely significant that a year prior to when he began writing *Descent*, when Darwin was only planning "a little essay on the Origin of Mankind," he would write in a letter to Wallace that "I still strongly think... that sexual selection has been the main agent in forming the races of Man" [31, p. 109].

Of further significance are the differences Darwin [30] drew between natural selection and sexual selection. Even though he states that "in most cases it is scarcely possible to distinguish between the effects of natural selection and sexual selection" (p. 257), he also states that sexual selection "acts in a less rigorous manner than natural selection" (p. 278), since natural selection is a matter of "life or death at all ages" and sexual selection rarely results in death and only begins its operation at the age of reproduction. In addition, sexual selection has no limits to the process unless when checked by natural selection, so that the products of natural selection, which are about "the external conditions of life," are "rather more perfect" (p. 279) and sexual selection will be "dominated by natural selection for the general welfare of the species" (p. 296). This is not to say that sexual selection cannot produce genuine adaptations in Darwin's view; it can, but they are not the kind of adaptations that would distinguish species, not only because, as he puts it in the *Origin* [5], "no one dreams of separating them [the two sexes]" (p. 424), given that male and female are not separate populational entities and sexual dimorphism is inherited from common parents, but because modifications produced by sexual selection, at best, are adaptations only in a secondary sense, since they do not (or generally do not) contribute to the survival of the organism and its fit in what Darwin repeatedly calls "places" in "the economy of nature" (e.g., pp. 81, 315).

This brings us to a process related to natural and sexual selection and a case related to Darwin on the races of man, namely, Darwin on domestic pigeons. In the *Origin* [5] Darwin states that "Altogether at least a score of pigeons might be chosen, which if shown to an ornithologist, and he were told that they were wild birds, would certainly, I think, be ranked by him as well-defined species. Moreover, I do not believe that any ornithologist would place the English carrier, the short-faced tumbler, the runt, the barb, pouter, and fantail in the same genus" (pp. 22–23). This is similar to Darwin's statement above on what a non-evolutionary naturalist would conclude about the races of man based on specimens "without any further information," that he would conclude that they represented "good and true species." But the case of pigeon breeds introduces some interesting differences, and not only that it was not a hot-button politically-charged topic.

In response to a request from T.H. Huxley for written sources on domestic breeding so as better to defend Darwin in a lecture he was preparing, Darwin in his first reply letter states that he knows of "no one Book" and that he "found it important associating with fanciers & breeders" [29, p. 404]. Less than three weeks later he wrote Huxley again supplying an "enclosure" containing some direct quotations from two books on pigeon breeding by John Matthews Eaton, a prize-winning pigeon breeder and also a friend of Darwin's. In those quotations Eaton refers to the different breeds of domestic pigeons as different "species" (p. 429). Darwin remarks in his letter that if Huxley could see the drawings which Darwin himself has, then Huxley "would have grand display of extremes of diversity" (p. 428). Darwin therefore had a good reason to follow Eaton and others and designate the different breeds of pigeons as different species. But he did not do this. Instead, both in his letter to Huxley and in the *Origin* [5], Darwin repeatedly refers to the different breeds as all one "species" (pp. 20–28), a practice he would continue in *Variation* [32 I, chs. V and VI].

The reason for this cannot simply have been that Darwin [5] thought that all the domestic breeds were derived from a common stock, the rock pigeon, *Columba livia* (p. 23). If that were his reason then he also would have had to designate primroses and cowslips as a single species, because, as we have seen above, he thought they were all derived from a common stock, too. The reason, it turns out, is because primroses and cowslips were evolved by natural selection whereas domestic pigeons were evolved by artificial selection, and there is a substantial difference in product between the two processes. As Darwin puts it in the *Origin* (pp. 83–86), natural selection acts "on every shade of constitutional difference" and produces adaptations, which are for "the good of the being," whereas artificial selection acts only on "external and visible characters," and "only" for the good of the breeder. Moreover nature has tens of thousands and ultimately millions of years for the operation of natural selection, whereas the wishes and efforts of man are "fleeting," his time is "short," so that "how poor will his products be, compared with those accumulated by nature during whole geological periods." Thus, for Darwin, "nature's productions should be far 'truer' in character than man's productions" and "should be infinitely better adapted to the most complex conditions of life." Elsewhere in the *Origin* he states that natural selection "is as immeasurably superior to man's feeble efforts, as the works of Nature are to those of Art" (p. 61). Similarly in *Variation* [32 I], just before summarizing his two chapters on pigeons, Darwin states that "It is not likely that characters selected by the caprice of man should resemble differences preserved under natural conditions either from being of direct service to each species, or from standing in correlation with other modified and serviceable structures" (p. 233). In short, Darwin did not call different pigeon breeds different "species," although it clearly would have aided his argument for evolution in the eyes of many, because he did not want to equivocate on the term "adaptation." Natural selection, and natural selection alone, produces genuine, real adaptations—based on the original meaning of the term in natural history as a possessive, namely, a complex (not simple) heritable trait or behavior that benefits the organism that has it in what Darwin [5] calls the "struggle for life" or "struggle for existence" (pp. 61–62).[1] Artificial selection, on the other hand, does not produce real adaptations (although it could in principle), which is why primroses and cowslips are different species in Darwin's view but not the different breeds of domestic pigeon.

Sports or monstrosities provide another interesting test category, cases of saltation (sudden origin), which some of Darwin's fellow naturalists classified as new species but which in every case Darwin refused to accept as such. In *Variation* [32 I] he discusses the Himalayan breed of rabbit, which is born albino but develops brownish-black ears, nose, feet, and upper tail, and which if kept separate breeds true. Darwin, however, rejects the status of *Lepus nigripes* as a distinct species, arguing that its distinctive characters are a matter of "reversion, supervening at different periods of growth and in different degrees, either to the original black or to the original albino parent variety" (p. 115).

Another example is the "'japanned' or 'black-shouldered'" form of peafowl, of which there were a number of cases in Great Britain, which bred true, and at least one "high authority" named it a distinct species, *Pavo nigripennis*. In *Variation* [32 I], however, Darwin refuses to call it a "species," but instead a "strongly marked variety or 'sport', which tends at all times and in many places to reappear" (p. 307). In a letter written a number of years earlier, Darwin wagers that it will "prove a variety,—hardly more surprising in its origin, than the so-called Himalayan rabbit" [25, p. 193].

---

1 Darwin's concept of adaptation included two important additions. One is that an organism can have an adaptation that benefits it very little or not at all but that fully benefitted its ancestors. He calls these structures and organs "rudimentary" (pp. 450–456). The other important addition is that in social animals some adaptations do not benefit the organism that has it but instead the social group, the "community," of which the organism is a part (pp. 87, 202–203).

A similar case is a variant of *Begonia frigida* that had arisen by saltation in the Royal Botanic Gardens at Kew. William Henry Harvey, the professor of botany at Trinity College, Dublin, wrote about it as a case against Darwin, not only as a distinct species but, if it had been discovered in nature, as possibly the type of a distinct order. In his letters on the topic, Darwin refers to such monstrosities as distinct "forms" [33, p. 102] but never as "species." As he puts it in a letter to Asa Gray on the topic of adaptation by sudden variation, "There seems to me in almost every case too much, too complex, & too beautiful adaptation in every structure to believe in its sudden production" (p. 317). The "almost every case" only indicates that Darwin was not dogmatic on the topic. The same view was elaborated by Darwin a few years earlier in *Natural Selection* [16]: "I cannot believe that in a state of nature new species arise from changes of structure in old species so great & sudden as to deserve to be called monstrosities.... Nor can I believe that structures could arise from any sudden & great change of structure (excepting possibly in rarest instances) so beautifully adapted as we know them to be, to the extraordinarily complex conditions of existence against which every species has to struggle" (p. 319). In all of this the connection between species and adaptations should not be lost, regardless of whether Darwin thought it *possible* that a genuine adaptation could arise by saltation instead of by natural selection.

A related test category for Darwin on species is speciation by hybridization. In the *Origin* [5] Darwin tells us that "The view generally entertained by naturalists is that species, when intercrossed, have been specially endowed with the quality of sterility, in order to prevent the confusion of all organic forms" (p. 245). We should therefore not expect to find naturalists among Darwin's contemporaries who claimed that a new species had arisen by hybridization. And yet there were some who accepted fertility between species, in various degrees, and who even called distinct hybrids "good species" if they bred true. The most famous case prior to Darwin's time is that of Linnaeus, whose experiences with what he named *Peloria*, a new genus which he believed was produced in nature by hybridization, changed his view on species and spawned among his fellow naturalists the purported discovery of a large number of hybrid species [34, p. 148]. They were not contemporaries of Darwin, and yet there were new purported cases of hybrid species in Darwin's day. Perhaps the most famous was the exceedingly beautiful *Bryanthus erectus*, an intergeneric hybrid originally produced by the Scottish nurseryman James Cunningham in 1841. The plant was regularly cultivated in British gardens within the decade, and Cunningham's discovery was mentioned in the commemoration of his death in 1851 in *Gardeners' Chronicle and Agricultural Gazette* of the same year, a journal to which Darwin had long subscribed. Two of Darwin's correspondents during 1863 and 1864 wrote to Darwin about their experiments in breeding the hybrid [35, pp. 18, 79, 81, 98, 353–354, 386; 36, pp. 82, 94]. Darwin did indeed include the name of the purported species in his list of plants for his planned "hothouse" at Down for "experimental purposes" [35, p. 747], which was built in February of 1863. But as Burkhardt *et al.* [36] note, Darwin in one of his reply letters "evidently expressed doubt about Scott's reference to *Bryanthus* as a bigeneric hybrid" (p. 84n11). The species name is not to be found in any of Darwin's published writings, and except for his "hothouse" list it is not to be found in any of his extant correspondence. At any rate, from what we have already seen, it would be surprising to find

Darwin claim, or accept, that a new species had arisen by hybridization. In fact we should not be able to find a single case, given his belief in the origin of species by means of natural selection.

The above is a case of negative evidence, but the theory itself can be tested, as all theories of history can be tested, namely, retrodictively, by making predictions against the past. My prediction was and continues to be that one cannot find anywhere in Darwin's writings where he calls a hybrid form a new species. I was therefore shocked when Ghiselin [9] states that Darwin, in one of his later papers [37], calls a hybrid form a new species. As Ghiselin puts it, Darwin in his paper on primroses and cowslips "demonstrates that the intermediate form, or oxlip, is a sterile hybrid, and supports this inference by showing that the oxlip occurs where the parent species are present, but not otherwise. The third species is shown to be sterile when crossed with the others, and to be distinct in morphology and in geographical range" (p. 100). If Ghiselin is right, I thought, that Darwin called the oxlip a "species," then my whole theory on Darwin on species is shot to pieces! The paper by Darwin was not one I had read, since it was not included in the collection of Darwin's papers by Barrett [38]. Upon examining the actual paper [37], however, I was relieved, for I found that nowhere in that paper did Darwin call the oxlip a "species." The three species Darwin in that paper refers to are not the primrose, cowslip, and oxlip, but rather the common primrose (*Primula vulgaris*), the common cowslip (*Primula veris*), and the Bardfield oxlip (*Primula elatior*), the latter which Darwin argues "is not a hybrid," and he adds that all three "have as good a right to receive distinct specific names as have, for instance, the ass, quagga, and zebra" (p. 451). The common oxlip Darwin simply refers to as a "hybrid," a taxonomic practice which Darwin continues, along with the three named "true species" above, in his *Different Forms of Flowers* [39, ch. II].

Our final example is taken from Darwin's *Orchids* [40]. In that book Darwin states that "The object of the following work is to show that the contrivances by which orchids are fertilized, are as varied and almost as perfect as any of the most beautiful adaptations in the animal kingdom" (p. 1). In this work Darwin also does not always follow what his fellow naturalists called "species." A case in point is *Habernaria chlorantha*, which the distinguished botanist George Bentham, President of the Linnean Society from 1861 to 1874, had classified as a variety of *H. bifolia* in his *Handbook of British Flora*, published in 1858. In a letter Darwin wrote in 1861 to Bentham, he states that "I must think you are mistaken in ranking Hab. chlorantha as a var. of H. bifolia: the pollen-masses & stigma differ more than in most of the best species of Orchis" [41, p. 185]. In *Orchids* [40] Darwin mentions that Bentham "and some other botanists" rank *H. chlorantha* as a mere variety (p. 73), but Darwin explicitly calls it a "species," not only because it is among "the most wonderful cases of adaptation which has ever been recorded" (p. 44), but also because it is distinguished from *H. bifolia* in "the stations inhabited," and no matter if "these two forms be hereafter proved to graduate into each other" (p. 73).

## 6. Darwin's species concept and its implications

The modern entomologist Hugh Paterson thought he found in Darwin a precursor of his own species concept, which he called the Recognition Species Concept and which he based on distinct fertilization mechanisms evolved by natural selection. Looking to the first line of Darwin's *Orchids*, which I quoted above, Paterson [42] claims that Darwin there "expresses a view in perfect agreement with the recognition concept" (p. 25). And a few years earlier he would write [43], looking now at Darwin's *Origin*, that "Although Darwin's views were inadequate in detail, for he did not realize that species and varieties were qualitatively distinct, I, nevertheless, accept that they are philosophically ancestral to the recognition concept" (p. 275). Paterson, unfortunately, by a selectivity of evidence, missed seeing that Darwin did not distinguish species only by fertilization mechanisms produced by natural selection but by any and all adaptations produced by natural selection.

Many years earlier the pioneering geneticist T.H. Morgan [44], in a paper devoted to the topic of adaptations, complained that "to-day, accepting evolution,... it is notorious that, by systematists, specific distinctions rest in many cases on differences that have no adaptive significance whatever" (p. 203). Morgan himself says, "from this time forward when I speak of the origin of species I mean the origin of the adaptive characters of species" (p. 204). Surprisingly, both of these statements are taken in the context of a discussion on Darwin's *Origin*, not just on Darwin's theory of natural selection operating on chance variations (Morgan was now over his anti-Darwinism phase), but also on the meaning of the title of Darwin's book. As Morgan puts it, "Darwin's famous book is entitled 'The Origin of *Species*' but his theory of natural selection explains the *adaptations* of living things." What is remarkable is that Morgan failed to realize that his own species concept in his paper comes from Darwin's *Origin*.

Morgan and Paterson are examples of scholars who, unlike the scholars whose very recent works were examined in the second section above, "came close but no cigar," to use a common metaphor. It is time at last for the cigar, for Darwin's species concept, and what we have seen in the previous two sections should give us sufficient confidence that we now have it. Accordingly, using modern language, we may express Darwin's species concept as: *A species is a primarily horizontal population of organisms in the Tree of Life united by common descent and distinguished from other species by at least one organismic or group adaptation.*

Of course, it is not enough to simply have the cigar. We should not simply want to look at it, we should want to smoke it, too. Or to use a better metaphor, we should want to see if "light will be thrown" on Darwin's related activities. I shall now argue that Darwin's species concept, as understood above, reveals a wider consistency (though not perfect) with Darwin's writings than hitherto realized.

For a start, light is thrown on the title of Darwin's most famous book, *On the Origin of Species by Means of Natural Selection*. It was not misnamed such that it should have been *"On the Unreality of Species as Shown by Natural Selection"* [45, p. 143]. The book was properly named because, according to Darwin, (i) natural selection is the one and only process that produces those wonders of biology rightly called "adaptations," and (ii) adaptations rightly distinguish one species from another.

Nor can one turn to Darwin's original title for the *Origin* in order to obfuscate the matter, which was *An Abstract of An Essay on the Origin of Species and Varieties Through Natural Selection* [29, p. 270]. The title certainly does not imply that varieties are basically the same as species, for if natural selection and only natural selection produces adaptations it does not follow that natural selection does nothing else. The title is simply consistent with Darwin's view that varieties are "incipient species" and that, as we have seen above, only species are distinguished by adaptations. And it is certainly significant that in a letter to John Murray, dated September 10, 1859, Darwin wrote that he wanted to omit the word "Varieties" from the title, mainly because of an objection raised by "a friend," and added that "The case of Species is the real important point" (p. 331).

In a related matter, light is also thrown on a curious passage that comes at the end of the Introduction to the *Origin* [5], a passage that has misled many, in which Darwin says he is "convinced that Natural Selection has been the main but not exclusive means of modification" (p. 6). "Modification" simply means heritable change. Darwin allowed for many processes which "modify" a species, not only natural selection but also use and disuse, the direct effects of climate, and correlation of growth. But that does not mean he was not a good Darwinian. The latter three causes (along with some others) are discussed together in a chapter Darwin titled "Laws of Variation" (ch. V), not "Laws of Adaptation." But just as important, it needs to be recognized that adaptation is a much narrower concept than modification, and it is a recurring theme in Darwin's writings that only natural selection produces adaptations. All throughout the *Origin* it is clear that the job of natural selection is to explain, not merely the modification of species, but, as he puts it in the Introduction, "that perfection of structure and coadaptation which most justly excites our admiration" (p. 3). Darwin's project, after all, was basically the same as Paley's *Natural Theology*, which Darwin read and greatly admired back when he was a student at Cambridge, a point he emphasizes in his correspondence [29, p. 338] and much later in his *Autobiography* [46, p. 59]. But it was the adaptations themselves, not Paley, that really mattered. As Darwin puts it in his *Autobiography*, "I had always been much struck by such adaptations, and until these could be explained it seemed to me almost useless to endeavour to prove by indirect evidence that species have been modified" (p. 119). The fundamental point is that nowhere in the *Origin* does Darwin credit any explanation for adaptations other than the process of natural selection.

It should also now be clear that Darwin did not really believe that species and varieties were basically the same, even though he made statements in the *Origin* [5] that have the contrary appearance, such as that "there is no essential [or 'fundamental'] distinction between species and varieties" (pp. 276, 278), passages that have fooled many. What is clear from Darwin's *practice* is that species in his view, but not varieties, are distinguished by adaptations. And what makes a variety an "incipient species" is whether any part of it is being worked on by natural selection in the direction of an adaptation. The key is heritable variation. As Darwin puts it in a letter to Hooker, "if a variation be not inherited, it is of no significance to us" [29, p. 297].[2] One might base a variety on a variation that is incapable of being inherited (or if so capable, that is not profitable to the organism), but it would not be an "incipient species" in Darwin's view (let alone a species).

The focus on adaptations also explains "borderline cases" in Darwin's writings. Varieties, as he repeatedly puts it in the *Origin* [5], are "incipient species" (e.g., pp. 52, 111), and given that the evolution of an adaptation is a very slow process, one taking hundreds if not thousands of generations, borderline cases at any horizontal level in the Tree are only to be expected, cases where expert naturalists—Darwinians at that—cannot determine whether closely allied forms should be classified as varieties or as species. But this should not take away from the fact that for Darwin there are clear cases when an organismic structure or behavior is an adaptation and when it is not.

Darwin's focus on adaptations also throws light on what he says about genealogy and classification, that, as we have seen above from the *Origin* [5], "all true classification is genealogical; that community of descent is the hidden bond which naturalists have been unconsciously seeking" (p. 420). Darwin does not say this because he was a proto-cladist, motivated by producing an objective *method* of classification, one without an appeal to similarity, nor does he say this because he was motivated by the *logic* of individuals as opposed to classes. His motive, instead, was *biological*, based on the fact (as he understood it) that adaptations do not arise in organisms spontaneously but are only transmitted by generation, by the reproduction of one organism by another. It is logically and biologically possible that the same adaptation could arise independently in different lineages, but it still for Darwin is the case that it is mainly by descent that adaptations get shared, even more so that a suite of adaptations get shared. The *origin* of adaptations, of course, is by slow, gradual, cumulative natural selection according to Darwin, but the *transmission* of adaptations is only by reproduction.[3] Hence, given that species are the basal units of taxonomy and that it is only the adaptations in organisms that rightly distinguish for Darwin one species from another, the only legitimate conclusion for Darwin was that classification must be "genealogical." But it would not be exclusively genealogical, for if a species acquires a new adaptation in its evolution (or loses one) without any branching, it would not be numerically the *same* species, a consequence that would make its way up the various taxonomic (manmade) ranks. All of this, and this alone, explains why Darwin in the *Origin* uses the phrase "the *vera causa* of community of descent" (p. 159).

The term *vera causa*, meaning "true cause," brings us to another important point. The science of Darwin's time was based on the *vera causa* ideal, both for scientific explanation and for scientific classification. This was made especially evident in Sir John Herschel's *Preliminary Discourse on the Study of Natural Philosophy* [47], which Darwin had read at least twice (shortly before his *Beagle* voyage and then again shortly after his first reading of Malthus) and which he highly regarded along with its author, who many considered the greatest scientist

---

2 The point is likewise made in the Origin [5]: "Any variation which is not inherited is unimportant for us" (p. 12). To this should be added a further point that is made: "unless profitable variations do occur, natural selection can do nothing" (p. 82).

3 As Darwin loosely puts in the Origin [5], "the chief part of the organisation of every being is simply due to inheritance" (p. 199).

and philosopher of science of his time [48]. The principal aim of natural science, Herschel argues, is to discover laws of nature, which themselves are the effects of real forces in nature, which Herschel [47] following Newton calls *"vera causæ"* (esp. pp. 18, 76, 91, 104–105, 144). According to Herschel it is only laws of nature that scientifically explain natural phenomena. But Herschel also argues that scientific classification is based on laws. He devotes an entire chapter to scientific classification, in which he distinguishes between "artificial and natural systems of classification in general" (p. 143n). Two of the fields he discusses are botany and mineralogy, each notorious for their "artificial systems" of lumpers and splitters, and it was surely not lost on Darwin that Herschel throughout his book elaborates on the case of mineralogy, which did not have *real* "mineral species" and hence scientific classification until the discovery of the "laws of crystallography" (pp. 123, 139–140, 183, 239–243, 290–296). Accordingly, although Darwin in the *Origin* [5], surprisingly, does not explicitly call natural selection a *vera causa*, he does nevertheless call it a "power" (pp. 61, 109, 205, 454) and a "law" (pp. 244, 472, 489–490), and he clearly expresses his belief that the evolution of species by natural selection is a true theory (pp. 457–458, 481). Little wonder, then, that Darwin would stress in his correspondence following the publication of the *Origin* that he thought of natural selection as a "veracausa" because of its power to explain a variety of phenomena [33, pp. 76, 84, 102, 123]. And little wonder also that Darwin felt it a "great blow & discouragement" when he heard that Herschel had rejected his theory as "the law of higgledy-piggledy" [29, p. 423]. In arguing for *the origin of species by means of natural selection,* Darwin in the *Origin* [5] had attempted to solve "that mystery of mysteries, as it has been called by one of our greatest philosophers [Herschel]" (p. 1) by bringing *both* the origin *and* the classification of species into the scientific domain. Arguing for evolution by natural selection, therefore, did not involve in his mind simply following what his fellow naturalists *called* "species."

## 7. Deeper implications of Darwin's species concept

Further questions still remain, questions that are more controversial. First, was Darwin a species category realist? Well if he in fact had a realist species concept then it would probably make more sense to categorize him as a species category realist than not. It all depends on what one means by "species category realist." Having a realist species concept might not be a sufficient condition, but I would certainly think it is a necessary one. What is interesting is that if we use Ereshefsky's [14] "minimum threshold" (p. 413) for species category realism, then Darwin as understood in the present chapter would have to be categorized as a species category realist. First, for Darwin not only most but *all* of the taxa *clearly* categorized by him as species share a common feature, namely, a distinguishing organismic or group adaptation (presumably also he would accept a unique set of adaptations). Second, that feature helps us to understand the nature of the members of the category, in that adaptations are not produced spontaneously but only over many generations by the process of natural selection and are transmitted only via reproduction. Third and finally, the feature that distinguishes members of the species category, viz., adaptations, distinguishes those members from members of other categories, such as the members of genera (the species that make up a genus are not distinguished by *species* adaptations) or the members of ecological categories (members of the predator category, for example, are not distinguished by one or more adaptations that define "predator").

This raises a further question: Is Darwin's species concept seriously flawed because it is parasitic on the concept of adaptation? Some might think so. As Gregory Radick [49] points out, "Now historians have thrown doubt on the naturalness of the Darwinian kind 'adaptation'" (p. 162). They take it to be a social construction—"inseparable from Britain in the age of complex machines and counter-revolutionary theology" (pp. 153–154)—and nothing more. I suspect that most if not all modern biologists would reject this view, not just a few such as Richard Dawkins. (Radick himself makes Darwin's theorizing on species parasitic on the *vera causa* ideal embodied in Whig ambitions for British science and society.) As touched upon in the opening section, I find this kind of contextualist history of science, epitomized by the books by Desmond and Moore [2, 3], not only more suggestive than substantive, but more argumentation by an ideologically-driven cloud of detail than argumentation by logic and evidence including inference to the best explanation. The work ethic is not in dispute here, only the philosophy behind it. The problem is not simply the self-imposed problem found in history of science in general, what the historian Mary Winsor [50] calls the "Taboo Problem," the reluctance to deal with the evidence itself for whether any of the science they study is true (pp. 240–241). The problem, more specifically, is the dogmatic allegiance to a kind of deterministic reasoning that in logic is known as *cum hoc ergo propter hoc*. No scientist is a genius, according to this kind of contextualist history of science, no scientist is ahead of his or her time in anything, no scientist really sees into nature. But these historians themselves somehow manage to transcend their milieu and see into the nature of science and scientists!

What should be disturbing is not that Darwin's species concept is based on the concept of adaptation, but that Darwin's species concept is not at all compatible with some of what he explicitly says about species *per se*, such as that "It is really laughable to see what different ideas are prominent in various naturalists minds, when they speak of 'species'.... It all comes, I believe, from trying to define the indefinable" [51, p. 309] and that "we shall at least be freed from the vain search for the undiscovered and undiscoverable essence of the term species" [5, p. 485]. Other passages on species in the *Origin* alone [5], such as those on "amount of difference" (pp. 57, 485), on characters "constant and distinct" (pp. 47, 484), and on "well-marked and permanent varieties" (p. 133) or "strongly marked and fixed varieties" (p. 155), can perhaps be harmonized with Darwin's species concept elucidated in this chapter. But not the other passages above, the passages, most interestingly, that have gained for Darwin the longstanding reputation of being a species nominalist.

The time would seem ripe, then, for proceeding to the (in itself) undesirable terrain of *dissimulation hypotheses* for Darwin, possibly even *radical* dissimulation. The phrase "dissimulation hypothesis" comes from scholarship on "the father of modern philosophy," René Descartes. There is strong evidence suggesting, for example, that Descartes was not sincere with his proofs of the existence of God [52], and that he was deceptive in the ordering of his three proofs of God's existence [53]. But science, too, has its fair share of dissimulation, rang-ing from the manipulation of data in the pursuit of truth to outright fraud in the pursuit of self-aggrandizement. Sir Isaac Newton, for example, in the second edition of his *Principia*, claimed greater mathematical precision for his work on gravity, sound, and the precession of the equinoxes than his work warranted, while his attack on Leibniz for plagiarizing calcu-lus verged on the neurotic [54]. Robert Millikan, who received a Nobel Prize in 1923 for his discovery of the unit amount of charge in an electron, selectively edited data for his publish-ed reports as shown in his private notebooks [55]. Much more recently, the groundbreaking work of Woo Suk Hwang and colleagues on the creation of human embryonic stem cell lines was shown to be largely fabricated, resulting not only in retracted articles by *Science* but in legal charges [56]. Examples abound [57], though clearly one should not want to throw out the proverbial baby with the bathwater.

In the case of Darwin, assuming that the analysis presented in this chapter is basically cor-rect, we can only speculate as to his motive or motives for withholding his real species con-cept. Perhaps it was because he thought that an explicit nominalistic definition would help to either break through or bypass the psychological barrier of what Wollaston [58], in char-acterizing the core of the "general" concept of species, called the "axiom" of special creation and non-transmutation (p. 133). Or perhaps Darwin chose to bypass defining "species" for mainly linguistic rather than psychological reasons, simply in order to better communicate his theory of evolution [8, p. 266]. And then perhaps Darwin felt that he would be lowering himself by adding to what he called the "really laughable" babble of species concepts among "systematic naturalists," something he complained about in the letter to Hooker quoted from above [51, p. 309] and repeated in *Natural Selection* [16, p. 98].

And then perhaps his motive changed with time and circumstances. In the years following the publication of the *Origin*, the *evolution* of species became widely accepted in scientific cir-cles, but not his proposed *mechanism*, natural selection. As time passed he might then have viewed bringing out his species concept as pointless. Revealing is his statement, written in 1863, that "Whether the naturalist believes in the views given by Lamarck, or Geoffroy St.-Hillaire, by the author of the 'Vestiges,' by Mr. Wallace and myself, or in any other such view, signifies extremely little in comparison with the admission that species have descend-ed from other species and have not been created immutable" [38 II, p. 81]. The same point is found in his letters of the same year [35, pp. 36, 40]. Darwin himself, no doubt, continued to believe in the truth of his theory of natural selection, but perhaps what mattered more to him now was the acceptance of the evolution of species *per se* than the evolution of species *by means of* natural selection. This is a remarkable thought. What it means for our topic is that if establishing the product (evolution) had become primary and the process (natural se-lection) secondary, then making the primary proof (for evolution) hinge on a questionable species concept, which depended on the secondary proof (for natural selection), might not have seemed wise. And certainly with regard to natural selection Darwin was prescient, since even as *a* mechanism of species evolution natural selection was not generally accepted in his lifetime and only began to achieve wide acceptance in biology in the 1930s [59].

## 8. Conclusion

We shall probably never have a convincing theory on Darwin's motive. But no matter. For the great many, myself included, who admire Darwin as one of the greatest scientists of all time, and also as an English gentleman, the idea that he might have intentionally misrepresented his species concept (whatever his motive) "may not be a cheering prospect," but the evidence must be followed wherever it leads, no less in history of science than in natural science. That Darwin did intentionally mislead in this matter does not take away from his scientific accomplishments, but it does bring him down from the heights of the virtuous gods of knowledge and makes him appear much more human, even as a scientist.

## Author details

David N. Stamos

Address all correspondence to: dstamos@yorku.ca

York University, Department of Philosophy, Toronto, Ontario, Canada

## References

[1] Lightman, Bernard, ed. (1997). *Victorian Science in Context*. Chicago: University of Chicago Press.

[2] Desmond, Adrian, and Moore, James (1991). *Darwin*. London: Penguin Books.

[3] Desmond, Adrian, and Moore, James (2009). *Darwin's Sacred Cause*. London: Penguin Books.

[4] Ashcraft, Richard (1986). *Revolutionary Politics & Locke's* Two Treatises of Government. Princeton: Princeton University Press.

[5] Darwin, Charles (1859). *On the Origin of Species by Means of Natural Selection*. London: John Murray.

[6] Ospovat, Dov (1981). *The Development of Darwin's Theory: Natural History, Natural Theology, and Natural Selection, 1838–1859*. Cambridge: Cambridge University Press.

[7] Wilkins, John S. (2009). *Species: A History of the Idea*. Berkeley: University of California Press.

[8] Beatty, John (1985). "Speaking of Species: Darwin's Strategy." In David Kohn, ed. (1985). *The Darwinian Heritage*. Princeton: Princeton University Press, pp. 265–281.

[9]  Ghiselin, Michael T. (1969). *The Triumph of the Darwinian Method*. Berkeley: University of California Press.

[10]  Sloan, Phillip R. (2009). "Originating Species: Darwin on the Species Problem." In Michael Ruse and Robert J. Richards, eds. (2009). *The Cambridge Companion to the "Origin of Species."* Cambridge: Cambridge University Press, pp. 67–86.

[11]  Ellis, Mark W., and Wolf, Paul G. (2010). "Teaching 'Species'." *Evolution: Education and Outreach* 3, pp. 89–98.

[12]  Mayden, R.L. (1997). "A Hierarchy of Species Concepts: the Denouement in the Saga of the Species Problem." In M.F. Claridge, *et al.*, eds. (1997). *Species: The Units of Biodiversity*. London: Chapman & Hall, pp. 381–424.

[13]  Mallet, James (2010). "Why Was Darwin's View of Species Rejected by Twentieth Century Biologists?" *Biology & Philosophy* 25, pp. 497–527.

[14]  Ereshefsky, Marc (2010). "Darwin's Solution to the Species Problem." *Synthese* 175, pp. 405–425.

[15]  Schweitzer, Albert (1933). *Out of My Life and Thought: An Autobiography*. Antje Bultmann Lemke, trans. New York: Holt, Rinehart and Winston.

[16]  Stauffer, R.C., ed. (1975). *Charles Darwin's Natural Selection, Being the Second Part of His Big Species Book Written From 1856 to 1858*. Cambridge: Cambridge University Press.

[17]  McOuat, Gordon R. (2001). "Cataloguing Power: Delineating Competent Naturalists and the Meaning of Species in the British Museum." *British Journal for the History of Science* 34, pp. 1–28.

[18]  Wittgenstein, Ludwig (1953). *Philosophical Investigations*. G.E.M. Anscombe, trans. Oxford: Basil Blackwell. 3rd ed. (1958).

[19]  Stamos, David N. (1999). "Darwin's Species Category Realism." *History and Philosophy of the Life Sciences* 21 (2), pp. 21–70.

[20]  Stamos, David N. (2007). *Darwin and the Nature of Species*. Albany: SUNY.

[21]  Padian, Kevin (1999). "Charles Darwin's Views of Classification in Theory and Practice." *Systematic Biology* 48 (2), pp. 352–364.

[22]  Darwin, Charles (1854). *A Monograph on the Sub-Class Cirripedia, the Balanidæ*. London: The Ray Society.

[23]  Burkhardt, Frederick, and Smith, Sydney, eds. (1988). *The Correspondence of Charles Darwin, Volume 4, 1847–1850*. Cambridge: Cambridge University Press.

[24]  Kohn, David. (2009). "Darwin's Keystone: The Principle of Divergence." In Michael Ruse and Robert J. Richards, eds. (2009). *The Cambridge Companion to the "Origin of Species."* Cambridge: Cambridge University Press, pp. 87–108.

[25]  Burkhardt, Frederick, *et al.*, eds. (1997). *The Correspondence of Charles Darwin, Volume 10, 1862*. Cambridge: Cambridge University Press.

[26]  Lyell, Charles (1832). *Principles of Geology*. Vol. II. London: John Murray.

[27]  Watson, Hewett C. (1845). "Report of an Experiment Which Bears Upon the Specific Identity of the Cowslip and Primrose." *The Phytologist* 2, pp. 217–219.

[28]  Watson, Hewett C. (1845). "On the Theory of 'Progressive Development,' Applied in Explanation of the Origin and Transmutation of Species." *The Phytologist* 2, pp. 108–113, 140–147, 161–168, 225–228.

[29]  Burkhardt, Frederick, and Smith, Sydney, eds. (1991). *The Correspondence of Charles Darwin, Volume 7, 1858–1859*. Cambridge: Cambridge University Press.

[30]  Darwin, Charles (1871). *The Descent of Man, and Selection in Relation to Sex*. Two Volumes. London: John Murray.

[31]  Burkhardt, Frederick, *et al.*, eds. (2005). *The Correspondence of Charles Darwin, Volume 15, 1867*. Cambridge: Cambridge University Press.

[32]  Darwin, Charles (1868). *The Variation of Animals and Plants Under Domestication*. Two Volumes. 2nd ed. (1875). London: John Murray.

[33]  Burkhardt, Frederick, *et al.*, eds. (1993). *The Correspondence of Charles Darwin, Volume 8, 1860*. Cambridge: Cambridge University Press.

[34]  Glass, Bentley (1959). "Heredity and Variation in the Eighteenth Century Concept of the Species." In Bentley Glass, *et al.*, eds. (1950). *Forerunners of Darwin: 1745–1859*. Baltimore: Johns Hopkins, pp. 144–172.

[35]  Burkhardt, Frederick, *et al.*, eds. (1999). *The Correspondence of Charles Darwin, Volume 11, 1863*. Cambridge. Cambridge University Press.

[36]  Burkhardt, Frederick, *et al.*, eds. (2001). *The Correspondence of Charles Darwin, Volume 12, 1864*. Cambridge: Cambridge University Press.

[37]  Darwin, Charles (1869). "On the Specific Difference between *Primulaveris, P. vulgaris*, and *P. elatior*; and on the Hybrid Nature of the common Oxlip." *Journal of the Linnean Society (Botany)* 10, pp. 437–454.

[38]  Barrett, Paul H., ed. (1977). *The Collected Papers of Charles Darwin*. Two Volumes. Chicago: University of Chicago Press.

[39]  Darwin, Charles (1877). *The Different Forms of Flowers on Plants of the Same Species*. 2nd ed. (1884). London: John Murray.

[40]  Darwin, Charles (1862). *On the Various Contrivances by which Orchids are Fertilised by Insects*. 2nd ed. (1877). London: John Murray.

[41]  Burkhardt, Frederick, *et al.*, eds. (1994). *The Correspondence of Charles Darwin, Volume 9, 1861*. Cambridge: Cambridge University Press.

[42] Paterson, H.E.H. (1985). "The Recognition Concept of Species." In E.S. Vrba, ed. (1985) *Species and Speciation.* Pretoria. Transvaal Museum, pp. 21–29.

[43] Paterson, H.E.H. (1982). "Darwin and the Origin of Species." *South African Journal of Science* 78, pp. 272–275.

[44] Morgan, T.H. (1910). "Chance or Purpose in the Origin and Evolution of Adaptation." *Science* 31 (789), pp. 201–210.

[45] Sober, Elliott (1993). *Philosophy of Biology.* Boulder, CO: Westview Press.

[46] Barlow, Nora, ed. (1958). *The Autobiography of Charles Darwin 1809–1882.* London: Collins.

[47] Herschel, John F.W. (1830). *Preliminary Discourse on the Study of Natural Philosophy.* London: Longman, Rees, Brown, & Green.

[48] Cannon, Walter F. (1961). "John Herschel and the Idea of Science." *Journal of the History of Ideas* 22 (2), pp. 215–239.

[49] Radick, Gregory (2003). "Is the Theory of Natural Selection Independent of its History?" In Jonathan Hodge and Gregory Radick, eds. (2003). *The Cambridge Companion to Darwin.* Cambridge: Cambridge University Press, pp. 143–167.

[50] Winsor, Mary P. (2001). "The Practitioner of Science: Everyone her Own Historian." *Journal of the History of Biology* 34, pp. 229–245.

[51] Burkhardt, Frederick, and Smith, Sydney, eds. (1990). *The Correspondence of Charles Darwin, Volume 6, 1856–1857.* Cambridge: Cambridge University Press.

[52] Loeb, Louis E. (1986). "Is There Radical Dissimulation in Descartes' *Meditations?*" In AmélieOksenbergRorty, ed. (1986). *Essays on Descartes'* Meditations. Berkeley: University of California Press, pp. 243–270.

[53] Stamos, David N. (1997). "The Nature and Relation of the Three Proofs of God's Existence in Descartes' *Meditations.*" *Auslegung* 22 (1), pp. 1–37.

[54] Westfall, Richard S. (1973). "Newton and the Fudge Factor." *Science* 179, pp. 751–758.

[55] Franklin, Allen (1981). "Millikan's Published and Unpublished Data on Oil Drops." *Historical Studies in the Physical Sciences* 11, pp. 185–201.

[56] Franzen, Martina, *et al.* (2007). "Fraud: Causes and Culprits as Perceived by Science and the Media. Institutional Changes, Rather than Individual Motivations, Encourage Misconduct." *EMBO Reports* 8 (1), pp. 3–7.

[57] Broad, William, and Wade, Nicholas (1982). *Betrayers of the Truth: Fraud and Deceit in the Halls of Science.* New York: Simon & Schuster.

[58] Wollaston, Thomas Vernon (1860). "On the Origin of Species by Means of Natural Selection." *Annals and Magazine of Natural History* 5 (3rd ser.), pp. 132–143.

[59]  Bowler, Peter J. (1988). The Non-Darwinian Revolution: Reinterpreting a Historical Myth. Baltimore: Johns Hopkins.

# Permissions

The contributors of this book come from diverse backgrounds, making this book a truly international effort. This book will bring forth new frontiers with its revolutionizing research information and detailed analysis of the nascent developments around the world.

We would like to thank Igor Ya Pavlinov, for lending his expertise to make the book truly unique. He has played a crucial role in the development of this book. Without his invaluable contribution this book wouldn't have been possible. He has made vital efforts to compile up to date information on the varied aspects of this subject to make this book a valuable addition to the collection of many professionals and students.

This book was conceptualized with the vision of imparting up-to-date information and advanced data in this field. To ensure the same, a matchless editorial board was set up. Every individual on the board went through rigorous rounds of assessment to prove their worth. After which they invested a large part of their time researching and compiling the most relevant data for our readers. Conferences and sessions were held from time to time between the editorial board and the contributing authors to present the data in the most comprehensible form. The editorial team has worked tirelessly to provide valuable and valid information to help people across the globe.

Every chapter published in this book has been scrutinized by our experts. Their significance has been extensively debated. The topics covered herein carry significant findings which will fuel the growth of the discipline. They may even be implemented as practical applications or may be referred to as a beginning point for another development. Chapters in this book were first published by InTech; hereby published with permission under the Creative Commons Attribution License or equivalent.

The editorial board has been involved in producing this book since its inception. They have spent rigorous hours researching and exploring the diverse topics which have resulted in the successful publishing of this book. They have passed on their knowledge of decades through this book. To expedite this challenging task, the publisher supported the team at every step. A small team of assistant editors was also appointed to further simplify the editing procedure and attain best results for the readers.

Our editorial team has been hand-picked from every corner of the world. Their multi-ethnicity adds dynamic inputs to the discussions which result in innovative outcomes. These outcomes are then further discussed with the researchers and contributors who give their valuable feedback and opinion regarding the same. The feedback is then collaborated with the researches and they are edited in a comprehensive manner to aid the understanding of the subject.

Apart from the editorial board, the designing team has also invested a significant amount of their time in understanding the subject and creating the most relevant covers. They scrutinized every image to scout for the most suitable representation of the subject and create an appropriate cover for the book.

The publishing team has been involved in this book since its early stages. They were actively engaged in every process, be it collecting the data, connecting with the contributors or procuring relevant information. The team has been an ardent support to the editorial, designing and production team. Their endless efforts to recruit the best for this project, has resulted in the accomplishment of this book. They are a veteran in the field of academics and their pool of knowledge is as vast as their experience in printing. Their expertise and guidance has proved useful at every step. Their uncompromising quality standards have made this book an exceptional effort. Their encouragement from time to time has been an inspiration for everyone.

The publisher and the editorial board hope that this book will prove to be a valuable piece of knowledge for researchers, students, practitioners and scholars across the globe.

# List of Contributors

**Igor Ya. Pavlinov**
Zoological Museum, Moscow Lomonosov State University, Moscow, Russia

**Richard A. Richards**
University of Alabama, Tuscaloosa, Alabama, USA

**Kirk Fitzhugh**
Research & Collections Branch, Natural History Museum of Los Angeles County, Los Angeles, CA, USA

**Larissa N. Vasilyeva**
Institute of Biology and Soil Science, Far East Branch of the Russian Academy of Sciences, Vladivostok, Russia

**Steven L. Stephenson**
Department of Biological Sciences, University of Arkansas, Fayetteville, USA

**Victor Prokhorovich Shcherbakov**
Institute of Problems of Chemical Physics, Moscow Region, Russia

**James T. Staley**
Department of Microbiology, University of Washington, USA

**Richard L. Mayden**
Department of Biology, Laboratory of Integrated Biodiversity, Conservation, and Genomics, Saint Louis University, St. Louis, Missouri, USA

**Arley Camargo**
Unidad de Diversidad, Sistemática y Evolución, Centro Nacional Patagónico, Consejo Nacional de Investigaciones Científicas y Técnicas, Puerto Madryn, Chubut, Argentina

**Jack Jr. Sites**
Department of Biology and Bean Life Science Museum, Brigham Young University, Provo, Utah, USA

**V. S. Friedmann**
Moscow State University, Faculty of Biology, Moscow, Russia

**David N. Stamos**
York University, Department of Philosophy, Toronto, Ontario, Canada

Printed in the USA
CPSIA information can be obtained
at www.ICGtesting.com
JSHW011453221024
72173JS00005B/1057

9 781632 395740